Solid Mechanics and Its Applications

Volume 283

The fundamental questions arising in mechanics are: Why?, How?, and How much? The aim of this series is to provide lucid accounts written by authoritative researchers giving vision and insight in answering these questions on the subject of mechanics as it relates to solids. The scope of the series covers the entire spectrum of solid mechanics. Thus it includes the foundation of mechanics; variational formulations; computational mechanics; statics, kinematics and dynamics of rigid and elastic bodies; vibrations of solids and structures; dynamical systems and chaos; the theories of elasticity, plasticity and viscoelasticity; composite materials; rods, beams, shells and membranes; structural control and stability; soils, rocks and geomechanics; fracture; tribology; experimental mechanics; biomechanics and machine design. The median level of presentation is the first year graduate student. Some texts are monographs defining the current state of the field; others are accessible to final year undergraduates; but essentially the emphasis is on readability and clarity.

Springer and Professors Barber and Klarbring welcome book ideas from authors. Potential authors who wish to submit a book proposal should contact Dr. Mayra Castro, Senior Editor, Springer Heidelberg, Germany, email: mayra.castro@springer.com

Indexed by SCOPUS, Ei Compendex, EBSCO Discovery Service, OCLC, ProQuest Summon, Google Scholar and SpringerLink.

Jiang-Bo Bai • Tian-Wei Liu • Nicholas Fantuzzi

Folding Mechanics of Deployable Composite Structures

Springer

Jiang-Bo Bai
School of Transportation Science and Engineering
Beihang University
Beijing, China

Tian-Wei Liu
School of Automation Science and Electrical Engineering
Beihang University
Beijing, China

Nicholas Fantuzzi
DICAM Department
University of Bologna
Bologna, Italy

ISSN 0925-0042 ISSN 2214-7764 (electronic)
Solid Mechanics and Its Applications
ISBN 978-3-031-99253-7 ISBN 978-3-031-99254-4 (eBook)
https://doi.org/10.1007/978-3-031-99254-4

This Springer imprint is published by the registered company Springer Nature Switzerland AG
The registered company address is: Gewerbestrasse 11, 6330 Cham, Switzerland

Preface 1

This book is based on the latest research advances in large-deformation flexible composite structures and the authors' recent works in this field. Comprising 12 chapters, it systematically introduces the progress in the study of large-deformation flexible composite structures. The book provides a concise introduction to fundamental theories, including the energy method, variational principles, and composite mechanics. It further explores analytical models for various typical folding modes of large-deformation flexible composite structures, covering compressive folding for composite tape springs, bending folding for composite tape springs, compressive folding for composite helical structures, coiling folding for lenticular deployable composite booms, coiling folding for tubular deployable composite booms, and bistable deployable composite booms. Additionally, it offers an in-depth discussion on finite element analysis and optimization strategies for such structures. The book also summarizes advancements in engineering applications and the challenges currently faced in the field.

This book serves as a valuable resource for undergraduate and graduate students in solid mechanics, composite mechanics, composite structural design, and aerospace engineering. It also provides a useful reference for researchers engaged in the mechanical design of composite structures, particularly those working on smart materials and structures, which have become an emerging research focus in recent years.

Beijing, China	Jiang-Bo Bai
Beijing, China	Tian-Wei Liu
Bologna, Italy	Nicholas Fantuzzi

Preface 2

Large-deformation flexible composite structures are a class of engineering structures based on flexible composite materials that can undergo significant geometric deformations while maintaining or recovering specific configurations under certain conditions. These structures integrate both structural and functional attributes, offering advantages such as high packing efficiency, excellent mechanical properties, lightweight characteristics, and strong design flexibility. In the aerospace field, they are commonly referred to as foldable or deployable composite structures, while in aviation applications, they are known as morphing or shape-adaptive composite structures. Although different professions may have different terminologies, the fundamental characteristics and essential mechanisms remain the same. To ensure clarity and consistency, the authors have chosen to adopt the unified term "large-deformation flexible composite structures" throughout this book. In recent years, these structures have been extensively studied and widely applied in engineering fields such as space deployable structures, morphing wings, foldable architecture, flexible electronic devices, energy harvesters, and adaptive structures.

The authors have been engaged in fundamental and applied research on advanced composite materials and their structures for many years, accumulating extensive experience in the development of large-deformation flexible composite structures. This includes material properties, analytical modeling, numerical simulations, manufacturing processes, experimental validation, as well as design of engineering prototypes. Some of these research achievements have already been successfully implemented in major engineering projects.

This book encapsulates the authors' research efforts in the folding mechanics of large-deformation flexible composite structures, keeping pace with the latest advancements and presenting the most recent research findings and development trends in this field. The book is organized around this central theme and explores it systematically and in depth. It begins with an overview of the research progress in large-deformation flexible composite structures, providing readers with a broad perspective on the field and the current state of research. Then, it briefly introduces fundamental theories such as the energy method, variational principles, and basic composite mechanics, laying the foundation for subsequent discussions. The book then delves into the analytical modeling of various typical folding modes, including compressive folding for composite tape springs, bending folding for composite tape springs, compressive folding for composite helical structures, coiling folding for lenticular deployable composite booms, coiling folding for tubular deployable composite booms, and bistable deployable composite booms. Furthermore, it presents finite element analysis and optimization design strategies, enriching the analytical framework. Finally, the book summarizes recent engineering applications and offers insights into the challenges ahead.

Many of the topics discussed in this book originate from the authors' extensive engineering practice, experience, and reflections. The analytical models presented are highly practical, addressing the well-known challenges of traditional finite element methods in handling large-deformation geometric nonlinear problems, such as poor convergence and high computational costs. Unlike numerical simulations, analytical or semi-analytical models can deliver near-instantaneous results with minimal input parameters, offering superior computational efficiency. This advantage makes them particularly effective in solving practical engineering problems, which is a key motivation behind the authors' research. In aerospace engineering, where project timelines are often urgent and design optimization through extensive finite element analysis is infeasible, the computational methods introduced in this book enable rapid design iteration, aligning with development schedules, a benefit the author has personally experienced.

This book is a summary of the authors' most significant work before turning 40, a milestone filled with deep emotions and reflections. It is astonishing to look back and realize that nearly two decades have been dedicated to large-deformation flexible composite structures. The journey of scientific research has been as winding as life itself, often filled with obstacles and moments of despair, where it seemed like all paths were blocked. Yet, it was through these struggles that the most profound experiences and meanings were uncovered, reinforcing the commitment to long-term perseverance and staying true to one's original aspirations. Looking ahead, one may wonder: Dare to ask where is the road? It is under our feet.

Due to inevitable limitations in expertise and scope, this book may contain inaccuracies or omissions, and the authors sincerely welcome critical feedback and constructive suggestions from readers.

Beijing, China — Jiang-Bo Bai
Beijing, China — Tian-Wei Liu
Bologna, Italy — Nicholas Fantuzzi

Contents

Chapter 1
Progress in Thin-Walled Deployable Composite Structures

Abstract This chapter focuses on the research progress of deployable composite structures in aerospace applications. Driven by the rapid development of aerospace technology, there is an urgent demand for high-performance space deployable structures. Elastic strain energy-driven large-deformation flexible composite structures have become a research frontier due to their advantages such as high packing efficiency and light weight. The chapter systematically reviews deployable composite hinges (including tape springs, integral hinges, and combined hinges), booms (tubular, non-circular, and bistable types), reflectors (taco-shaped, petal-shaped, umbrella-shaped, and warp-rid types), and composite helical structures, covering design, modeling, fabrication, optimization, and applications. It also outlines the book's structure, which systematically introduces theoretical models, computational methods, numerical simulations, optimization designs, and engineering applications of large-deformation flexible composite structures.

1.1 Introduction

With the rapid development of aerospace technology, the demand for large-scale spacecraft missions has been increasing, leading to an urgent need for high-performance, highly stable deployable space structures, such as ultra-large space antennas, solar sails, flexible robotic arms, and solar power stations. These structures are stowed in a compact form before launch, greatly reducing the space they occupy, allowing more spacecraft to be loaded into the limited storage capacity of launch vehicles. Once in space, they can deploy according to a predefined procedure, forming the required large-scale structures to complete various complex tasks. Thus, deployable space structures solve the conflict between multifunctionality, large size, and limited storage capacity of launch vehicles. According to Web of Science data, the number of papers on deployable space structures has grown exponentially since 1999, with over 2300 relevant papers published by 2023, as shown in Fig. 1.1a. This indicates the tremendous potential of deployable space structures

J.-B. Bai et al., *Folding Mechanics of Deployable Composite Structures*, Solid Mechanics and Its Applications 283,
https://doi.org/10.1007/978-3-031-99254-4_1

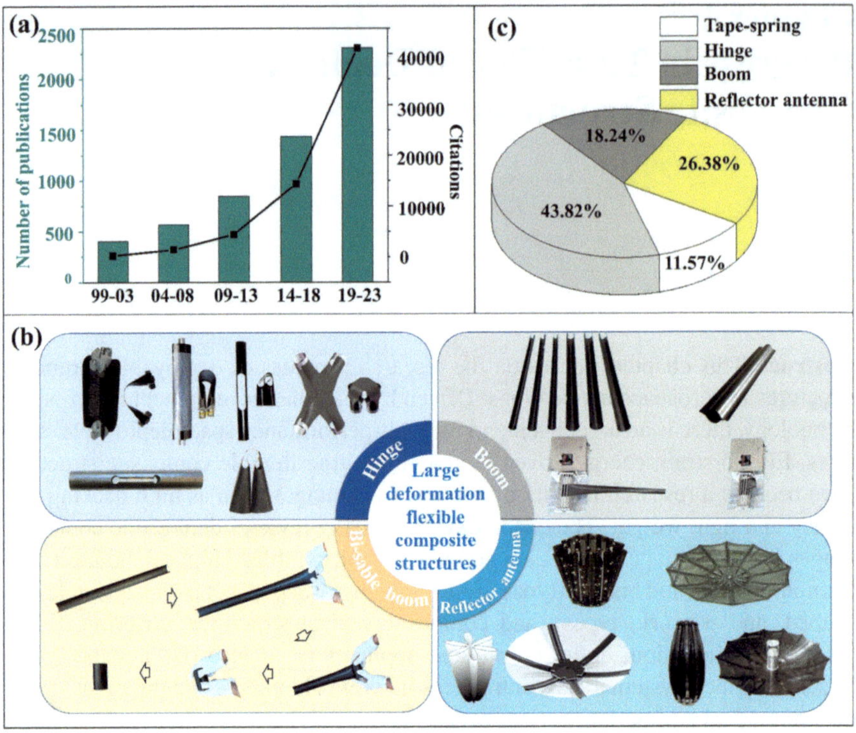

Fig. 1.1 Keyword analysis and publication trend of large-deformation flexible composite structures: (**a**) Number of publications and citations over the past 25 years. (**b**) Typical space applications. (**c**) Main types of structural configurations

in advancing space technology, gaining increasing recognition from researchers. The emerging body of research shows that more and more people are recognizing and utilizing the capabilities of deployable space structures to address various challenges [1].

Currently, deployable space structures can be categorized into four main types: mechanical, inflatable, shape-memory-driven, and elastic strain energy-driven structures [1–4]. Among these, elastic strain energy-driven large-deformation flexible composite structures feature an integrated structure-function design, utilizing the strain energy stored during folding to achieve elastic deployment. Compared to other traditional types of deployable space structures, elastic strain energy-driven large-deformation flexible composite structures offer advantages such as simplicity, high integration, excellent mechanical properties, high volume-to-stowage ratio, high deployment precision, and light weight [5–8]. Therefore, they have become a research hotspot and frontier in the field of aerospace composite structures in recent years. To date, typical elastic strain energy-driven large-deformation flexible composite structures mainly include deployable composite tape spring, hinges, booms, and reflectors, as shown in Fig. 1.1b. According to Web of Science data, Fig. 1.1c provides a comprehensive overview of the percentage distribution of the four main types of elastic strain energy-driven large-deformation flexible composite structures, highlighting the diversification of designs to meet specific needs. This section reviews the latest developments in the design, modeling, fabrication, optimization, and application of large-deformation flexible composite structures for space applications, as shown in Fig. 1.2.

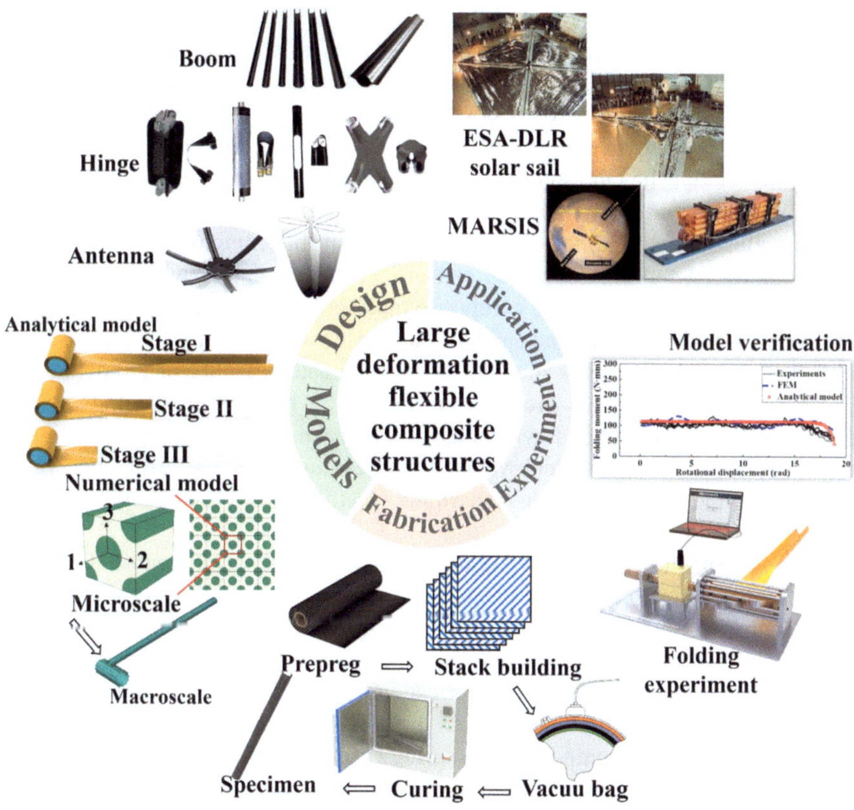

Fig. 1.2 Overview of design, models, fabrication, experimental verification, and applications of large-deformation flexible composite structures

1.2 Deployable Composite Hinges

1.2.1 Deployable Composite Tape Springs

Inspired by the concept of a tape measure, researchers have developed a structure called a deployable composite tape spring, which can store and release elastic strain energy [9, 10]. The deployable composite tape spring is a thin shell with a curved cross section, typically exhibiting uniform curvature and a cross-sectional angle smaller than 180°. Due to its curved cross section being capable of flattening, the deployable composite tape spring is well suited for folding and can achieve self-deployment through the stored strain energy. The folding of the deployable composite tape spring can be categorized into two modes: when the longitudinal and transverse curvatures have the same direction, it is referred to as "equal-sense folding," whereas when they have opposite directions, it is referred to as "opposite-sense folding," as shown in Fig. 1.3a. Regardless of the folding mode employed, the deployable composite tape spring possesses the ability to reversely rotate and lock, enabling important functionalities such as actuation, reverse rotation, and locking. When evaluating the folding behavior of the deployable composite tape spring, the following parameters are of primary concern: (1) peak moment, which refers to the critical moment the deployable composite tape spring can withstand when experiencing buckling instability and serves as a measure of its ability to resist external disturbances—a higher peak moment indicates greater stability of the deployable composite tape spring; (2) propagation moment, which represents the moment exerted during the initial deployment stage after folding deformation, reflecting its capability to actuate space deployable structures; and (3) stress level—accurately determining the stress level helps determine whether failure occurs and is used for strength design.

Figure 1.3b illustrates the relationship between folding moment and bending angle for the deployable composite tape spring under both equal-sense and opposite-sense folding conditions. In the case of opposite-sense folding, starting from point O, the folding moment M shows a linear variation and gradually increases within a small range of bending angles. The corresponding configuration exhibits a smooth curvature along the longitudinal axis. As the bending angle increases, the cross section of the deployable composite tape spring gradually becomes flatter until reaching the peak moment, which represents the maximum value of the folding moment. Subsequently, due to the structural buckling, the folding moment decreases rapidly (AB). In this configuration, only the middle portion of the deployable composite tape spring undergoes deformation due to folding, while the rest remains straight (shown in Fig. 1.3a, c). Once buckling occurs, theoretically larger bending angles do not alter the folding moment (BC), which is defined as the steady-state moment M_+^*. During the stable phase, only the arc length of the folding region is affected by the variation in bending angle. In the deploying process of the opposite-sense folding configuration, the first part of the curve (CB) coincides with the loading path. However, at point B, the red curve remains unchanged until reaching point D and

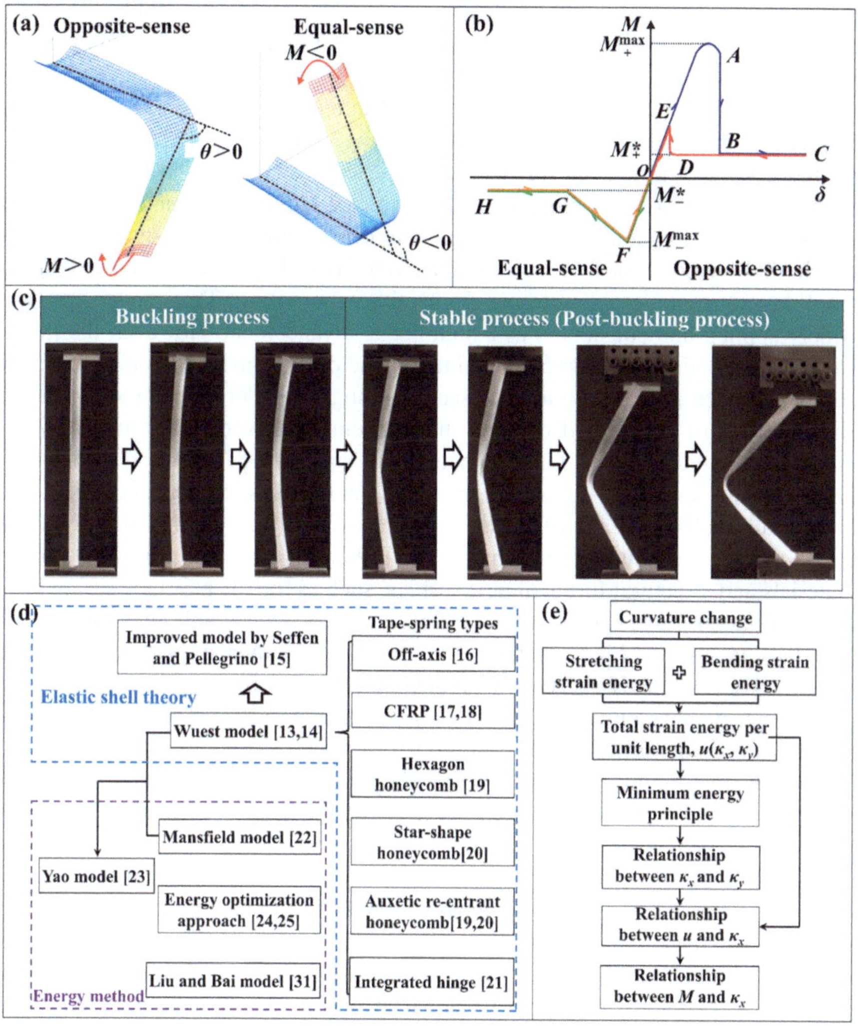

Fig. 1.3 Folding behavior of the deployable composite tape spring: (**a**) Illustration of the senses of folding and sign convention [11]. (**b**) Analytical evolution of the folding moment with the applied bending angle [11]. (**c**) Opposite-sense folding process [12]. (**d**) Analytical system of analytical models. (**e**) The modeling process by Liu and Bai [34]

then returns to the loading curve at point E. It subsequently linearly recovers back to point O (EO). In the equal-sense folding process, starting from O again, the folding moment shows a linear relationship with the bending angle (OF), exhibiting the same stiffness as the opposite-sense folding. The folding moment typically reaches its maximum value at relatively small bending angles, known as the peak moment during the equal-sense folding process. Then, similar to the opposite-sense folding, the post-buckling behavior (GH) is defined by the steady-state moment M_{-}^{*}. Finally, it is generally assumed that the unloading path of the equal-sense folding theoretically matches the loading path, although it may not be an exact match in practice [11, 12].

This section aims to introduce analytical models for predicting the folding behavior of the deployable composite tape spring. Currently, there are two analytical models (the elastic shell method and the energy method) used to describe the folding behavior of the deployable composite tape spring, particularly the folding moment versus longitudinal curvature relationship. Based on the small-deflection elastic shell theory, Wuest [13, 14] first derived the folding moment versus longitudinal curvature curve for isotropic deployable tape spring. This model has been widely applied and considered as a significant advancement. However, the model has some limitations, such as underestimating the peak moment and only predicting deployable tape spring with a cross-sectional angle smaller than 1 radian. Seffen and Pellegrino [15] pointed out that the moment balance equation in the Wuest model is valid only for the flattened state of the tape spring. To extend the applicability to tape springs with larger cross-sectional angles, Seffen and Pellegrino numerically solved the modified integral limits of the moment balance equation. Consequently, the Wuest model has been further developed to apply to various materials and structural forms of tape springs, such as the off-axis tape spring [16], deployable composite tape spring [17, 18], hexagon honeycomb tape spring [19], star-shape tape spring [20], auxiliary reentrant honeycomb tape spring [19, 20], and integrated tape spring [21]. These analytical models for different structural forms of tape springs do not significantly differ in modeling approach compared to the Wuest model but consider different material properties.

Shell energy minimization is another widely used method for predicting the folding moment versus curvature relationship of deployable tape springs. Mansfield [22] proposed an analytical model for predicting the folding behavior of isotropic deployable tape springs under large-deflection conditions. This model considers both stretching and bending strain energy, establishes a total strain energy expression based on the curvature variation of the deployable tape spring during the folding process, and obtains the quantitative relationship between folding moment and longitudinal curvature through variational minimization. In addition, Mansfield derived an analytical model for the torsional buckling of the tape spring, i.e., buckling occurs under equal-sense bending load without end effects. Yao et al. [23], combining the Wuest model and the Mansfield model, based on Calladine shell theory and von Karman thin plate large deflection bending theory, derived the expression for the folding moment versus curvature relationship of the deployable composite tape spring using the energy method. This model considers the influence of transverse bending, longitudinal bending, and longitudinal tension on the elastic strain energy. Recently, Knott [24] and Shore [25] proposed new energy optimization analytical models for analyzing the folding moment versus curvature relationship of the deployable composite tape spring. One key feature of this model is the use of a single path variable to describe the cross section of the tape spring. Comparative analyses have shown that the predictions of this method are in good agreement with the Wuest and Mansfield models [25]. The aforementioned analytical models have made significant progress in predicting the folding behavior of the deployable composite tape spring, especially in terms of the folding moment versus curvature relationship. However, accurately predicting the folding behavior of these structures relies on determining their geometric configuration during the folding process. Iqbal et al. [26] hypothesized that the cross section of the deployable composite tape spring remains circular during the folding deformation. This hypothesis has also been validated by other literature [27–30]. Inspired by this work, Liu and Bai [31] developed a novel analytical model for predicting the folding performance of deployable composite tape springs based on the same assumption. The model encompasses geometric configurations, relationship between folding moment and longitudinal curvature, and stress levels. Compared to the Wuest model [13, 14], which is based on elastic shell theory, and the Yao model [23], which employs the energy method, this new model demonstrates higher predictive accuracy. Detailed discussions are provided in subsequent sections. The analytical modeling approach for the folding behavior of the deployable composite tape spring is summarized in Fig. 1.3d, which presents the analytical framework.

In addition, by changing the boundary conditions and loading methods of the deployable composite tape spring, axial compression folding deformation can also be achieved. This folding method of the deployable composite tape spring can be applied to structures such as hinge [32] and cabin [33]. However, there are relatively few analytical modeling studies on this folding method. Liu and Bai [34] proposed an analytical model based on the Euler-Bernoulli beam theory to predict the load versus displacement curve, shape function, and stress level during the axial compression folding of the deployable composite tape spring. The study showed that the load versus displacement curve exhibits a three-stage nonlinear behavior similar to bending folding behavior, and the model was validated through experimental and numerical analysis [33]; the modeling flowchart is shown in Fig. 1.3e.

Based on the above summary, it can be concluded that the elastic shell method and energy method show a trend of evolution in the space application to the deployable composite tape spring. The focus of research lies in the folding behavior of the deployable composite tape spring, particularly the folding moment versus longitudinal curvature relationship. Existing analytical models can predict nonlinear characteristics similar to experiments, but there is a common issue of underestimating the peak moment. Currently, most analytical models mainly predict the opposite-sense folding behavior of the deployable composite tape spring since both bending and torsion occur during the equal-sense folding deformation [13–15]. Therefore, establishing an analytical model that accurately predicts the equal-sense folding behavior of the deployable composite tape spring is one of the key areas for future research.

1.2.2 Integral Deployable Composite Hinges

Due to simplicity and ease of use, the deployable composite tape spring has been widely employed in the design of space deployable structures. However, they exhibit certain limitations in terms of mechanical properties, such as relatively low bending stiffness and load-bearing capacity. To overcome these issues, integrated deployable composite hinges (DCHs) have been developed. An integrated deployable composite hinge is a cylindrical shell structure with symmetric openings and an integral hinge mechanism that enables 180° folding functionality (shown in Fig. 1.4a) [35, 36]. Due to the structural integrity, it possesses features like easy fabrication and fewer connection points. However, this type of hinge experiences stress concentration during folding, and repeated folding can lead to damage or failure. In 2003, the European Space Agency (ESA) launched the Mars Advanced Radar for Subsurface and Ionosphere Sounding (MARSIS) aboard the Mars Express spacecraft, which was the first radar antenna employing an integrated deployable composite hinge (shown in Fig. 1.4b) [37]. This marked the inaugural application of an integrated deployable composite hinge in space, generating significant interest among researchers and driving further research and broader applications of this hinge. The folding

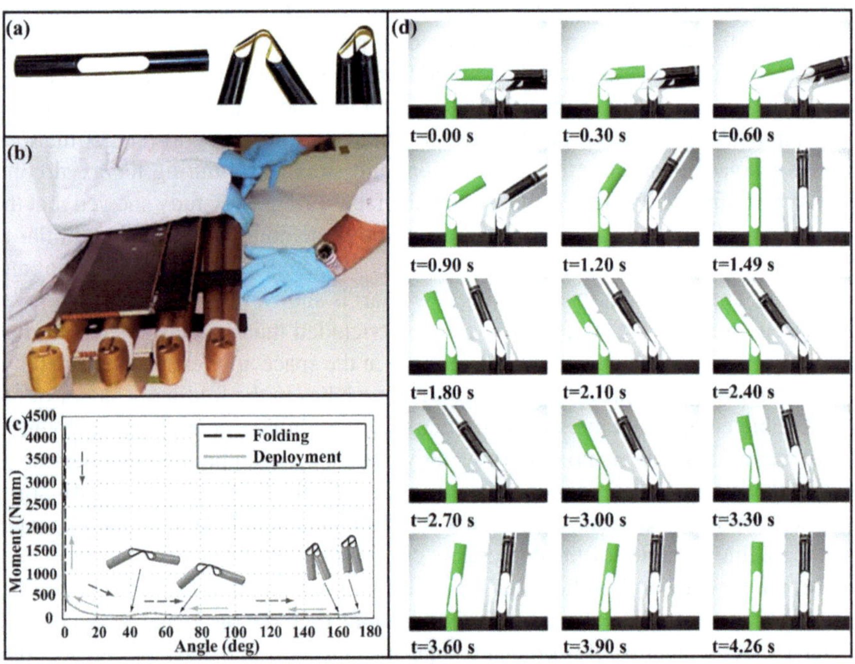

Fig. 1.4 Folding behavior of integrated deployable composite hinges: (**a**) Deployed, partially folded and fully folded configuration [36]. (**b**) MARSIS [38]. (**c**) Folding moment-bending angle profile for folding and deployment simulation [44]. (**d**) Comparison of snapshots from experiment and simulation [49]

behavior of integrated deployable composite hinges that needs to be focused is similar to that of the deployable composite tape spring, including folding moment-bending angle curve, peak moment, propagation moment, damage or failure, etc. [38–47]. Folding moment versus bending angle curve of integrated deployable composite hinges also exhibit a typical three-stage characteristic [44], as shown in Fig. 1.4c. It is worth noting that most of the aforementioned studies are based on experimental and numerical methods because the geometric nonlinearity and shell contact during the folding deformation of integrated deployable composite hinges make analytical modeling challenging. In addition to the folding deformation, the dynamic impact response during the rapid deployment process of integrated deployable composite hinges is another crucial consideration. Some literature has numerically modeled the deployment process of integrated deployable composite hinges and used high-speed cameras to measure the varying angle versus time (shown in Fig. 1.4d) [44, 46–49].

Integrated deployable composite hinges typically experience maximum stress (i.e., the critical location) near the end of the slot [44, 46, 47], as shown in Fig. 1.5a, which is a stress concentration area. To address this issue, Mallikarachchi and Pellegrino [50–52], as well as Mobrem et al. [37], introduced circular cutouts at both ends of the slot in the integrated deployable composite hinge (shown in Fig. 1.5b). The folding and deployment process of this new design was investigated through experiment and numerical simulation. The research revealed that the circular cutouts effectively reduce the stress level. However, in certain design configurations, failure may occur at the location of the circular cutout, as shown in Fig. 1.5c. To further reduce the folding radius and stored strain energy in the integrated deployable composite hinge, Sakovsky et al. [35, 53] proposed a dual-matrix integrated deployable composite hinge (illustrated in Fig. 1.5d). The folding and deployment process of this hinge was studied using experiment and numerical simulation, as shown in Fig. 1.5e. The predicted geometric configurations matched well with the actual configurations. It is worth noting that the aforementioned integrated deployable composite hinges are all based on a straight tube configuration. Ferraro and Pellegrino [54, 55] introduced a novel integrated deployable composite hinge with an angled shape, where two thin-walled cylinders intersect at a 90-degree angle (shown in Fig. 1.5f).

The mentioned researchers primarily conducted comprehensive studies on the folding and deployment behavior of symmetrically open integrated deployable composite hinges using experimental and numerical methods. However, they also identified stress concentration issues in these hinges during the folding process, which could lead to damage or failure upon repeated folding. By optimizing the shape and parameters of the cutout, the reliability of integrated deployable composite hinges can be improved. Future research can further explore the application of integrated deployable composite hinges in complex space environments and enhance their bending stiffness and load-carrying capacity through optimized designs. Additionally, attention should be given to hinge durability and damage prediction, along with exploring their potential application in complex folding mechanisms.

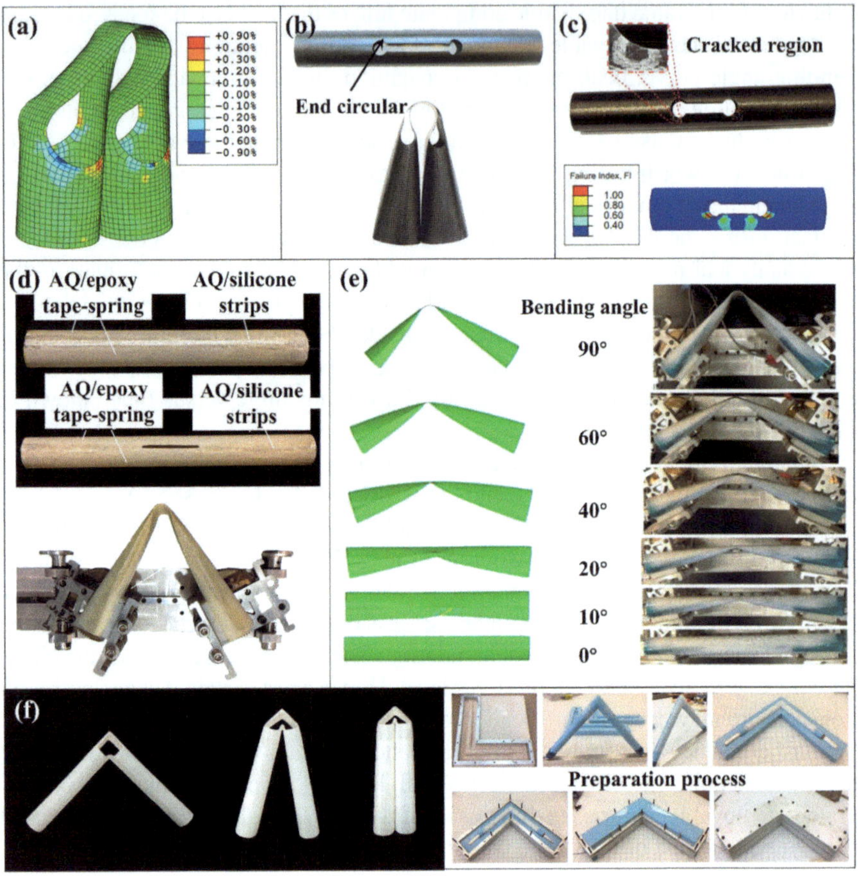

Fig. 1.5 Folding behavior of other types of integral deployable composite hinges: (**a**) Strain contour plot in the folding state [47]. (**b**) Integral deployable composite hinges with end circular [50]. (**c**) Failure at the end circular position after folding [52]. (**d**) Dual-matrix integral deployable composite hinge and folding apparatus [53]. (**e**) Comparison of experimentally observed and simulated deployments of dual-matrix integral deployable composite hinge [53]. (**f**) Folding process of deployable composite hinge with diagonal geometry and its manufacturing process with silicone molding [54]

1.2.3 Combined Deployable Composite Hinges

Combined deployable composite hinges have also gained increasing attention. They are typically composed of multiple deployable composite tape springs arranged in opposite or back-to-back configurations. These hinges offer higher bending stiffness and load-carrying capacity compared to deployable composite tape spring alone. Furthermore, they can be structurally and material optimized to meet specific application requirements. Echter et al. [56] and Pellegrino et al. [57, 58] proposed an opposite configuration for combined deployable composite hinges. In this design, two opposing deployable composite tape springs are secured together using end components, as shown in Fig. 1.6a, b. Francis et al. [59, 60] and Fosness [61] developed a back-to-back deployable composite hinge (also known as an X-hinge), as depicted in Fig. 1.6c. When deployed, this hinge aims to have equal bending stiffness in two directions perpendicular to the longitudinal axis of the hinge. Other combined spring-based deployable hinges have also received attention [62–70]. Figure 1.6d–g shows several representative examples of such hinges. In 2009, the United States Air Force Research Laboratory [71, 72] proposed a cross-shaped deployable composite hinge for use in mesh reflector antennas (shown in Fig. 1.6h). When combined with support rods, this hinge can be utilized in the design of various one-dimensional and two-dimensional structures.

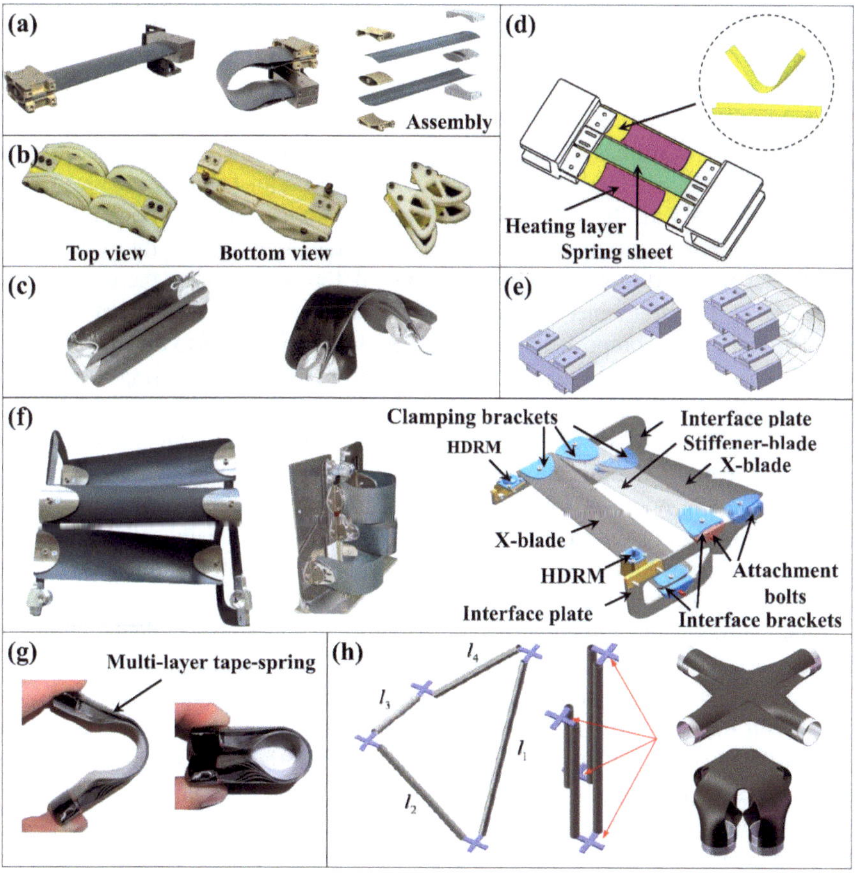

Fig. 1.6 Folding behavior of combined deployable composite hinges: (**a**) CAD model of equal-type hinge [57]. (**b**) Tape spring rolamite hinge [58]. (**c**) X-shape hinge [61]. (**d**) Hybrid intelligent hinge [63]. (**e**) Combined hinge with four tape springs [65]. (**f**) Ultralight deployment mechanism [69]. (**g**) Multilayer tape spring hinge [70]. (**h**) Four-cornered hinge [71]

1.3 Deployable Composite Booms

1.3.1 Tubular Deployable Composite Booms

Similar to the deployable composite tape spring, tubular deployable composite boom (DCB) is a thin-shell structure with uniform transverse curvature [73]. However, the cross-sectional angle is usually larger than 180°, enabling it to undergo coiling and folding deformation. The tubular deployable composite boom can be folded in two ways: equal-sense folding and opposite-sense folding, as shown in Fig. 1.7a, b. The tubular deployable composite boom can be not only compactly packaged through folding but also restored to its initial configuration by releasing the stored strain energy during the folding process. It serves as fundamental components for constructing space deployable structures, offering advantages such as lightweight, high stiffness, and repeatability of deployment [74]. Additionally, it is a crucial structure for space applications such as deployable mesh reflector antennas, deployable solar sails, thin-film antennas, and space probes [75, 76]. Regarding the folding behavior of the tubular deployable composite boom, the following parameters are of primary concern: (1) Folding moment: this refers to the moment required to coil the tubular deployable composite boom from its deployed state using an auxiliary mechanism. Accurate prediction of the folding moment is crucial for the design of the auxiliary mechanism. (2) Stored strain energy: this refers to the strain energy stored in the structure due to its deformation during the folding process. The stored strain energy significantly affects the deployment behavior of the tubular deployable composite boom. (3) Stress level: this refers to the magnitude of stress experienced by the tubular deployable composite boom during the folding process. It is used for strength design considerations.

In 1961, Klein designed the thin-walled storable tubular extendible member (STEM) for the communication antenna of Canada's first satellite, the Alouette. This deployable boom is folded and stored in corresponding constraints before launch while being fixed to the satellite. After launch, it gradually extends outward until fully deployed through motor control. The successful application of the Alouette satellite laid the foundation for the development of tubular deployable booms. Subsequently, many scholars established various analytical models to describe the folding behavior of tubular deployable composite booms. Early studies focused on using isotropic materials to fabricate deployable booms. Based on the plate shell theory, Rimrott et al. [77–80] conducted extensive analysis on various mechanical behavior of deployable booms, including bending stiffness, torsional stiffness, ploy zone length, stress distribution, and stored strain energy. Iqbal et al. [26] proposed an analytical model to predict the stored strain energy of the tubular bistable deployable composite boom (Bi-DCB) during the folding deformation process, considering the influence of stretching strain energy and bending strain energy. This model has been widely used and developed into various versions to analyze the stored strain energy, folding moment, and stress level of the tubular deployable composite boom. For example, Wang et al. [81] developed the Iqbal's model and

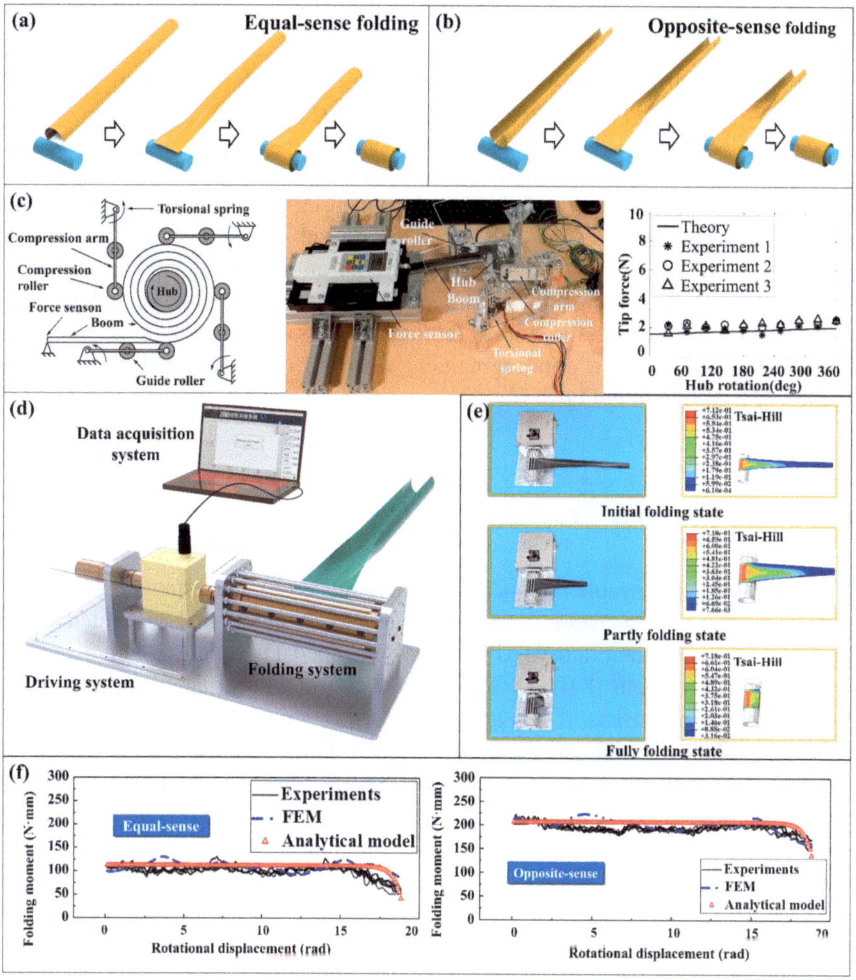

Fig. 1.7 Folding behavior of the tubular deployable composite boom: (**a**) Equal-sense folding. (**b**) Opposite-sense folding. (**c**) Folding and deploying experimental setup and tip force during blossoming [81]. (**d**) Folding machine. (**e**) Folding deformation processes of the tubular deployable composite boom through experiments and numerical simulations. (**f**) Folding moment versus rotational displacement curves in the folding processes

used the energy method to model the folding and deploying process of the tubular deployable composite boom. The model was used to predict the tip force that the boom can withstand before blossom occurs, as shown in Fig. 1.7c. In addition, experimental setup was developed to realize the folding and deploying process of the tubular deployable composite boom. The experimental results were compared with the predictions of the analytical model, and they showed good agreement. Liu and Bai [82] considered the boundary conditions and geometric configurations of the tubular deployable composite boom under realistic working conditions and developed a folding machine to achieve equal-sense and opposite-sense folding processes of the tubular deployable composite boom. This folding machine can measure the driving moment during the deformation process of the tubular deployable composite boom, including folding system, driving system, and data acquisition system, as shown in Fig. 1.7d. The folding behavior of the tubular deployable composite boom was analyzed through experiments and numerical simulations, and the predicted folding moment-rotational displacement curve showed good agreement with the experiments, as shown in Fig. 1.7e. In subsequent work, based on the energy method and two classical failure criteria, an analytical model was established to predict the folding moment and stress level of the tubular deployable composite boom throughout the entire folding process. The predictions using the analytical model were in good agreement with the numerical and experimental results [83], as shown in Fig. 1.7f.

1.3.2 No-Circular Cross-Section Deployable Composite Booms

In recent years, deployable booms have undergone continuous innovation in terms of their cross-sectional forms, attracting widespread attention. The United States Air Force Research Laboratory (AFRL) has developed a triangular rollable and collapsible (TRAC) boom [84, 85]. The cross section of the triangular rollable and collapsible boom consists of two arcs connected on one side, forming two curved edges and a flat belly plate, as shown in Fig. 1.8a. This boom has been used in three different solar sail missions, namely NanoSail-D [86], The Planetary Society's LightSail-1 [87], and LightSail-2 [88]. In all three cases, the triangular rollable and collapsible booms were made of metal. Recent studies have shown that the metal triangular rollable and collapsible boom is sensitive to temperature gradients. In space, when one edge faces the sun while the other remains in shadow, significant end deflections can occur [89]. Therefore, in recent years, extensive research has been conducted on the triangular rollable and collapsible boom made of carbon fiber-reinforced composites. The preparation of composite triangular rollable and collapsible boom involves a secondary bonding process. First, two half-sections are cured, and then they are bonded together. Figure 1.8b shows the fabrication process of the triangular rollable and collapsible boom. Lecler [90] proposed an analytical model to predict the buckling load of the triangular rollable and collapsible boom under pure bending and designed a buckling test setup (shown in Fig. 1.8c). Buckling modes and locations have been investigated through experiments, numerical simulations, and analytical models [91] (shown in Fig. 1.8d). Lecler et al. [92–94] and Luo [95] analyzed stress concentration and material failure in the ploy zone of the triangular rollable and collapsible boom using analytical and numerical methods. Inspired by the design of the TRAC boom, Yang et al. [96, 97] proposed the concepts of N-shaped and M-shaped deployable composite booms (shown in Fig. 1.8e, f) and studied the strain energy stored during the folding process as well as the peak bending moment.

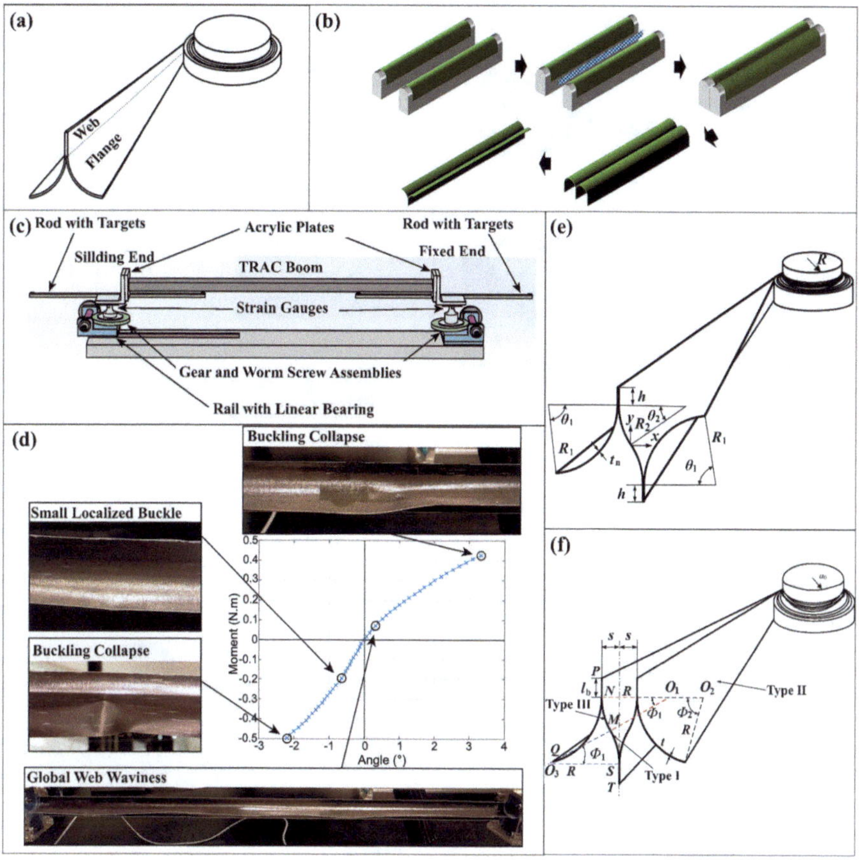

Fig. 1.8 No-circular deployable composite booms: (**a**) Triangular rollable and collapsible boom [84]. (**b**) Manufacturing process of triangular rollable and collapsible boom [90]. (**c**) Bending experimental setup (the sliding end is mounted on a rail to allow longitudinal translation) [90]. (**d**) Buckling modes under equal-sense and opposite-sense moments [91]. (**e**) N-shaped deployable composite boom [96] (**f**). M-shaped deployable composite boom [97]

It is worth noting that a key characteristic of the aforementioned booms is their open cross section, which typically exhibits lower torsional stiffness. In 1999, the German Aerospace Center (DLR) first used carbon fiber-reinforced composites to fabricate a lenticular deployable composite boom [98], consisting of two axially symmetric Ω-shaped carbon fiber composite shells, as shown in Fig. 1.9a, b. Due to its closed cross section, the lenticular deployable composite boom has higher torsional stiffness. There are numerous space applications for the lenticular deployable composite boom, including rollable rib antennas developed by Oxford Space Systems Limited (OSS), solar sail spacecraft by the ESA and DLR, and solar sail spacecraft used by NASA in deep space science and exploration missions [99]. Bai et al. conducted analytical analyses on various key performance aspects of the lenticular deployable composite boom, including nonlinear mechanical response

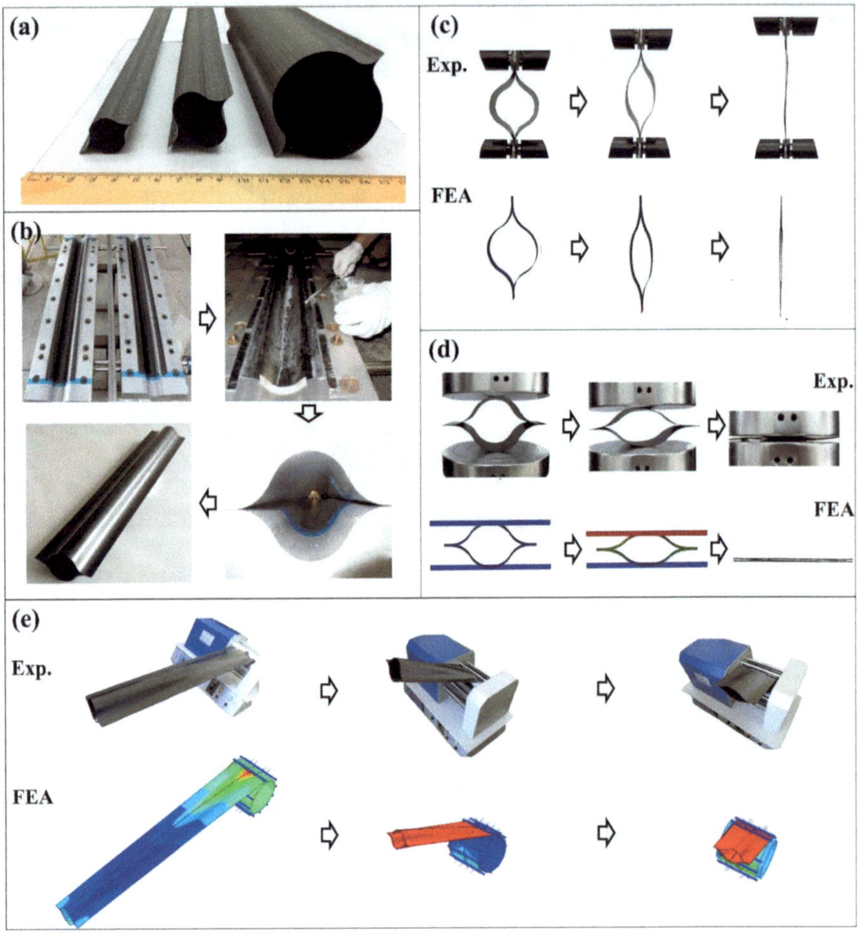

Fig. 1.9 Lenticular deployable composite booms [106]: (**a**) The samples. (**b**) Manufacturing process. (**c**) Tensile deformation process determined by the Exp. and FEA. (**d**) Compressive deformation process determined by the Exp. and FEA. (**e**) Coiling deformation process determined by the Exp. and FEA

during tensile and compressive flattening processes [100–102], neutral cross-section position in the flattened state [103], in-plane strain and interlaminar shear stress during folding [104], and folding moment and stored strain energy [105, 106]. Figure 1.9c–e, respectively, depicts the tensile, compressive, and folding deformations of the lenticular deployable composite boom determined by experiments and numerical simulations. The mechanical behavior of the lenticular deployable composite boom mainly focuses on buckling, stiffness, and thermal effect in space environment, such as local buckling under axial compressive loading [107] (shown in Fig. 1.10a), global Euler buckling [108], nonlinear buckling and post-buckling

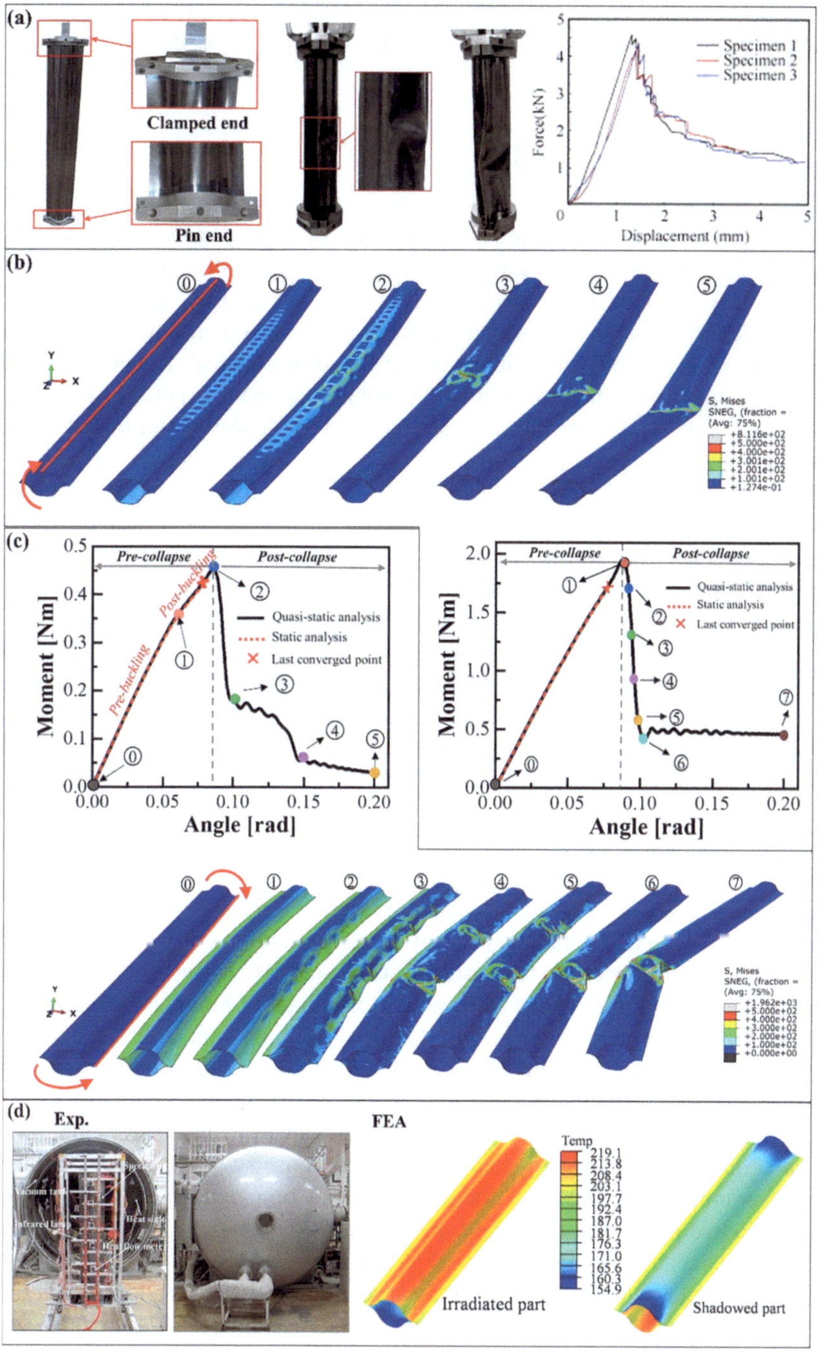

Fig. 1.10 Buckling and thermal behavior of the lenticular deployable composite boom: (**a**) Subjected to axial compression [107]. (**b**) Subjected to a pure bending load about x-axis [109]. (**c**) Subjected to a pure bending load about y-axis [110]. (**d**) Temperature field and thermal deformation in extreme environment [111]

under pure bending conditions [109, 110], and temperature field and thermal deformation in extreme environment [111] (as shown in Fig. 1.10b–d).

1.3.3 Bistable Deployable Composite Booms

The aforementioned deployable composite booms typically have only one deployed stable state, while the folded state usually possesses high strain energy and cannot stay stable without external constraints. Therefore, mechanisms are required to control the deployment process, which increases system mass and the risk of deployment failure. By employing suitable material and mechanical designs, deployable composite booms can be designed with bistable functionality [112, 113], meaning they possess two stable equilibrium states in both the deployed and folded configurations (shown in Fig. 1.11a). Bistable deployable composite booms can maintain a folded stable state without any restraining devices and exhibit lower strain energy in the folded stable state, demonstrating more controllable self-deployment characteristics [114–116]. As a result, bistable deployable composite booms hold significant potential for a wide range of applications.

Since the initial proposal of the tubular bistable deployable composite boom by British scholar Daton Lovett in 1996, Cambridge University has been at the forefront of related research in this field. Various analytical models have been developed to predict the geometrical configuration, stored strain energy, and stress levels of the tubular bistable deployable composite boom in the folded stable state. Figure 1.11b provides an overview of the evolution and development of the most representative analytical models. Based on the classical laminate theory (CLT) and principle of minimum potential energy, Iqbal et al. [26] successfully established a simple linear elastic analytical model and derived expressions for the strain energy as a function of longitudinal and transverse curvatures, as well as the cross-sectional angle. This model can be used to predict the cross-sectional radius of the tubular bistable deployable composite boom in the folded stable state. However, the bistable behavior exhibited by symmetric and antisymmetric laminates cannot be distinguished by this model. Galletly and Guest [27] further developed a "beam model" based on the Iqbal model, considering all possible deformation modes, including torsion, shear, and stretch-bend coupling effects. The analytical predictions are consistent with numerical simulations but show significant discrepancies with experimental results. In subsequent studies, Galletly and Guest [28] proposed a "shell model" that abandoned assumptions about cross-sectional shapes. Compared to the "beam model," the predictive performance of the "shell model" is essentially the same, but still presents noticeable deviations from experimental results. This difference may be attributed to plastic deformation of the matrix at high strains, while the material is assumed to be linear elastic throughout in the analytical models. Additionally, Guest and Pellegrino [29] introduced a two-parameter analytical model assuming a uniformly distributed deformation of the shell, with zero Gaussian curvature at all times. This model can determine the cross-sectional radius of the tubular bistable

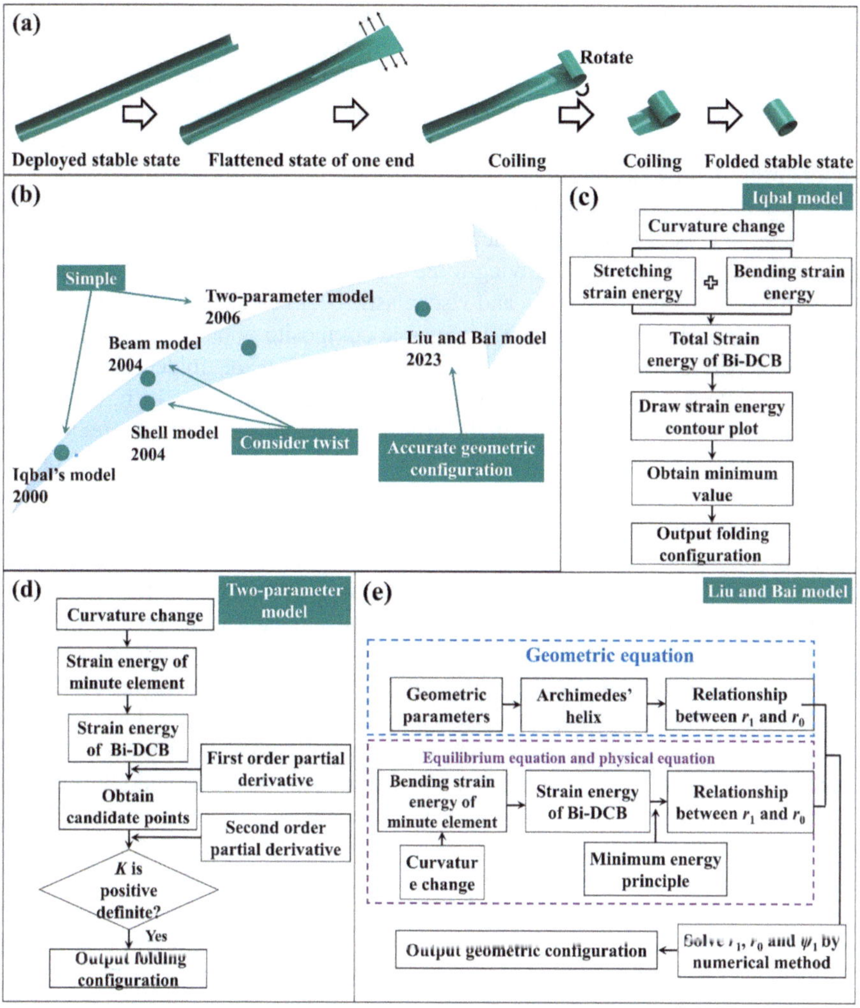

Fig. 1.11 Tubular bistable deployable composite booms: (**a**) Bistable deformation process. (**b**) Evolution and development process of analytical models. (**c**) Modeling process of Iqbal model. (**d**) Modeling process of two-parameter model. (**e**) Modeling process of Liu and Bai model

deployable composite boom in the folded stable state, and its predictions align with those of the aforementioned three analytical models. However, these four analytical models still cannot accurately predict the true geometrical configuration of longer tubular bistable deployable composite boom in the folded stable state, as their cross-sectional geometry more closely resembles an Archimedes' helix rather than a circle. Therefore, based on the works of Iqbal et al. [26] and Guest et al. [29], Liu and Bai [117, 118] introduced more precise geometric models to describe the folded stable state (arbitrary length and number of coils) of tubular bistable deployable composite booms. An analytical model was established to predict the folding behavior, including stored strain energy, geometric configuration, and stress levels of the tubular bistable deployable composite boom. Based on this, Liu and Bai further analyzed the effect of temperature on the folded stable state through analytical modeling and numerical simulation [119]. Figure 1.11c–e illustrates the modeling process of the representative Iqbal model, two-parameter model, and improved model proposed by Liu and Bai.

The deployment behavior of bistable deployable composite booms, which are of particular interest for space applications, needs to be carefully examined. This includes stability [120, 121], driving force [122], deployment speed [123], deployment deformation process [124], and viscoelastic behavior [125–141]. Additionally, the ploy region length of bistable deployable composite booms should be considered, which refers to the length of the region between the folded state and the deployed state. The length of the ploy region can range from a few inches to several feet and has a significant impact on the efficiency of folding. Furthermore, the ploy region can potentially weaken the final structure, necessitating additional support structures for reinforcement. Yang et al. [142] hypothesized that longitudinal and transverse curvatures are the primary factors influencing the ploy region and derived a function describing the variation of energy with curvature along its length. They found that as the curvature approaches the final deployed curvature, the total energy gradually increases. This model provides a convenient way to predict the ploy region length and curvature distribution. Based on the analytical model proposed by Rimrott et al. for isotropic tubular deployable boom, Wang and colleagues [143–145] theoretically predicted the ploy region length of bistable deployable composite booms. Rakow and Reedy [146] compared these two analytical models through experiments and numerical simulations and discussed the parameters influencing the ploy region length. These studies highlight the importance of understanding and predicting the behavior of the ploy region in order to optimize the design and performance of bistable deployable composite booms.

Apart from the tubular bistable deployable composite boom, several researchers have utilized the analytical model established by Guest and Pellegrino [29] along with numerical simulation to design various other configurations of bistable deployable composite booms, for instance, the parabolic bistable deployable composite boom [147–149], the bistable deployable composite boom with initial longitudinal curvature [150, 151], the bistable deployable composite boom with flat region [152], the lenticular bistable deployable composite boom [153–155], and bistable triangular rollable and collapsible boom [156, 157].

1.4 Deployable Composite Reflectors

Deployable composite reflector antennas are a common type of antenna in the space field. It is made of composite materials and possesses high flexibility and toughness. Deployable composite reflector antennas have been used for wide applications in spacecraft. For instance, they can be used as a high-gain antenna in communication satellites, a radar antenna in earth observation, a microwave radiometer in remote sensing, and a radio telescope in scientific exploration. Additionally, deployable composite reflector antennas can also be utilized for adaptive deformation and vibration control of spacecraft. Based on the folding and deploying methods, deployable composite reflector antennas can be classified into the taco-shaped type, petal-shaped type, umbrella-shaped type, and wrap-rib type.

1.4.1 Taco-Shaped Deployable Composite Reflectors

The taco-shaped deployable composite reflector antenna, also known as the spring-back reflector antenna, is a common type of antenna in the space field. The reflector of this antenna is made of composite materials and features a secondary backplane and support structure to provide deployment stiffness and accuracy. During launch, a restraint device is used to maintain the folded configuration. Once in orbit, the restraints are released, and the antenna automatically deploys. The taco-shaped deployable composite reflector has been successfully applied to communication antennas on the satellite by the Canadian Space Agency (CSA) [158], as shown in Fig. 1.12a. However, the deployment stiffness of the taco-shaped deployable composite reflector is relatively low. To address this issue, Tan et al. proposed a method of locally reinforcing the reflector [159, 160], as shown in Fig. 1.12b. Research has shown that local reinforcement significantly increases the stiffness and accuracy of the reflector without adding mechanical complexity to the system. This type of reflector meets the stringent requirements of high-precision communication while offering advantages such as low cost, high-dimensional accuracy, and ease of folding. For Ku-band missions, Türkmen et al. designed a full-sized off-axis reflector and considered different reinforcement schemes (shown in Fig. 1.12c) [161]. The study reveals that this design has advantages in terms of mass, natural frequency, number of components, production, and assembly. Subsequent work involved the validation of its performance through testing the modal behavior, folding, dynamic deployment, and surface accuracy of prototype samples [162]. Soykasap et al. developed a taco-shaped deployable composite reflector based on composite tape springs as the supporting structure [163]. The study demonstrates that design variables such as initial curvature of the tape springs, projected length, and three-dimensional folding angle have significant influences on its bending behavior. However, further research is needed to investigate the effects of the connection between the tape spring and reflector on the entire folding and deploying processes. Figure 1.12d illustrates the folding and deploying processes of this reflector. Tan et al. analyzed a taco-shaped deployable composite reflector with curved reinforcing ribs [164]. It is found that by utilizing curved reinforcing ribs, local instability can be prevented, the maximum packing force required for the reflector can be increased, and its bending stiffness can be improved. Soykasap et al. presented a design scheme for an X-band satellite taco-shaped deployable composite reflector, which was experimentally validated for folding and deployment functionality

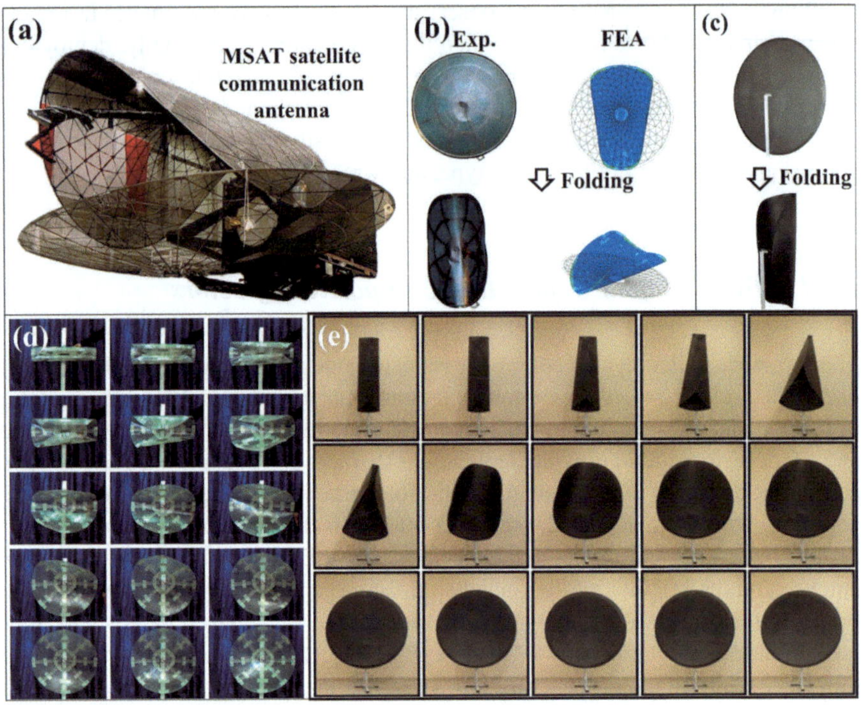

Fig. 1.12 Taco-shaped deployable composite reflectors: (**a**) Spring-back reflector in deployed and folded configurations (Courtesy of the Canadian Space Agency http://www.space.gc.ca) [158]. (**b**) Stiffened spring-back reflector [160]. (**c**) Predesign reflector made of three-ply plain weave carbon/epoxy ([0°/45°/0°]) [161]. (**d**) Ultrathin shell deployable reflector based on curved large tape springs [163]. (**e**) Dynamic deployment of self-deployable flexible composite shell reflector [165]

(shown in Fig. 1.12e) [165]. It should be noted that this type of deployable composite reflector also has some limitations, such as difficulties in controlling dimensional accuracy, lower folding efficiency, and potential for sudden release of stored strain energy causing deployment shocks, thus restricting its applications.

1.4.2 Petal-Shaped Deployable Composite Reflectors

The petal-shaped deployable composite reflector antenna is typically composed of a reflector and a feed source at the focal point. The reflector features a petallike shape; hence, it is referred to as a petal-shaped reflector [166] (shown in Fig. 1.13a). This type of reflector exhibits high mechanical complexity to achieve the required accuracy during the folding and deployment process. To meet the requirements of deployable antennas with aperture sizes ranging from 0.3 m to 1 m, Stilesh and Garrett [167] proposed a radial segmented winding scheme, which divides the flexible shell surface into multiple segments along a spiral path. Through the design of

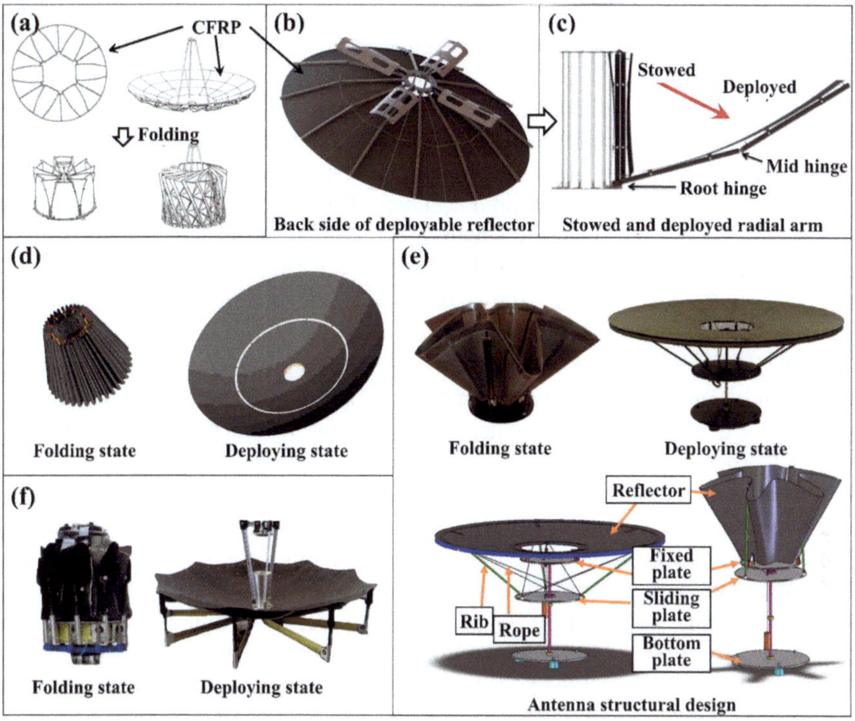

Fig. 1.13 Petal-shaped deployable composite reflectors: (**a**) Early reflector schematic diagram [166]. (**b**) Back side of multi-arm radial composite reflector [169]. (**c**) Folded and deployed multi-arm radial composite reflector [169]. (**d**) Flexible precision reflector [170]. (**e**) Breadboard reflector [171]. (**f**) Origami deployable reflector [172]

the spiral shell surface, each segment remains purely curved in the folded configuration, avoiding the shape accuracy variation caused by shear loads. To achieve synchronization during the deployment process, a scissor-hinge-like connection scheme is employed along the circumferential edge of each segment. However, due to the weak stiffness of a single shell segment, significant deformations can occur under the influence of gravity; thus, the prototype was only subjected to deployment verification tests on a forming mold. Enhancing the stiffness of the individual shell segments and controlling the deployment behavior are important issues to be addressed for petal-shaped deployable composite reflectors. Reynolds et al. [168] designed a highly compact petal-shaped deployable composite reflector that can be stored within a 1 U CubeSat. They developed a radial segmented winding method and utilized constraints to limit the reflector, resulting in the desired load characteristics of the planar structure. Footdale et al. [169] developed a multi-arm radial composite (MARCO) reflector to reduce costs (shown in Fig. 1.13b). This reflector employs segmented composite shell bodies as the main part and a simplified assembly technique to lower the costs. The deployed diameter of this reflector is 4 m, and it can be stowed within a cylindrical container with a diameter of 0.66 m and a height of 1.1 m (shown in Fig. 1.13c). Experimental validation was conducted to demonstrate that the shape and gravity-induced deformations of the reflector meet the accuracy requirements for surface profiles.

Petal-shaped reflector antennas offer high gain, high directivity, wideband characteristics, excellent interference resistance, and a high signal-to-noise ratio. However, such reflectors require highly complex mechanical structures for folding and deploying to achieve the desired accuracy. The use of mechanical components introduces thermal deformations, which further complicate system design. To effectively address the limitations of petal-shaped reflector antennas, Harris Corporation has developed a novel type of petal-shaped deployable composite reflector antenna, also known as a flexible precision reflector antenna [170] (shown in Fig. 1.13d). This flexible precision reflector is composed of thin composite shells and can be folded in an umbrellalike folded manner. By dividing the reflector into multiple panels, a larger diameter can be achieved while ensuring high precision and flatness. Harris Corporation also developed a lightweight and structurally simple flexible precision reflector antenna with a diameter of 0.9 m (shown in Fig. 1.13e), which operates in frequencies above 40 GHz [171]. The feasibility of this design approach has been validated through the fabrication and testing of a petal-shaped reflector antenna model. Dufour et al. [172] reported a new concept for a flexible precision reflector antenna, as shown in Fig. 1.13f. Compared to traditional fixed reflector antennas, this deployable design offers smaller folding dimensions and lighter weight. Additionally, the use of origami-inspired techniques enables higher surface accuracy, thereby improving RF performance. The reflector is suitable for antennas ranging from 0.5 m to 2 m in diameter and can operate at high frequencies in the Ka-band, providing more options for communication and scientific missions.

1.4.3 Umbrella-Shaped Deployable Composite Reflectors

Umbrella-shaped deployable composite reflector antennas are characterized by low areal density and compact stowage volume. They can operate in a high-frequency range and achieve a large deployment diameter. Compared to traditional flexible reflectors, umbrella-shaped deployable composite reflectors can withstand larger stowage strains without creep and maintain high accuracy with fewer ribs (shown in Fig. 1.14a) [173].

To meet the requirements of spaceborne synthetic aperture radar (SAR) antennas, Schmid and Barho [174] proposed an umbrella-shaped deployable composite reflector scheme with a 3-m aperture. The reflector is assembled from 30 thin-shell panels, each with a thickness of approximately 0.4 mm and a surface accuracy of around 0.5 mm. The weight of each panel is approximately 200 g. Folding and deploying of the panels are achieved by arranging radial support ribs on the back side. When folded, the back-support ribs rotate around the axis, causing the shells to tilt and undergo pure bending, allowing for overlapping of adjacent panels and improving stowage efficiency. For example, the stowage diameter of a 3-m aperture

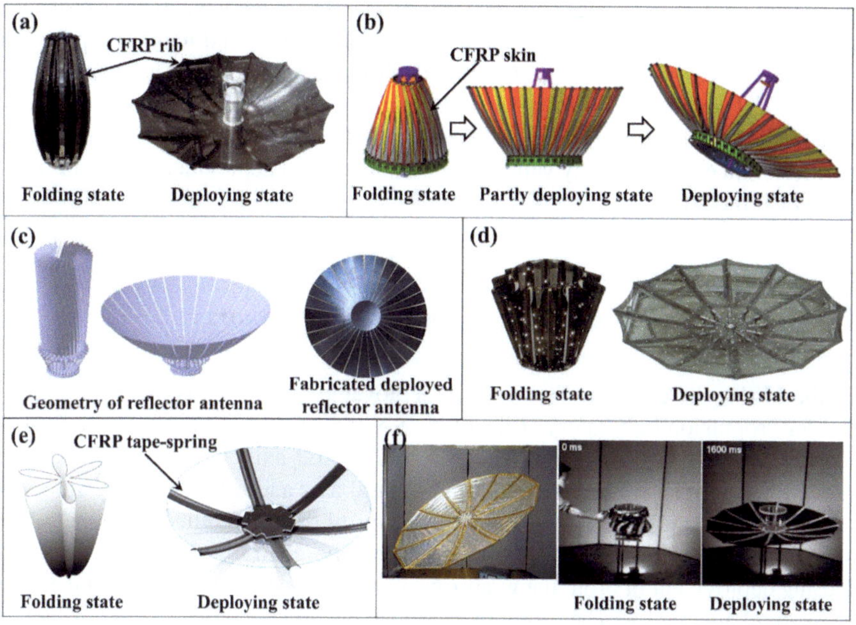

Fig. 1.14 Umbrella-shaped deployable composite reflectors: (**a**) 1-m-diameter FlexRS®-I SUTRA antenna in deployed and folded states [173]. (**b**) 3-m unfurlable composite skin reflector [174]. (**c**) Deployable composite reflector for SAR [175]. (**d**) Flexible shell-membrane reflector [176]. (**e**) Deployable reflector with a continuous composite thin-shell and six tape springs [177]. (**f**) Collapsible rib-tensioned surface reflector [178]

reflector is less than 1 meter, with a weight of approximately 10 kg. After deployment, the panels are individually formed and combined to form the reflector, with a gap of about 0.5 mm between panels, avoiding the problem of reduced surface accuracy due to mismatched dimensions (shown in Fig. 1.14b). This scheme is suitable for X-band frequencies and above, with an aperture range between 3 m and 6 m. Yoon et al. [175] investigated an umbrella-shaped deployable composite reflector (shown in Fig. 1.14c), which offers advantages such as light weight, low loss, high precision, and stability. They also studied the radiation pattern, gain, and polarization variation effects of the antenna in the X-band by considering the changes in effective electrical properties affected by fiber direction. Datashvili et al. [176] designed a mechanically reconfigurable umbrella-shaped deployable composite reflector (shown in Fig. 1.14d) by simplifying support ribs. This reflector can achieve larger diameters and operating frequencies with high surface accuracy. Klimm and Kwok [177] proposed a concept for an umbrella-shaped deployable reflector based on composite tape springs. The composite tape springs store strain energy during deployment and guide the development direction while simplifying the structure. The composite tape springs in this approach closely follow the shape of the reflector and can independently recover their original shape (shown in Fig. 1.14e). Pellegrino [178] presented a design for a novel multifunctional deployable reflector. The reflector consists of a central extendable hub, several composite tape springs connected to the hub, ribs for support, and tensioned precision membrane, as shown in Fig. 1.14f. For a deployable reflector with a 10 m diameter and a focal length of 7.8 m, a surface accuracy of approximately 2 mm can be achieved. Additionally, they introduced a proof-of-concept design with a 1.5-m diameter and experimentally validated the accuracy and deployment behavior of the analytical model.

1.4.4 Wrap-Rib Deployable Composite Reflectors

In the 1970s, the concept of a wrap-rib deployable reflector was first proposed by the Jet Propulsion Laboratory (JPL) of NASA and Lockheed Martin Space System Company [168, 179]. It mainly consists of a central axis, wrap-ribs, and a mesh reflector. The wrap-ribs are hinged to the central axis and extend radially along the axis to support and maintain the parabolic shape of the mesh reflector. In the folded state, the wrap-ribs are tightly wound around the central axis in one direction. During deployment in orbit, the wrap-ribs rely on their own elastic potential energy to rebound and deploy. Therefore, the wrap-rib deployable reflector has advantages such as high stowage efficiency, simple folding and deployment control mechanisms, and high reliability.

The wrap-rib deployable reflector was first installed on the ATS-6 satellite launched in 1974 (shown in Fig. 1.15a). The reflector consisted of 48 wrap-ribs,

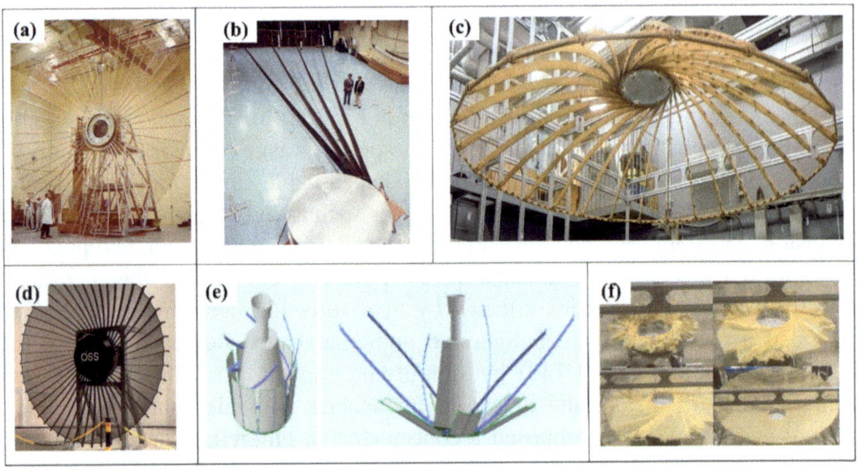

Fig. 1.15 Wrap-rib deployable composite reflectors: (**a**) $D = 9.5$-m deployable composite reflector fabricated in the 1970s [180]. (**b**) $D = 55$-m deployable composite reflector fabricated in the 1980s [181]. (**c**) Prototype of wrap-rib deployable composite reflector [182]. (**d**) Deployed 2.7-m OSS wrap-rib deployable composite reflector [182]. (**e**) Conceptual mechanism [187]. (**f**) Deployment test [187]

with a central axis diameter of 2.0 m and a height of 0.45 m. When deployed, the effective aperture of the reflector was 9.1 m. The complete structure weighed approximately 60 kg [180]. Following this, the U.S. Department of Defense successfully developed a series of wrap-rib deployable reflectors. For example, a partially folded sector of a 55-m reflector structure was successfully deployed on the ground in 1984 (shown in Fig. 1.15b) [181]. Urata et al. [182, 183] developed a 3.6-m-diameter wrap-rib deployable reflector for earth observation, as depicted in Fig. 1.15c. OSS designed a wrap-rib deployable reflector antenna for X-band SAR in low earth orbit, as shown in Fig. 1.15d [184–186]. This antenna has a diameter range from 2.7 m to 5 m and utilizes a mesh reflector surface. It includes a central axis, multiple lenticular deployable composite booms to support the reflector, and a deployable secondary composite reflector and supporting pole. Urate et al. [187] designed a novel wrap-rib deployable reflector, and Fig. 1.15e, f illustrate the schematic and experimental process of the reflector deployment, respectively.

However, wrap-rib deployable reflectors suffer from poor stiffness, vibration resistance, and accuracy. To improve the surface accuracy, a large number of wrap-ribs and a central axis with a corresponding volume are required. This significantly increases the weight of the reflector and makes it difficult to fold, making it unable to meet the requirements for larger apertures.

1.5 Composite Helical Structures

Due to the excellent ability of storing and releasing elastic strain energy, composite helical structures (CHSs) are indispensable machine elements in many engineering applications. Taking the recent examples, the helical structure can be used as a deployable antenna to receive and transmit signals in the aerospace field [188, 189], and it can also be used as a spring in the automobile field to play the role of shock-absorbing function [190, 191]. Traditional helical structures are generally made of metal materials. The manufacturing process of metal helical structures is simple, but it does not satisfy requirements of lightweight design due to the heavy weight. Using composite helical structure instead of metal helical structures can effectively reduce the weight of helical structures and take into account excellent mechanical properties. A large number of experimental, analytical, and numerical investigations have been carried out on helical structure stiffness [192, 193], strength [194, 195], and compressive stability [196, 197]. It is possible to effectively optimize and improve these properties with the aid of some analytical or numerical models.

Many investigations have been conducted to deal with basic mechanical properties of helical structures [198–202]. There are many analytical models for calculating the compressive stiffness and compressive strength of helical structures. In these models, the change of geometric parameters of helical structures with the increase of compressive load is not considered. In 1963, Wahl [203] deduced analytical expressions of basic mechanical properties (i.e., stiffness, strength, compressive stability) of helical structures based on the energy method. Even today, Wahl's work is still used by many scholars and helical structure manufacturers. With the aid of Wahl model, Mehdi et al. [204] and Budan et al. [205], respectively, analyzed compressive behavior of three different fiber-reinforced composite helical structures. The results showed that the weight and stiffness of helical structures can be effectively optimized by changing the type of material [206]. Patil et al. [207–209] calculated the compressive stiffness and compressive stability of cylindrical helical structures based on Wahl model. Compared with the experimental results, it was shown that helical structures were more prone to instability with the increase of the helix angle under compressive load. In the following work, Patil proposed an analytical instability equation for conical helical structures and verified that conical helical structures can effectively solve the problem of compressive instability. Choi and Choi [210] established an analytical model for predicting the compressive stiffness of unidirectional reinforced composite helical structure and analyzed the influences of ply orientation and helix diameter on compressive stiffness of composite helical structure. The predictions were compared with the experimental results and numerical simulations to evaluate the validity of the analytical model. Xiong et al. [211] proposed an analytical model for predicting the compressive stiffness of composite helical structure with hollow circular cross section and discussed the factors affecting the compressive stiffness of composite helical structure. The predictions have a good agreement with the experimental data. Gobbi and Mastinu [212] established mathematical models for predicting compressive stiffness, strength, global stability, local stability, and vibrations of composite helical structure with circular cross section. The predictions correlated well with the experimental results. In addition, they also optimized parameters of composite helical structure using genetic algorithm to obtain optimal mechanical properties. It is noteworthy that all the above models only consider the influence of internal twisting moment on deformation of helical structures under compressive load and neglect the influences of axial force, shear force, and bending moment. Thus, these models are only effective for predicting the compressive behavior of the helical structures with small helix angle (i.e., $\alpha \leq 5°$). Furthermore, since these models do not consider geometric nonlinearity effect, it is impossible to predict mechanical properties of helical structures within large-deformation process (i.e., geometric nonlinearity). In fact, when the helical structures are subjected to external load, geometric parameters continuously change with the increase of external load.

In practice, few researchers have considered the influence of geometric parameter variations (i.e., geometric nonlinearity) when studying the mechanical properties of helical structures. Babenko [213], based on the variational principle, developed an analytical model to predict the compressive performance of helical structures while accounting for geometric nonlinearity. The computational results indicated that only the relationship between helix diameter and helix angle is approximately linear, while other geometric parameters exhibit significant nonlinear relationships with the helix angle. However, this analytical model neglects the effects of boundary conditions and is applicable only to helical structures with a large number of effective coils. Savi [214] and Toi [215] observed pronounced nonlinear behavior in the stiffness and load versus displacement curves of shape memory alloy helical structures through tensile and compressive experiments. Numerical simulations showed good agreement with experimental results, revealing that geometric nonlinearity and constitutive relations significantly influence the tensile performance of helical structures. However, the reconstruction of finite element models becomes highly time-consuming due to variations in the geometric parameters of the helical structures. Hashemi [216], based on Wahl model, proposed a numerical model applicable to both small and large deformations of helical structures under external loading. The numerical results were in good agreement with experimental data. However, obtaining sufficient initial input parameters through experiments and computations remains a challenging task. Goncharenko [217] introduced an analytical model to predict the tensile performance of helical structures while considering geometric nonlinearity. The study demonstrated a notable nonlinear behavior during the transition of helical structures from a free state to a straightened state. While the nonlinear tensile performance of helical structures during large deformations can be explained by geometric nonlinearity, the model neglects the influence of material nonlinearity. Consequently, its predictions deviate from experimental results when applied to tensile performance analysis of metallic helical structures prone to plastic deformation.

From current studies, it is evident that most analytical and numerical models have been developed specifically for metallic helical structures. Therefore, it is necessary to establish more accurate and efficient analytical models that account for geometric nonlinearity to predict the compressive performance of composite helical structures, particularly for composite helical structures with large helix angles (i.e., $\alpha > 5°$).

1.6 Arrangement

Large-deformation flexible composite structures, characterized by structure-function integrated features, can utilize the strain energy stored during their folding process to achieve elastic deploying. These structures have garnered widespread attention in space applications. Building on existing research, significant progress has been observed in this field concerning deformation mechanisms, modeling, design, optimization, and applications [1]. This book, themed "Folding Mechanics of Large-Deformation Flexible Composite Structures," systematically introduces the research achievements in theory, experiments, and simulations. The proposed analytical models, computational methods, and experimental approaches are highly practical and can effectively solve the nonlinear geometric issues of large deformations in engineering practice.

This book is divided into 12 chapters, with the following content arrangement:

Chapter 1 introduces the research progress on large-deformation flexible composite structures both domestically and internationally, providing a brief overview of the content of this book.

Chapters 2 and 3 summarize the fundamental theories, including the energy method, variational principles, and basic mechanics of composite materials, which serve as critical analytical references for the modeling work presented in this book.

Chapter 4 constructs geometric equations for the axial compressive folding deformation of composite cylindrical shells based on the classical Euler-Bernoulli beam model. These equations form the basis of an analytical model for predicting load versus displacement curves, shape functions, and stress levels. Samples were fabricated, and axial compressive folding tests were conducted to validate the analytical model. The effects of various geometric parameters on axial compressive folding performance were also analyzed.

Chapter 5 develops an analytical model for predicting the bending folding performance of composite tape springs using the Newton-Raphson method and the principle of minimum energy. This model encompasses key metrics such as geometric configuration and relationship between folding moment and longitudinal curvature. The model considers the contributions of tensile and bending strain energy to the total strain energy. Its predictive accuracy was validated by comparisons with three classical models and experimental tests.

Chapter 6 establishes an analytical model for composite helical structures, incorporating geometric nonlinear effects. Using the energy method, the load versus displacement relationship was derived, and the compressive stiffness was obtained through numerical fitting. The composite strength criterion was applied to derive the analytical expression for compressive strength. The model's accuracy was validated, and the effects of geometric factors such as helix angle and helix diameter on compressive performance were discussed.

Chapter 7 formulates geometric equations describing the folding deformation performance of lenticular deployable composite booms. By employing equilibrium equations and functional relationships, an analytical model predicting the load versus displacement relationship for folding deformation was proposed. Furthermore, analytical formulas for folding moments and the critical coiling radius were derived. Samples were fabricated using vacuum bagging and subjected to tension, compression, coiling, and folding tests, which confirmed the predictive accuracy of the analytical model.

Chapter 8 develops geometric equations for the coiling and folding performance of tubular deployable composite booms. An analytical model for folding moment versus rotational angle curves during coiling and folding was proposed based on equilibrium equations and energy principles. The strength criteria were used to derive analytical expressions for stress levels. Samples were fabricated and subjected to large-deformation experiments, validating the analytical model's predictions. Finally, the effects of geometric parameters on coiling and folding performance were analyzed.

Chapter 9 formulates geometric equations describing the folded stable state of tubular bistable deployable composite booms. Two analytical models for predicting bistable performance were derived using the minimum energy principle and bistability criteria. Stress-level expressions were developed based on the composite strength criterion, and the models' accuracy was validated through bistability experiments. The effects of geometric parameters on bistable performance were also analyzed.

Chapter 10 introduces numerical simulations for analyzing the mechanical performance of large-deformation flexible composite structures. Taking the deployable composite booms as a typical example, accurate finite element models were established based on actual geometric conditions and application scenarios. The large-deformation mechanism of the structures was revealed, and corresponding folded stable state mechanical behavior was effectively predicted. The influence of various parameters was systematically discussed.

Chapter 11 integrates the analytical models with surrogate modeling and evolutionary algorithms to perform multi-objective optimization designs for typical deployable composite booms. For deployable composite booms, the optimization achieved the minimization of the ultimate coiling radius, folding moment, and linear density while maximizing the sectional moment of inertia. For tubular deployable composite booms, the optimization minimized the average folding moment and maximized the bending stiffness. For bistable deployable composite booms, the optimization minimized the linear density and the longitude-sectional polar radius at the end point while maximizing the average strain energy and natural frequency. All optimization cases considered optimal combinations of design variables, with evolutionary algorithms used to identify the best designs that met performance requirements.

Chapter 12 reviews the space applications and development history of large-deformation flexible composite structures in aerospace. It also highlights future challenges and development opportunities for these structures.

References

1. Liu TW, Bai JB, Fantuzzi N (2024) Thin-walled deployable composite structures: a review. Prog Aerosp Sci 146:100985
2. Ramazanova RD, Alifanov MO (2021) Promising manned spacecraft for a long-duration flight with a flexible inflatable shell. J Space Safety Eng 8(4):259–265
3. Li Y, Li M, Dai F (2022) Twisting control strategy of bistable composite shell with initial curvature. Thin-Walled Struct 181:110021
4. Ji C, Liu J, Wu C et al (2023) Dynamic analysis and parametric optimization of telescopic tubular mast applied on solar sail. Chinese J Mech Eng 36(1):1–12
5. Morozov EV, Lopatin AV, Shatov AV et al (2024) Composite anisogrid lattice toroidal shell: application to a load-carrying rim of the spacecraft reflectarray antenna. Comp Struct 331:117860
6. Yang H, Feng J, Liu YB et al (2022) Design and kinematic analysis of a large deployable mechanism of the parabolic cylindrical antenna with multi tape-spring hinges. J Mech Eng 58(3):75–83
7. An N, Jia Q, Jin H et al (2022) Multiscale modeling of viscoelastic behaviour of unidirectional composite laminates and deployable structures. Mater Des 219:110754
8. Zhao P, Wu C, Li Y (2023) Design and application of solar sailing: a review on key technologies. Chinese J Aeronaut 36(5):125–144
9. Seffen KA (2020) Folding a ridge-spring. J Mech Phys Solids 137:103820

10. Seffen KA (2001) On the behaviour of folded tape-springs. J Appl Mech 68(3):369–375
11. Dewalque F, Collette JP, Bruls O (2016) Mechanical behaviour of tape springs used in the deployment of reflectors around a solar panel. Acta Astronaut 123:271–282
12. Kwok K, Pellegrino S (2013) Folding, stowage, and deployment of viscoelastic tape springs. AIAA J 51(8):1908–1918
13. Wuest W (1954) Einige anwendungen der theorie der zylinderschale. J Appl Math Mech/Z Angew Math Mech 34(12):444–454
14. Panovko YG, Gubanova II (1965) Stabilities and oscillations of elastic systems: paradoxes, fallacies and new concepts. Consultants Bureau, New York
15. Seffen KA, Pellegrino S (1999) Deployment dynamics of tape-springs. Proc Royal Soc Lond Series A: Math Phys Eng Sci 455(1983):1003–1048
16. Soykasap Ö (2007) Analysis of tape-spring hinges. Int J Mech Sci 49(7):853–860
17. Yee JCH, Soykasap O, Pellegrino S (2004) Carbon fibre reinforced plastic tape-springs. In: 45th AIAA/ASME/ASCE/AHS/ASC structures, structural dynamics and materials conference, 2004, p 1819
18. Zuo Y, Jin G, Xie P (2018) Calculative and experimental study of the CFRP tape-spring. J Mech Sci Technol 32:3603–3609
19. Yang Y, Wang F, Liu J (2022) Application of honeycomb structures in key components of space deployable structures. Adv Mater Sci Eng:4756272
20. Yang Y, Wang F, Liu J (2023) The exact solution of the bending moment in the folding process of negative Poisson's ratio honeycomb tape-spring and multi-objective optimization design. Aerospace 10(5):390
21. Yee JC, Pellegrino S (2005) Composite tube hinges. J Aerosp Eng 18(4):224–231
22. Mansfield EH (1973) Large-deflexion torsion and flexure of initially curved strips. Proc Royal Soc Lond A Math Phys Sci 334(1598):279–298
23. Yao XF, Ma YJ, Yin YJ et al (2011) Design theory and dynamic mechanical characterization of the deployable composite tube hinge. Sci China Phys Mech Astron 54:633–639
24. Knott G, Viquerat A (2018) Curved bistable composite slit tubes with positive Gaussian curvature. AIAA J 56(4):1679–1688
25. Shore J, Viquerat A, Richardson G et al (2020) An energy optimisation approach to modelling tape-spring behaviour. In: AIAA Scitech 2020 Forum, Orlando, Florida, January 6–10, 2020
26. Iqbal K, Pellegrino S, Daton-Lovett A (1998) Bistable composite slit tubes. In: IUTAM-IASS symposium on deployable structures. Cambridge, UK, September 6–9, 1998
27. Galletly DA, Guest SD (2004) Bistable composite slit tubes. I. A beam model. Int J Solids Struct 41(16–17):4517–4533
28. Galletly DA, Guest SD (2004) Bistable composite slit tubes. II. A shell model. Int J Solids Struct 41(16–17):4503–4516
29. Guest SD, Pellegrino S (2006) Analytical models for bistable cylindrical shells. Proc Royal Soc A: Math Phys Eng Sci:839–854
30. Zhang Z, Wu H, He X et al (2013) The bistable behaviours of carbon-fiber/epoxy anti-symmetric composite shells. Comp Part B: Eng 47:190–199
31. Liu TW, Bai JB, Fantuzzi N (2024) New insights into folding tape-spring. Comp Struct 341:118225
32. Boesch C, Pereira C, John R et al (2008) Ultra light self-motorized mechanism for deployment of light weight spacecraft appendages. In: Proceedings of 39th aerospace mechanisms symposium, Huntsville, Alabama, May 7–9, 2008
33. Liu TW, Bai JB (2022) Folding behaviour of a deployable composite cabin for space habitats-part 1: experimental and numerical investigation. Comp Struct 302:116244
34. Liu TW, Bai JB (2022) Folding behaviour of a deployable composite cabin for space habitats-Part 2: analytical investigation. Comp Struct 297:115929
35. Sakovsky M, Pellegrino S (2019) Closed cross-section dual-matrix composite hinge for deployable structures. Comp Struct 208:784–795

36. Fernandes P, Sousa B, Marques R et al (2021) Influence of relaxation on the deployment behaviour of a CFRP composite elastic-hinge. Comp Struct 259:113217
37. Mobrem M, Adams DS (2009) Deployment analysis of the lenticular jointed antennas onboard the mars express spacecraft. J Spacecraft Rockets 46(2):394–402
38. Bowen AG, Zucco G, Weaver PM (2021) Morphing of symmetric cross-ply cylindrical shells by minimising the Brazier moment: optimised hinge folding. Thin-Walled Struct 158:107122
39. Echter MA, Gillmer SR, Silver MJ et al (2020) A multifunctional high strain composite (HSC) hinge for deployable in-space optomechanics. Smart Mater Struct 29:105010
40. Lan X, Liu Y, Lv H et al (2009) Fiber reinforced shape-memory polymer composite and its application in a deployable hinge. Smart Mater Struct 18(2):024002
41. Soykasap Ö (2009) Deployment analysis of a self-deployable composite boom. Comp Struct 89(3):374–381
42. Warren P, Dobson B, Hinkle J et al (2005) Experimental characterization of lightweight strain energy deployment hinges. In: 46th AIAA/ASME/ASCE/AHS/ASC structures, structural dynamics and materials conference, Austin, Texas, April 18–21, 2005
43. Silver M, Hinkle J, Peterson L (2004) Modeling of snap-back bending response of doubly slit cylindrical shells. In: 45th AIAA/ASME/ASCE/AHS/ASC structures, structural dynamics and materials conference, Palm Springs, California, April 18–21, 2004
44. Mallikarachchi HMYC, Pellegrino S (2011) Quasi-static folding and deployment of ultrathin composite tape-spring hinges. J Spacecraft Rockets 48(1):187–198
45. Piovesan D, Zaccariotto M, Bettanini C et al (2016) Design and validation of a carbon-fiber collapsible hinge for space applications: a deployable boom. J Mech Robot 8(3):031007
46. Mallikarachchi H, Pellegrino S (2009) Deployment dynamics of composite booms with integral slotted hinges. In: 50th AIAA/ASME/ASCE/AHS/ASC structures, structural dynamics, and materials conference, Palm Springs, California, May 4–7, 2009
47. Mallikarachchi H, Pellegrino S (2008) Simulation of quasi-static folding and deployment of ultra-thin composite structures. In: 49th AIAA/ASME/ASCE/AHS/ASC structures, structural dynamics, and materials conference, Schaumburg, Illinois, April 7–10, 2008
48. Dewalque F, Schwartz C, Denoël V et al (2018) Experimental and numerical investigation of the nonlinear dynamics of compliant mechanisms for deployable structures. Mech Syst Signal Process 101:1–25
49. Mallikarachchi H, Pellegrino S (2014) Deployment dynamics of ultrathin composite booms with tape-spring hinges. J Spacecraft Rockets 51(2):604–613
50. Mallikarachchi H, Pellegrino S (2010) Optimized designs of composite booms with integral tape-spring hinges. In: 51st AIAA/ASME/ASCE/AHS/ASC structures, structural dynamics, and materials conference, Orlando, Florida, April 12–15, 2010
51. Mallikarachchi H, Pellegrino S (2011) Design and validation of thin-walled composite deployable booms with tape-spring hinges. In: 52nd AIAA/ASME/ASCE/AHS/ASC structures, structural dynamics and materials conference, Denver, Colorado, April 4–7, 2011
52. Mallikarachchi HM (2011) Thin-walled composite deployable booms with tape-spring hinges. University of Cambridge, Cambridge
53. Sakovsky M, Pellegrino S, Mallikarachchi H (2016) Folding and deployment of closed cross-section dual-matrix composite booms. In: 3rd AIAA Spacecraft Structures Conference, San Diego, California, January 4–8, 2016
54. Ferraro S, Pellegrino S (2018) Self-deployable joints for ultra-light space structures. In: AIAA spacecraft structures conference, Kissimmee, Florida, January 8–12, 2018
55. Ferraro S, Pellegrino S (2019) Topology optimization of composite self-deployable thin shells with cutouts. In: AIAA Scitech 2019 Forum, San Diego, California, January 7–11, 2019
56. Echter MA, Silver MJ, D'Elia E et al (2018) Recent developments in precision high strain composite hinges for deployable space telescopes. In: AIAA spacecraft structures conference, Kissimmee, Florida, January 8–12, 2018
57. Pellegrino S, Kebadze E, Lefort T et al (2002) Low-cost hinge for deployable structures. University of Cambridge, Cambridge

58. Hogstrom K, Pellegrino S (2016) Methods for characterizing the reliability of deployable modules for large optical reflectors. In: 3rd AIAA spacecraft structures conference, San Diego, California, January 4–8, 2016
59. Francis W, Lake M, Mallick K et al (2003) Development and testing of a hinge/actuator using elastic memory composites. In: 44th AIAA/ASME/ASCE/AHS/ASC structures, structural dynamics, and materials conference, Norfolk, Virginia, April 7–10, 2003
60. Barrett R, Francis W, Abrahamson E et al (2006) Qualification of elastic memory composite hinges for spaceflight applications. In: 47th AIAA/ASME/ASCE/AHS/ASC structures, structural dynamics, and materials conference, Newport, Rhode Island, May 1–4, 2006
61. Fosness E, Carpenter B, Tupper M et al (2003) Development of low shock deployment devices for aerospace applications using shape memory and elastic memory composites. In: AIAA Space 2003 conference and exposition, Long Beach, California, September 23–25, 2003
62. Wang C, Wang Y (2018) The mechanical design of a hybrid intelligent hinge with shape memory polymer and spring sheet. Compos Part B 134:1–8
63. Hoffait S, Brüls O, Granville D et al (2010) Dynamic analysis of the self-locking phenomenon in tape-spring hinges. Acta Astronaut 66(7–8):1125–1132
64. Jeong JW, Yoo YI, Shin DK et al (2014) A novel tape spring hinge mechanism for quasi-static deployment of a satellite deployable using shape memory alloy. Rev Sci Instrum 85(2):025001
65. Givois D, Jacques S, Mazoyer T (2001) A low cost hinge for appendices deployment: design, test and applications. In: 9th European space mechanisms and tribology symposium, Liège, Belgium, September 19–21, 2001
66. Sicre J, Givois D, Emerit A (2005) Application of "MAEVA" hinge to myriade microsatellites deployments needs. In: 11th European space mechanisms and tribology symposium, Lucerne, Switzerland, September 21–23, 2005
67. Eigenmann M, Schmalbach M, Schiller M et al (2011) Ultra-light deployment mechanism (UDM) for sectioned large deployable antenna reflectors. In: Proceeding of the 14th European space mechanisms and tribology symposium, Constance, Germany, September 28–30, 2011
68. Zajac K, Schmidt T, Schiller M et al (2013) Verification test for ultra-light deployment mechanism for sectioned deployable antenna reflectors. In: 15th European Space mechanisms and tribology symposium, Noordwijk, Netherland, September 25–27, 2013
69. Warren PA, Silver MJ, Dobson BJ Jr et al (2013) Experimental characterization of deployable outer barrel assemblies for large space telescopes. In: UV/Optical/IR space telescopes and instruments: innovative technologies and concepts VI, San Diego, California, 2013
70. Kim DY, Choi HS, Lim JH et al (2020) Experimental and numerical investigation of solar panels deployment with tape spring hinges having nonlinear hysteresis with friction compensation. Appl Sci 10(21):7902
71. Straubel M, Block J, Sinapius M et al (2011) Deployable composite booms for various gossamer space structures. In: 52nd AIAA/ASME/ASCE/AHS/ASC structures, structural dynamics and materials conference, Denver, Colorado, April 4–7, 2011
72. Murphey TW, Footdale JN. Deployable structures with quadrilateral reticulations. U.S. Patent 8, 384, 613. 2013-2-26
73. Wang SC, Shang HJ, Guo HW et al (2020) Mechanical feature analysis of a deployable thin-walled open cylindrical boom. Manned Spaceflight 26(2):159–165
74. Mchale C, Telford R, Weaver PM (2020) Morphing lattice boom for space applications. Compos Part B 202:108441
75. Seefeldt P, Grundmann JT, Hillebrandt M et al (2020) Performance analysis and mission applications of a new solar sail concept based on crossed booms with tip-deployed membranes. Adv Space Res 67(9):2736–2745
76. Takao Y, Mori O, Kawaguchi J (2021) Analysis and design of a spacecraft docking system using a deployable boom. Acta Astronaut 179:172–185
77. Rimrott FPJ (1980) Self-extension of an open section tube. Theory Shells:495–508

78. Rimrott FPJ, Fritzsche G (1998) Fundamentals of STEM mechanics. In: IUTAM-IASS symposium on deployable structures: theory and applications, Cambridge, UK, September 6–9, 1998
79. Rimrott FPJ (1965) Storable tubular extendible member. Mach Des 37(28):156–165
80. Rimrott FPJ (1966) Two secondary effects in bending of slit thin-walled tubes. J Appl Mech 33(1):75
81. Wang S, Schenk M, Jiang S et al (2020) Blossoming analysis of composite deployable booms. Thin-Walled Struct 157:107098
82. Liu TW, Bai JB, Fantuzzi N et al (2023) Folding behaviour of thin-walled tubular deployable composite boom for space applications: experiments and numerical simulation. Acta Astronaut 209:159–171
83. Liu TW, Bai JB, Fantuzzi N (2023) Analytical models for predicting folding behaviour of thin-walled tubular deployable composite boom for space applications. Acta Astronaut 208:167–178
84. Banik J, Murphey T (2010) Performance validation of the triangular rollable and collapsible mast. In: 24th Annual AIAA/USU Conference on Small Satellites, Logan, Utah, August 9–12, 2010
85. Leclerc C, Pellegrino S (2017) Ultra-thin composite deployable booms. In: Proceedings of IASS Annual Symposia, Hamburg, Germany, September 25–27, 2017
86. Johnson L, Whorton M, Heaton A et al (2011) NanoSail-D: a solar sail demonstration mission. Acta Astronaut 68(5–6):571–575
87. Biddy C, Svitek T (2012) LightSail-1 solar sail design and qualification. In: Proceedings of the 41st aerospace mechanisms symposium, Pasadena, California, May 16–18, 2012
88. Betts B, Spencer DA, Nye B et al (2017) Lightsail 2: controlled solar sailing using a CubeSat. In: The 4th International symposium on solar sailing, Kyoto, Japan, January 17–20, 2017
89. Stohlman OR, Loper ER (2016) Thermal deformation of very slender triangular rollable and collapsible booms. In: 2016 AIAA Scitech conference, San Diego, California, January 4–8, 2016
90. Leclerc C (2020) Mechanics of ultra-thin composite coilable structures. California Institute of Technology, Pasadena
91. Leclerc C, Pellegrino S (2020) Nonlinear elastic buckling of ultra-thin coilable booms. Int J Solids Struct 203:46–56
92. Leclerc C, Pedivellano A, Pellegrino S (2018) Stress concentration and material failure during coiling of ultra-thin TRAC booms. In: 2018 AIAA Spacecraft structures conference, Kissimmee, Florida, January 8–12, 2018
93. Leclerc C, Pellegrino S (2019) Reducing stress concentration in the transition region of coilable ultra-thin-shell booms. In: AIAA Scitech 2019 Forum, San Diego, California, January 7–11, 2019
94. Hasanyan AD, Leclerc C, Pellegrino S (2020) Interface failure analysis of triangular rollable and collapsible (TRAC) booms. In: AIAA Scitech 2020 Forum, Orlando, Florida, January 6–10, 2020
95. Luo W, Pellegrino S (2022) Propagating instabilities in coilable booms. In: AIAA Scitech 2022 Forum, San Diego, California, January 3–7, 2022
96. Yang H, Lu F, Guo H et al (2019) Design of a new N-shape composite ultra-thin deployable boom in the post-buckling range using response surface method and optimization. IEEE Access 7:129659–129665
97. Yang H, Guo H, Wang Y et al (2020) Analytical solution of the peak bending moment of an M boom for membrane deployable structures. Int J Solids Struct 206:236–246
98. Herbeck L, Leipold M, Sickinger C et al (2000) Development and test of deployable ultra-lightweight CFRP-booms for a solar sail. In: Spacecraft structures, materials and mechanical testing, Noordwijk, Netherlands, November 29–December 1, 2000
99. Yoshiro OGI, Reveles J, Fraux V, et al. (2019) Deployable wrap-rib assembly. U.S. Patent Application. 2019-11-28

100. Bai JB, Chen D, Xiong JJ et al (2017) A corrugated flexible composite skin for morphing applications. Compos Part B 131:134–143
101. Bai JB, Chen D, Xiong JJ et al (2019) A semi-analytical model for predicting nonlinear tensile behaviour of corrugated flexible composite skin. Compos Part B 168:312–319
102. Liu TW, Bai JB, Li SL et al (2023) Large deformation and failure analysis of the corrugated flexible composite skin for morphing wing. Eng Struct 278:115463
103. Wang LW, Bai JB, Shi Y (2021) Simplified analytical model for predicting neutral cross-section position of lenticular deployable composite boom in tensile deformation. Materials 14(24):7809
104. Bai JB, Xiong JJ, Gao JP et al (2013) Analytical solutions for predicting in-plane strain and interlaminar shear stress of ultra-thin-walled lenticular collapsible composite tube in fold deformation. Comp Struct 97:64–75
105. Liu TW, Bai JB, Fantuzzi N (2022) Folding behaviour of the thin-walled lenticular deployable composite boom: analytical analysis and many-objective optimization. Mech Adv Mater Struct:1–19
106. Bai JB, Chen D, Xiong JJ et al (2019) Folding analysis for thin-walled deployable composite boom. Acta Astronaut 159:622–636
107. Bai JB, Xiong JJ (2014) Temperature effect on buckling properties of ultra-thin-walled lenticular collapsible composite tube subjected to axial compression. Chinese J Aeronaut 27(5):1312–1317
108. Yao Z, Li D, Zhu S et al (2021) Analysis of buckling characteristics and parameter influence of composite thin-walled lenticular boom structures. E3S Web Conf EDP Sci 233:04009
109. Jia Q, An N, Ma X et al (2021) Exploring the design space for nonlinear buckling of composite thin-walled lenticular tubes under pure bending. Int J Mech Sci 207:106661
110. Jia Q, An N, Ma X et al (2022) A dynamic finite element procedure for bending collapse of composite thin-walled lenticular tubes. Comp Struct 287:115364
111. Bai JB, Shenoi RA, Xiong JJ (2017) Thermal analysis of thin-walled deployable composite boom in simulated space environment. Comp Struct 173:210–218
112. Yang LY (2019) Deployment behaviours of woven composite bistable cylindrical shells. Harbin Institute of Technology, Harbin
113. Shan JL (2015) Research on bistable cylindrical shells in view of energy principle. Xi'an University of Architecture And Technology, Xi'an
114. Zhang Z, Li Y, Yu X et al (2019) Bistable morphing composite structures: a review. Thin-Walled Struct 142:74–97
115. Schultz MR, Hulse MJ, Keller PN et al (2008) Neutrally stable behaviour in fiber-reinforced composite tape springs. Compos A: Appl Sci Manuf 39(6):1012–1017
116. Fernandez JM, Viquerat A, Lappas VJ et al (2014) Bistable over the whole length (BOWL) CFRP booms for solar sails. Adv Solar Sailing:609–628
117. Liu TW, Bai JB, Wang ZZ et al (2023) Determining folded stable state of bistable deployable composite boom: an improved two-parameter analytical model. Acta Astronaut 213:507–515
118. Liu TW, Bai JB, Fantuzzi N (2024) Analytical model for predicting folding stable state of bistable deployable composite boom. Chinese J Aeronaut 37(8):460–469
119. Liu TW, Bai JB, Xi HT et al (2023) Temperature effect on large deformation of bistable deployable composite boom. Comp Struct 318:117090
120. Wu C, Viquerat A (2018) Computational and experimental study on dynamic instability of extended bistable carbon/epoxy booms subjected to bending. Comp Struct 188:347–355
121. Wu C (2017) Dynamic analysis of extended bistable reeled fibre-reinforced composite booms for space applications. University of Surrey, Guildford
122. Kajihara S, Yokozeki T, Aoki T (2023) Dimension effects on the stowability and self-deployment behaviour of CFRP bistable open-section semi-cylindrical beam. Comp Struct 307:116628
123. Wang S, Xu S, Lu L et al (2023) Roll-out deployment process analysis of a fiber reinforced polymer (FRP) composite tape-spring boom. Polymers 15(11):2455

124. Fujioka E, Yokozeki T, Watanabe A et al (2019) Analysis on temperature-dependent deployment behaviour of bi-stable composite rods. Adv Compos Mater 28(3):245–257
125. Yang C, Wang B, Zhong S et al (2022) On tailoring deployable mechanism of a bistable composite tape-spring structure. Compos Commun 32:101171
126. Mao H, Shipsha A, Tibert G (2017) Design and analysis of laminates for self-deployment of viscoelastic bistable tape springs after long-term stowage. J Appl Mech 84(7):071004
127. Mallol P, Mao H, Tibert G (2018) Experiments and simulations of the deployment of a bistable composite boom. J Spacecr Rocket 55(2):292–302
128. Shen Z, Liu C, Li H (2020) Viscoelastic analysis of bistable composite shells via absolute nodal coordinate formulation. Comp Struct 248:112537
129. Shahryarifard M, Golzar M, Tibert G (2021) Viscoelastic bistable tape spring behaviour modeled by a two-dimensional system with rigid links, springs and dashpots. Comp Struct 258:113181
130. Mao H, Ganga PL, Ghiozzi M et al (2017) Deployment of bistable self-deployable tape spring booms using a gravity offloading system. J Aerosp Eng 30(4):04017007
131. Brinkmeyer A, Pellegrino S, Weaver PM (2016) Effects of long-term stowage on the deployment of bistable tape springs. J Appl Mech 83(1):011008
132. Kwok K (2018) Viscoelastic analysis of stowage and quasi-static deployment of composite tape springs. In: 2018 AIAA Spacecraft Structures Conference, Kissimmee, Florida, January 8–12, 2018
133. Zhang Z, Li Y, Wu H et al (2018) Viscoelastic bistable behaviour of antisymmetric laminated composite shells with time-temperature dependent properties. Thin-Walled Struct 122:403–415
134. Shahryarifard MM, Golzar M, Tibert G (2021) Modeling bistable behaviours in exposure of thermo-viscoelastic conditions by link-spring-dashpot structures. Modares Mech Eng 21(9):589–600
135. Ding S, Sun M, Li Y et al (2022) Novel deployable panel structure integrated with thick origami and morphing bistable composite structures. Materials 15(5):1942
136. Lakshmi SS, Varsha M, Krishnaa GMS et al (2022) Thermo-structural analysis of deployable composite booms with slotted hinges for space applications. Mater Today Proc 56:3564–3570
137. Roh JH, Bae JS (2017) Softenable composite boom for reconfigurable and self-deployable structures. Mech Adv Mater Struct 24(8):698–711
138. Zhang Z, Xiong L, Sun M et al (2022) Bistable characteristics of deployable carbon fiber/bismaleimide resin composite shells in megathermal environments. Mech Adv Mater Struct:1–12
139. Iqbal K, Pellegrino S (2000) Bi-stable composite shells. In: 41st Structures, structural dynamics, and materials conference and exhibit, Atlanta, Georgia, April 3–6, 2000
140. Long Y, Rique O, Fernandez JM et al (2022) Simulation of the column bending test using an anisotropic viscoelastic shell model. Comp Struct 288:115376
141. Kwok K, Pellegrino S (2017) Micromechanics models for viscoelastic plain-weave composite tape springs. AIAA J 55(1):309–321
142. Yang LY, Tan HF, Cao ZS (2018) Modeling and analysis of the ploy region of bistable composite cylindrical shells. Comp Struct 192:347–354
143. Wang B, Seffen KA, Guest SD (2021) Folded strains of a bistable composite tape-spring. Int J Solids Struct 233:111221
144. Wang B, Seffen KA, Guest SD (2019) Shape of a bistable composite tape-spring in folding. In: AIAA Scitech 2019 Forum, San Diego, California, January 7–11, 2019
145. Seffen KA, Wang B, Guest SD (2019) Folded orthotropic tape-springs. J Mech Phys Solids 123:138–148
146. Rakow AS, Reedy R (2022) Predicting ploy length of high-strain composite slit-tubes and tape-springs. In: AIAA Scitech 2022 Forum, San Diego, California, January 3–7, 2022
147. Lee AJ, Fernandez JM, Daye JG (2022) Bistable deployable composite booms with parabolic cross-sections. In: AIAA Scitech 2022 Forum, San Diego, California, January 3–7, 2022

148. Knott GP (2018) New forms of bistable composite slit tube. University of Surrey, Guildford
149. Lee A, Fernandez JM (2018) Mechanics of bistable two-shelled composite booms. In: AIAA Spacecraft Structures Conference, Kissimmee, Florida, January, 8–12, 2018
150. Yao Y, Ambruso A, Fernandez JM et al (2023) A multifunctional bistable ultrathin composite boom for in-space monitoring of deployment dynamics. In: AIAA Scitech 2023 Forum, National Harbor, Maryland, January 23–27, 2023
151. Liu TW, Bai JB, Xi HT et al (2023) A curved bistable composite slit tube for deployable membrane structures. Comp Commun:101648
152. Liu TW, Bai JB, Xi HT et al (2023) Bistable behaviour of deployable composite boom with central flat region. Eng Struct 290:116357
153. Lee AJ, Fernandez JM (2019) Inducing bistability in collapsible tubular mast booms with thin-ply composite shells. Comp Struct 225:111166
154. Fernandez JM, Rose G, Stohlman OR et al (2018) An advanced composites-based solar sail system for interplanetary small satellite missions. In: 2018 AIAA Spacecraft Structures Conference, Kissimmee, Florida, January 8–12, 2018
155. Fernandez JM, Lee AJ (2019) Bistability in collapsible tubular mast booms. In: AIAA Scitech 2019 Forum, San Diego, California, January 7–11, 2019
156. Fernandez JM, Lee AJ (2018) Bistable collapsible tubular mast booms. In: International conference on advanced lightweight structures and reflector Antennas, Tbilisi, Georgia, September 19–21, 2018
157. Fernandez JM (2017) Advanced deployable shell-based composite booms for small satellite structural applications including solar sails. In: International symposium on solar sailing, Kyoto, Japan, January 17–20, 2017
158. Tan LT, Pellegrino S (2004) Ultra thin deployable reflector antennas. In: 45th AIAA/ASME/ASCE/AHS/ASC structures, structural dynamics & materials conference, Palm Springs, California, April 19–22, 2004
159. Tan LT, Pellegrino S (2012) Thin-shell deployable reflectors with collapsible stiffeners: experiments and simulations. AIAA J 50(3):659–667
160. Soykasap O, Tan LT (2011) High-precision offset stiffened springback composite reflectors. AIAA J 49(10):2144–2151
161. Türkmen D, Soykasap Ö, Karakaya Ş (2012) Development of full-scale ultrathin shell reflector. Int J Anten Propag 2012(1):829780
162. Soykasap O, Karakaya S, Kaplan E (2013) Self-deploying composite shell reflector antenna for ku-band missions: 1.5 m diameter demonstrator. In: 6th International conference on Recent Advances in Space Technologies (RAST), Istanbul, Turkey, June 12–14, 2013
163. Soykasap Ö, Karakaya Ş, Türkmen D (2012) Curved large tape springs for an ultra-thin shell deployable reflector. J Reinf Plast Comp 31(10):691–703
164. Tan LT, Soykasap O, Pellegrino S (2005) Design & manufacture of stiffened spring-back reflector demonstrator. In: 46th AIAA/ASME/ASCE/AHS/ASC structures, structural dynamics and materials conference, Austin, Texas, April 18–21, 2005
165. Soykasap O, Karakaya S, Gayretli A et al (2015) Preliminary design of deployable flexible shell reflector of an X-band satellite payload. In: 2nd AIAA spacecraft structures conference, Kissimmee, Florida, January 5–9, 2015
166. Mikulas MM, Lou MC, Withnell PR et al (1993) Deployable concepts for precision segmented reflectors. In: JPL D-10947, June, 1993
167. Stiles L, Garrett J (2010) Development of deployable aperture concepts for CubeSats. In: 51st AIAA/ASME/ASCE/AHS/ASC structures, structural dynamics, and materials conference, Orlando, Florida, April 12–15, 2010
168. Reynolds W, Murphey T, Banik J (2011) Highly compact wrapped-gore deployable reflector. In: 52nd AIAA/ASME/ASCE/AHS/ASC structures, structural dynamics and materials conference, Denver, Colorado, April 4–7, 2011

169. Footdale JN, Peterson M, Griffee JC (2019) Flexible composite shell design for the MARCO deployable reflector. In: AIAA Scitech 2019 Forum, San Diego, California, January 7–11, 2019
170. Allen B, Willer F, Harless I, et al. (2002) Compactly stowable thin continuous surface-based antenna having radial and perimeter stiffeners that deploy and maintain antenna surface in prescribed surface geometry. U.S. Patent 6,344,835. 2002-2-5
171. Keller P, Lake M, Codell D et al (2006) Development of elastic memory composite stiffeners for a flexible precision reflector. In: 47th AIAA/ASME/ASCE/AHS/ASC structures, structural dynamics, and materials conference, Newport, Rhode Island, May 1–4, 2006
172. Dufour L, Datashvili LS, Guinot F et al (2021) Origami deployable reflector antenna for cubesats. In: AIAA Scitech 2021 Forum, 2021, p 1703
173. Datashvili L, Maghaldadze N, Baier H et al (2017) Enabling technologies and architectures at LSS GmbH for European large deployable and reconfigurable reflector antennas. In: Deutsche Gesellschaft für Luft-und Raumfahrt-Lilienthal-Oberth, Munich, Germany, 2017
174. Schmid M, Barho R (2003) Development summary and test results of a 3 meter unfurlable CFRP skin antenna reflector. In: 10th European space mechanisms and tribology symposium, San Sebastián, Spain. September 24–25, 2003
175. Yoon SS, Lee JW, Lee TK et al (2018) Insensitivity characteristics in the dual polarization of deployable CFRP reflector antennas for SAR. IEEE Trans Antenna Propag 66(1):88–95
176. Datashvili L, Baier H, Wehrle E et al (2010) Large shell-membrane space reflectors. In: 51st AIAA/ASME/ASCE/AHS/ASC structures, structural dynamics, and materials conference, Orlando, Florida, April 12–15, 2010
177. Klimm W, Kwok K (2010) Surface accuracy of viscoelastic composite thin-shell deployable reflector antennas. In: AIAA Scitech 2020 Forum, Orlando, Florida, January 6–10, 2020
178. Pellegrino S (2002) Deployable membrane reflectors. In: 2nd World engineering congress, Sarawak, Malaysia, July 22–25, 2002
179. Arya M, Sauder JF, Hodges R et al (2019) Large-area deployable reflectarray antenna for CubeSats. In: AIAA Scitech 2019 Forum, San Diego, California, January 7–11, 2019
180. Murphey T (2009) Historical perspectives on the development of deployable reflectors. In: 50th AIAA/ASME/ASCE/AHS/ASC structures, structural dynamics, and materials conference, Palm Springs, California, May 4–7, 2009
181. Freeland RE, Helms RG, Mikulas MM (2006) The applicability of past innovative concepts to the technology for new extremely large space antenna/telescope structures. SAE Trans:160–173
182. Urata KN, Sri Sumantyo JT, Santosa CE et al (2018) Development of an L-band SAR microsatellite antenna for earth observation. Aerospace 5(1):120
183. Chandra M, Kumar S, Chattopadhyaya S et al (2021) A review on developments of deployable membrane-based reflector antennas. Adv Space Res 68(9):3749–3764
184. Angevain JC, Ihle A, Rodrigues G et al (2019) Large deployable spaceborne reflector antennas in Europe: progress status and perspectives. In: 13th European Conference on Antennas and Propagation, Krakow, Poland, March 31-April 5, 2019
185. Angevain J, Ihle A, Rodrigues G et al (2018) Present status and future outlook of the large deployable space borne reflector antennas in Europe. In: Proceedings of the 3rd International Conference, Tbilisi, Georgia, September 19–21, 2018
186. Κάνε T. (2022) Topology optimisation and structural analysis of a deployable antenna of umbrella type rib for 3D-printing
187. Urate KN, Sumantyo JTS, Santosa CE et al (2019) A compact C-band CP-SAR microsatellite antenna for earth observation. Acta Astronaut 159:517–526
188. Sproewitz T, Block J, Bager A et al (2011) Deployment verification of large CFRP helical high-gain antenna for AIS signals. In: Aerospace conference, Big Sky, Montana, March 5–12, 2011
189. Cheng Y, Chen L, Chen X (2015) A beam scanning method based on the helical antenna for space-based AIS. J Navig 68(1):52–70

190. Gzal M, Groper M, Gendelman O (2017) Analytical, experimental and finite element analysis of elliptical cross-section helical spring with small helix angle under static load. Int J Mech Sci 130:476–486
191. Wu JJ (2006) Study on the inertia effect of helical spring of the absorber on suppressing the dynamic responses of a beam subjected to a moving load. J Sound Vib 297(3–5):981–999
192. Yıldırım V (2016) Exact determination of the global tip deflection of both close-coiled and open-coiled cylindrical helical compression springs having arbitrary doubly symmetric cross-sections. Int J Mech Sci 115–116:280–298
193. Pawar HB, Desale DD (2018) Optimization of three wheeler front suspension coil spring. Procedia Manuf 20:428–433
194. Ekanthappa J, Basavarajappa SGSS, Shankar GS (2017) Fabrication & experimentation of the glass-epoxy helical spring reinforced with graphite powder. Mater Today Proc 4(10):11034–11038
195. Kaiser B, Pyttel B, Berger C (2011) VHCF-behaviour of helical compression springs made of different materials. Int J Fatigue 33(1):23–32
196. Kacar I, Yildirim V (2015) Free vibration/buckling analyses of non-cylindrical initially compressed helical composite springs. Mech Based Des Struct Mach 44(4):340–353
197. Crescenzo FD, Salvini P (2019) Two-dimensional discrete model for buckling of helical springs. Procedia Struct Integr 24:28–39
198. Dia V, Bujoreanu LG, Stanciu S et al (2008) Study of the shape memory effect in lamellar helical springs made from Cu-Zn-Al shape memory alloy. Mater Sci Eng A 481–482:697–701
199. Champion R, Champion WL (2011) Departure from linear mechanical behaviour of a helical spring. Math Comput Model 53(5–6):915–926
200. Pöllänen I, Martikka H (2010) Optimal re-design of helical springs using fuzzy design and FEM. Adv Eng Softw 41(3):410–414
201. Lee CY, Zhuo HC, Hsu CW (2009) Lateral vibration of a composite stepped beam consisted of SMA helical spring based on equivalent Euler-Bernoulli beam theory. J Sound Vib 324(1–2):179–193
202. Frikha A, Cartraud P, Treyssède F (2013) Mechanical modeling of helical structures accounting for translational invariance. Part 1: Static behaviour. Int J Solids Struct 50(9):1373–1382
203. Wahl AW (1963) Mechanical springs. McGraw-Hill Book Company, NewYork
204. Bakhshesh M, Bakhshesh M (2012) Optimization of steel helical spring by composite spring. Int J Multidiscipl Sci Eng 3(6):47–51
205. Budan DA, Manjunatha TS (2010) Investigation on the feasibility of composite coil spring for automotive applications. World Acad Sci Eng Technol 4(10):1035–1039
206. Chiu CH, Hwan CL, Tsai HS et al (2007) An experimental investigation into the mechanical behaviours of helical composite springs. Comp Struct 77(3):331–340
207. Patil RV, Reddy PR, Laxminarayana P (2014) Comparison of cylindrical and conical helical springs for their buckling load and deflection. Int J Adv Sci Technol 73:33–50
208. Patil RV, Reddy PR, Laxminarayana P (2015) Development of buckling equation for conical spring and its experimental verification. Am J Eng Technol Res 15(1):67–81
209. Patil RV, Reddy PR, Laxminarayana P (2015) Establishment of buckling equation for conical spring and its experimental verification. J Autom Appl Mech 3(1)
210. Choi BL, Choi BH (2015) Numerical method for optimizing design variables of carbon-fiber reinforced epoxy composite coil springs. Compos Part B 82:42–49
211. Xiong ZY, Song RX, Kang ZX et al (2015) Analysis on rigidity of composite helical spring and its influence factors. Eng Mech 32(9):216–221. and 228
212. Gobbi M, Mastinu G (2001) On the optimal design of composite material tubular helical springs. Meccanica 36(5):525–553
213. Babenko AY, Soltannia B, Mobarakeh PS (2018) Solving geometrically nonlinear problem on deformation of a helical spring through variational methods. Int J Mech Appl 8(1):69–79
214. Savi MA, Pacheco PMC, Garcia MS et al (2015) Nonlinear geometric influence on the mechanical behaviour of shape memory alloy helical springs. Smart Mater Struct 24(3):035012

215. Toi Y, Lee JB, Taya M (2004) Finite element analysis of superelastic, large deformation behaviour of shape memory alloy helical springs. Comput Struct 82(20–21):1685–1693
216. Hashemi YM, Kadkhodaei M (2018) The effects of geometric parameters under small and large deformations on dissipative performance of shape memory alloy helical springs. J Stress Anal 3(1):69–79
217. Goncharenko MV, Goncharenko VV (2009) Geometric nonlinearity of mechanical behaviour as a consequence of large deformations of springs. J Eng Phys Thermophys 82(6):1229–1234

Chapter 2
Energy Method and Variational Principles

Abstract This chapter introduces the fundamental concepts and solution methods of the energy method and variational principles. The energy method describes mechanical laws in terms of energy variation, while variational principles represent physical laws in a variational form, both transforming elastic mechanics problems into functional extremum problems. It elaborates on the definitions of functionals and variations, calculations of strain energy and complementary energy in elastic bodies, and the minimum potential energy principle, minimum complementary energy principle, and Hamilton's variational principle. Additionally, it presents two solution methods: the Rayleigh-Ritz method, which assumes displacement functions to solve for undetermined parameters based on the minimum potential energy principle, and the Galerkin method, a representative of weighted residual methods, which obtains approximate solutions by minimizing weighted residuals.

The energy method, also known as the energy principle, is a mechanical law expressed in terms of energy variation. Broadly speaking, among all possible states that satisfy certain constraints, the true state should make the energy attain an extremum (or stationary value). Variational principles represent physical laws in a variational form, stating that in all possible states of motion of a system that satisfy certain constraints, the true motion state results in a physical quantity (such as the potential energy functional) reaching an extremum (or stationary value) [1–4]. Generally, for elastic mechanics problems, the entire elastic system must be considered, and the energy relationships of the system must be accounted for. This involves establishing variational equations for some functionals, reducing the elastic mechanics problem to a variational problem of seeking the extremum (or stationary value) of the functional under given constraints.

There are several types of energy principles, which can be broadly divided into two categories: the basic energy principle and the generalized variational principle. The basic energy principle was introduced earliest, initially arising naturally from physical concepts, and is also referred to as the natural energy principle. It includes two fundamental principles: the minimum potential energy principle and the minimum complementary energy principle. The minimum potential energy principle uses displacement as the primary variable, while the minimum complementary

J.-B. Bai et al., *Folding Mechanics of Deployable Composite Structures*, Solid Mechanics and Its Applications 283,
https://doi.org/10.1007/978-3-031-99254-4_2

energy principle uses stress as the primary variable. In other words, the basic energy principle is the typical energy principle that takes a single type of variable (displacement $\boldsymbol{u}$ or stress $\boldsymbol{\sigma}$) as the basic variable. The generalized variational principle typically refers to variational principles other than the minimum potential energy principle and the minimum complementary energy principle [2], often using displacement $\boldsymbol{u}$, strain $\boldsymbol{\varepsilon}$, and stress $\boldsymbol{\sigma}$ as basic variables, such as second-class and third-class variational principles. Many books on variational principles exist, with representative works including the classical texts by Hu [3, 4], Qian [5], and Washizu [6]. This chapter provides a brief introduction to the basic knowledge related to the energy method and variational principles.

2.1 Functionals and Variations

For each value of the independent variable x within a certain domain, there is a corresponding value of the dependent variable y. This relationship between the independent and dependent variables is called a function, denoted as $y = y(x)$. Suppose the form of a function $y(x)$ changes to a new function $Y(x)$, as shown in Fig. 2.1. If y has a small increment corresponding to a fixed value of x, the change is expressed as

$$\delta y(x) = Y(x) - y(x) \tag{2.1}$$

where $\delta y(x)$ is an increment referred to as the variation of the function $y(x)$. The definition of variation is similar to that of a differential, and variations and differentials can be understood through the following comparison: (1) A differential refers to a situation where the function $y(x)$ remains unchanged, and the independent variable x has an infinitesimal increment dx. Then, dy is called the differential of the function, and dy = y'dx, where the symbol y' represents the first derivative of y with respect to x. (2) A variation refers to a situation where the independent variable x remains unchanged, and the function $y(x)$ has an infinitesimal increment δy. Then,

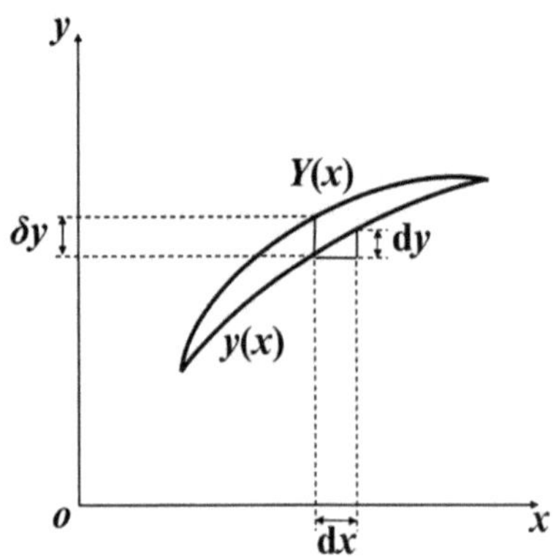

Fig. 2.1 Schematic diagram of the relationship between variation and differential

δy is called the variation of the function, and $\delta y = Y(x) - y(x)$. Additionally, the order of variations and differentials can be interchanged, i.e., $(\delta y)' = \delta(y')$.

If for every function $y(x)$ in a certain class of functions, there corresponds a value of a variable I, then the variable I is called a functional dependent on the function $y(x)$ and is denoted as $I = I[y(x)]$. Typically, functionals have the following form: $I\left[y(x)\right] = \int_a^b f\left(x,y,y'\right)\mathrm{d}x$. A functional is a type of function whose independent variable is a function. In simple terms, a functional is a function of a function. The integrand is generally a compound function of the independent variable x, the function $y(x)$, and its derivative y'. When the first-order variation of the functional $\delta I = 0$, the functional $I[y(x)]$ attains an extremum. Typically, problems concerning the extremum of a functional are classified as variational problems, while the variational principle primarily focuses on methods for solving functional extreme value problems.

2.2 The Energy of Elastic Body

Assume that during the loading process, the elastic body remains in equilibrium, meaning there is no change in kinetic energy, and thus no change in its nonmechanical energy. As a result, the decrease in potential energy from external forces (i.e., the work done by the external force) is entirely converted into strain energy, which is stored within the elastic body. Strain energy can be calculated by the work done by stress over the corresponding strain.

Assume that the elastic body is subjected to a uniform tensile stress σ_x in the x direction, with the corresponding linear strain ε_x. The strain energy per unit volume, or strain energy density, is given by

$$v_\varepsilon = \int_0^{\varepsilon_x} \sigma_x d\varepsilon_x \tag{2.2}$$

Strain energy density is a functional with strain components as its independent variables. It is represented in Fig. 2.2 as the area under the stress-strain curve.

The area in the upper-left part of the stress-strain curve in Fig. 2.2 is denoted as

$$v_c = \int_0^{\sigma_x} \varepsilon_x \mathrm{d}\sigma_x \tag{2.3}$$

where v_c represents the complementary strain energy per unit volume, also known as the complementary strain energy density. It is a functional with stress components as its independent variables, and in Fig. 2.2, it corresponds to the area of the rectangle $\sigma_x\varepsilon_x$, excluding the remaining portion of v_ε. When the stress-strain relationship of the elastic body is linear, since $\sigma_x = E\varepsilon_x$, or as seen from Fig. 2.2, we have

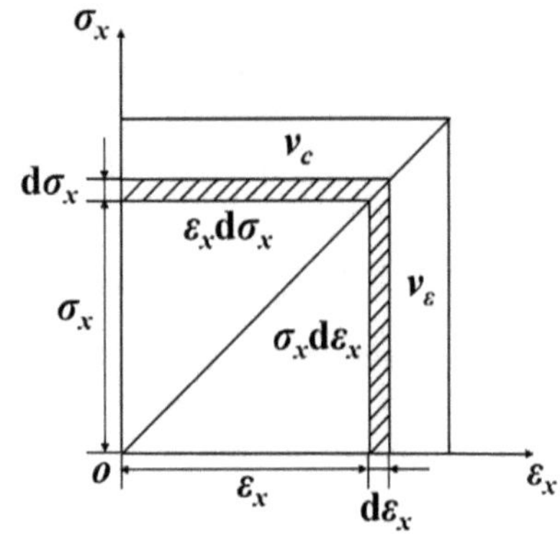

Fig. 2.2 Stress-strain curve

$$\begin{aligned} v_\varepsilon &= \int_0^{\varepsilon_x} \varepsilon_x \mathrm{d}\sigma_x = \frac{1}{2}\sigma_x\varepsilon_x \\ v_c &= \int_0^{\sigma_x} \sigma_x \mathrm{d}\varepsilon_x = \frac{1}{2}\varepsilon_x\sigma_x \end{aligned} \tag{2.4}$$

If the elastic body is subjected to all six stress components σ_x, σ_y, σ_z, τ_{yz}, τ_{xz}, τ_{xy}, the corresponding strain energy can be calculated using the method described above, by superimposing the strain energy densities for each stress component. The total strain energy density is given by

$$v_\varepsilon = \frac{1}{2}\left(\sigma_x\varepsilon_x + \sigma_x\varepsilon_x + \sigma_x\varepsilon_x + \tau_{yz}\gamma_{yz} + \tau_{xz}\gamma_{xz} + \tau_{xy}\gamma_{xy}\right) \tag{2.5}$$

Therefore, the total strain energy of the elastic body is

$$U_\varepsilon = \int_V v_\varepsilon \mathrm{d}V \tag{2.6}$$

Strain energy can be expressed as a functional with strain or stress components as independent variables, and the corresponding expressions can be derived using physical or geometric equations. Additionally, it can be shown that the strain energy density of the elastic body, when differentiated with respect to any stress component, yields the corresponding strain component. Similarly, the partial derivative of the strain energy density with respect to any strain component equals the corresponding stress component, as shown in Eq. (2.7):

$$\begin{cases} \dfrac{\partial v_\varepsilon}{\partial \varepsilon_x} = \sigma_x, \quad \dfrac{\partial v_\varepsilon}{\partial \varepsilon_y} = \sigma_y, \quad \dfrac{\partial v_\varepsilon}{\partial \varepsilon_z} = \sigma_z \\ \dfrac{\partial v_\varepsilon}{\partial \gamma_{yz}} = \tau_{yz}, \quad \dfrac{\partial v_\varepsilon}{\partial \gamma_{zx}} = \tau_{zx}, \quad \dfrac{\partial v_\varepsilon}{\partial \gamma_{xy}} = \tau_{xy} \end{cases} \tag{2.7}$$

Similarly, the complementary strain energy of the entire elastic body can be derived in a similar manner, and its expression is as follows:

$$V_c = \int_V v_c \mathrm{d}V \tag{2.8}$$

The expression for the complementary strain energy density v_c in the equation, when the stress-strain relationship is linear, is as follows:

$$v_c = \frac{1}{2}\left(\varepsilon_x\sigma_x + \varepsilon_y\sigma_y + \varepsilon_z\sigma_z + \gamma_{yz}\tau_{yz} + \gamma_{zx}\tau_{zx} + \gamma_{xy}\tau_{xy}\right) \tag{2.9}$$

Note that the complementary strain energy is a functional with stress components as its independent variables. Therefore, it can be proven that the rate of change of the elastic body's complementary strain energy density with respect to any stress component is equal to the corresponding strain component. The specific expression is as follows:

$$\begin{cases} \dfrac{\partial v_c}{\partial \sigma_x} = \varepsilon_x, & \dfrac{\partial v_c}{\partial \sigma_y} = \varepsilon_y, & \dfrac{\partial v_c}{\partial \sigma_z} = \varepsilon_z \\ \dfrac{\partial v_c}{\partial \tau_{yz}} = \gamma_{yz}, & \dfrac{\partial v_c}{\partial \tau_{zx}} = \gamma_{zx}, & \dfrac{\partial v_c}{\partial \tau_{xy}} = \gamma_{xy} \end{cases} \tag{2.10}$$

2.3 The Minimum Potential Energy Principle

Consider any elastic body in equilibrium under certain external forces, including body forces f_x, f_y, f_z and surface forces $\overline{f}_x, \overline{f}_y, \overline{f}_z$. Assume that there is a set of displacement components (u, v, w), which represent the actual displacement components of the elastic body, satisfying the equilibrium differential equations expressed in terms of displacement components, as well as the displacement boundary conditions and stress boundary conditions expressed in terms of displacement components. Now, assume that under the geometrical constraints of the elastic body, these displacement components have a small change, known as the virtual displacement or variation of displacement δu, δv, δw, resulting in a new set of displacements $u' = u + \delta u$, $v' = v + \delta v$, and $w' = w + \delta w$.

At this point, the work done by the external forces on the virtual displacements, that is, the virtual work, is

$$\begin{aligned} \delta V_\varepsilon &= \int_V \left(f_x\delta u + f_y\delta v + f_z\delta w\right)\mathrm{d}V \\ &+ \int_{S_\sigma} \left(\overline{f}_x\delta u + \overline{f}_y\delta v + \overline{f}_z\delta w\right)\mathrm{d}S \end{aligned} \tag{2.11}$$

where S is the boundary of the elastic body, and S_σ is the boundary of the applied external forces. This equation is the variational equation for displacement.

Assume that during the virtual displacement of the elastic body, there are no changes in temperature or velocity, meaning there is no change in thermal or kinetic energy. According to the law of conservation of energy, the increment of strain energy δU during the virtual displacement should be equal to the virtual work δV done by the external force on the virtual displacement, i.e., $\delta U = \delta V$:

$$\begin{aligned}\delta U_\varepsilon &= \int_V \left(f_x \delta u + f_y \delta v + f_z \delta w\right) \mathrm{d}V \\ &+ \int_{S_\sigma} \left(\overline{f}_x \delta u + \overline{f}_y \delta v + \overline{f}_z \delta w\right) \mathrm{d}S\end{aligned} \tag{2.12}$$

According to the variational principle, the operation of variation can be interchanged with that of definite integration, yielding

$$\delta U_\varepsilon = \delta \int_V v_\varepsilon \mathrm{d}V = \int_V \delta v_\varepsilon \mathrm{d}V \tag{2.13}$$

By considering the strain energy density v_ε as a function of the strain components and applying Eq. (2.7), we obtain

$$\begin{aligned}\delta U_\varepsilon &= \int_V \left(\frac{\partial v_\varepsilon}{\partial \varepsilon_x}\delta\varepsilon_x + \frac{\partial v_\varepsilon}{\partial \varepsilon_y}\delta\varepsilon_y + \frac{\partial v_\varepsilon}{\partial \varepsilon_z}\delta\varepsilon_z + \frac{\partial v_\varepsilon}{\partial \gamma_{yz}}\delta\gamma_{yz} + \frac{\partial v_\varepsilon}{\partial \gamma_{xz}}\delta\gamma_{xz} + \frac{\partial v_\varepsilon}{\partial \gamma_{xy}}\delta\gamma_{xy}\right)\mathrm{d}V \\ &= \int_V \left(\sigma_x\delta\varepsilon_x + \sigma_y\delta\varepsilon_y + \sigma_z\delta\varepsilon_z + \tau_{yz}\delta\gamma_{yz} + \tau_{xz}\delta\gamma_{xz} + \tau_{xy}\delta\gamma_{xy}\right)\mathrm{d}V\end{aligned} \tag{2.14}$$

Substituting Eq. (2.14) into Eq. (2.11),

$$\begin{aligned}&\int_V \left(f_x \delta u + f_y \delta v + f_z \delta w\right) \mathrm{d}V + \int_{S_\sigma} \left(\overline{f}_x \delta u + \overline{f}_y \delta v + \overline{f}_z \delta w\right) \mathrm{d}S \\ &= \int_V \left(\sigma_x\delta\varepsilon_x + \sigma_y\delta\varepsilon_y + \sigma_z\delta\varepsilon_z + \tau_{yz}\delta\gamma_{yz} + \tau_{xz}\delta\gamma_{xz} + \tau_{xy}\delta\gamma_{xy}\right)\mathrm{d}V\end{aligned} \tag{2.15}$$

This is known as the principle of virtual displacement, also referred to as the principle of virtual work. Thus, the principle of virtual displacement for an elastic body can be stated as follows: If an elastic body is in equilibrium under a known distribution of body forces and surface forces, then the virtual work done by these external forces in a virtual displacement process is equal to the virtual strain energy accumulated in the elastic body. Conversely, for any virtual displacement, if the external virtual work of the object is equal to its internal virtual work, the object must be in a state of equilibrium.

The constraints of the virtual displacement principle are the geometric equations and displacement boundary conditions within the domain. It can be proven that the principle of virtual work is equivalent to the equilibrium equations and force boundary conditions; by substituting the strain-displacement relations and displacement boundary conditions into the principle of virtual work, the equilibrium equations and force boundary conditions can be obtained.

According to Eq. (2.11), since the virtual displacement is infinitesimally small, the magnitude and direction of the external forces can be assumed to remain constant throughout the virtual displacement process. Therefore, the variation symbol within the integral can be brought outside the integral and simplified, resulting in

$$\delta\left[U_\varepsilon - \int_V \left(f_x u + f_y v + f_z w\right)\mathrm{d}V - \int_{S_\sigma} \left(\bar{f}_x u + \bar{f}_y v + \bar{f}_z w\right)\mathrm{d}S\right] = 0 \tag{2.16}$$

By setting,

$$V = \int_V \left(f_x u + f_y v + f_z w\right)\mathrm{d}V + \int_{S_\sigma} \left(\bar{f}_x u + \bar{f}_y v + \bar{f}_z w\right)\mathrm{d}S \tag{2.17}$$

where V represents the work done by external forces on the actual displacements u, v, and w. Assuming that the external force originates from a potential field, ($-V$) should equal the potential energy of the external force. Using V_p to denote this, where $u = v = w = 0$ represents the natural state with zero potential energy, we get

$$V_p = -\int_V \left(f_x u + f_y v + f_z w\right)\mathrm{d}V - \int_{S_\sigma} \left(\bar{f}_x u + \bar{f}_y v + \bar{f}_z w\right)\mathrm{d}S \tag{2.18}$$

Substituting back, the sum of the strain energy of the elastic body and the potential energy of the external forces is called the total potential energy of the elastic system, denoted by Π, which is given by

$$\delta\Pi = \delta\left(V_\varepsilon + V_p\right) = 0 \tag{2.19}$$

This is the minimum potential energy principle: For an elastic body in equilibrium under a given external force and satisfying displacement boundary conditions, the actual displacement configuration will make the total potential energy reach an extremum. Considering the second variation, $\delta^2\Pi > 0$ is obtained, proving that for a stable equilibrium state, this extremum is a minimum. Furthermore, due to the uniqueness of solutions in elasticity theory, the minimum of the total potential energy is indeed the absolute minimum.

The minimum potential energy principle further states that, among all possible virtual displacements, the virtual displacements that also satisfy equilibrium conditions correspond to the smallest structural potential energy variation. This principle not only provides the condition for the stationary value of potential energy but also, for linearly elastic structures, the second variation of the potential energy is always positive, which means the total potential energy of the structure achieves a minimum. Thus, the structure reaches the minimum potential energy in the actual displacement configuration.

2.4 The Minimum Complementary Energy Principle

Consider any elastic body in equilibrium under the action of external forces. The six stress components (σ_x, σ_y, σ_z, τ_{yz}, τ_{xz}, τ_{xy}) are the actual stress components, which satisfy the equilibrium differential equations, stress boundary conditions, and compatibility equations, and the corresponding displacements also satisfy the displacement boundary conditions. Now, assume that the body forces and the surface forces given by the stress boundary conditions remain unchanged, but the stress components undergo a small variation ($\delta\sigma_x$, …, $\delta\tau_{yz}$, …), known as virtual stresses or variations of stress, which change the stress components to $\sigma_x^{'} = \sigma_x + \delta\sigma_x, \ldots, \tau_{yz}^{'} = \tau_{yz} + \delta\tau_{yz}, \ldots$.

If strain energy is considered as a function of stress components, the variation of strain energy is given by

$$\delta V_c = \int_V \delta v_c \mathrm{d}V = \int_V \left(\frac{\partial v_c}{\partial \sigma_x} \delta\sigma_x + \cdots + \frac{\partial v_c}{\partial \tau_{yz}} \delta\tau_{yz} + \cdots \right) \mathrm{d}V \tag{2.20}$$

Substituting Eq. (2.10) into Eq. (2.20), we get

$$\delta V_c = \int_V \left(\varepsilon_x \delta\sigma_x + \cdots + \gamma_{yz} \delta\tau_{yz} + \cdots \right) \mathrm{d}V \tag{2.21}$$

Then, substituting the equilibrium equations and boundary conditions satisfied by the stress components, we can simplify the expression to

$$\delta V_c = \int_{S_u} \left(\overline{u} \delta p_x + \overline{v} \delta p_y + \overline{w} \delta p_z \right) \mathrm{d}S \tag{2.22}$$

This is the stress variation equation, where δp_x, δp_y, and δp_z represent the variations in the surface force components, known as virtual surface forces. The integral on the right-hand side of this equation is taken over the displacement boundary and represents the work done by the variation in surface forces on the actual given displacement. Therefore, due to the variation in stress, the variation of strain energy equals the work done by the virtual surface forces on the actual displacement.

As mentioned above, if the surface force is given on a certain part of the boundary, then the surface force on that part of the boundary cannot vary, so $\delta p_x = \delta p_y = \delta p_z = 0$, and the corresponding integral term on the right-hand side of Eq. (2.22) becomes zero. Similarly, if the given displacement is zero on a certain part of the boundary, the corresponding integral term on the right-hand side of Eq. (2.22) also becomes zero. Therefore, the integral on the right-hand side of the stress variation Eq. (2.22) should only be taken over the boundaries that satisfy the following conditions: the surface force is not given, and the given displacement is not equal to zero.

By applying Eq. (2.10), the corresponding principle of virtual stress can be derived:

$$\begin{aligned}\delta V_c &= \int_V \left(\begin{aligned} &\frac{\partial v_c}{\partial \sigma_x}\delta\sigma_x + \frac{\partial v_c}{\partial \sigma_y}\delta\sigma_y + \frac{\partial v_c}{\partial \sigma_z}\delta\sigma_z + \frac{\partial v_c}{\partial \gamma_{yz}}\delta\tau_{yz} \\ &+ \frac{\partial v_c}{\partial \gamma_{xz}}\delta\tau_{xz} + \frac{\partial v_c}{\partial \gamma_{xy}}\delta\tau_{xy} \end{aligned} \right) \mathrm{d}V \\ &= \int_V \left(\varepsilon_x \delta\sigma_x + \varepsilon_y \delta\sigma_y + \varepsilon_z \delta\sigma_z + \gamma_{yz}\delta\tau_{yz} + \gamma_{xz}\delta\tau_{xz} + \gamma_{xy}\delta\tau_{xy} \right) \mathrm{d}V\end{aligned} \tag{2.23}$$

Substituting into Eq. (2.22), it can be simplified to

$$\begin{aligned}&\int_{S_u} \left(\overline{u}\delta P_x + \overline{v}\delta Py + \overline{w}\delta P_z \right) \mathrm{d}V \\ &= \int_V \left(\varepsilon_x \delta\sigma_x + \varepsilon_y \delta\sigma_y + \varepsilon_z \delta\sigma_z + \gamma_{yz}\delta\tau_{yz} + \gamma_{xz}\delta\tau_{xz} + \gamma_{xy}\delta\tau_{xy} \right) \mathrm{d}V\end{aligned} \tag{2.24}$$

This is the principle of virtual stress. It states that if the elastic body is in equilibrium before the virtual stress occurs, then during the process of virtual stress, the work done by the virtual surface forces on the given displacement boundary is equal to the work done by the virtual stress on the strain within the elastic body.

It can be proven that the principle of virtual stress is equivalent to the geometric equation and displacement boundary conditions. In other words, substituting the equilibrium equations and force boundary conditions into the principle of virtual stress yields the geometric equations describing the strain-displacement relationship and displacement boundary conditions.

From the stress variation equation, the minimum complementary energy principle can be derived:

$$\begin{aligned}&\delta V_c - \int_{S_u} \left(\overline{u}\delta p_x + \overline{v}\delta p_y + \overline{w}\delta p_z \right) \mathrm{d}S = 0 \\ &\delta\left[V_c - \int_{S_u} \left(\overline{u}p_x + \overline{v}p_y + \overline{w}p_z \right) \mathrm{d}S \right] = 0\end{aligned} \tag{2.25}$$

where $V_c - \int_{S_u} \left(\overline{u}p_x + \overline{v}p_y + \overline{w}p_z \right) \mathrm{d}S$ represents the total complementary energy of the elastic body. It can be seen that, among all possible stress states that satisfy the equilibrium equations and stress boundary conditions, the actual stress state should make the total complementary energy of the elastic body an extremum. If second-order variations are considered, then $\delta^2 \left[V_c - \int_{S_u} \left(\overline{u}p_x + \overline{v}p_y + \overline{w}p_z \right) \mathrm{d}S \right] > 0$, and it can be proven that this extremum is a minimum; thus, the above conclusion is referred to as the minimum complementary energy principle.

Additionally, the minimum complementary energy principle can be further elaborated as follows: among all permissible stress states that satisfy the equilibrium equations and stress boundary conditions in the elastic body domain, the actual stress (i.e., the stress that satisfies the geometric equations and displacement boundary conditions) must make the total complementary energy a minimum. Conversely, the stress that makes the total complementary energy an extremum must be the actual stress.

From this analysis, it can be concluded that the actual stress, in addition to satisfying the equilibrium equations and stress boundary conditions, should also satisfy the compatibility equations, and the corresponding displacement should satisfy the displacement boundary conditions. Furthermore, the actual stress, in addition to satisfying the equilibrium differential equations and stress boundary conditions, also satisfies the stress variation equation. Moreover, through calculations, the compatibility conditions can also be derived from the stress variation equation. Therefore, it is evident that the stress variation equation can replace the compatibility conditions and displacement boundary conditions.

2.5 Hamilton's Variational Principle

Hamilton's principle is the extension of the principle of virtual displacement to dynamic problems. For static problems, the minimum potential energy principle can be used to obtain the static equilibrium equations and the boundary conditions for forces within the domain. For dynamic problems, Hamilton's variational principle can be used to derive the dynamic equilibrium equations and the boundary conditions for forces. Hamilton's variational principle has forms based on one variable and two variables. Here, a brief introduction to the one-variable form of Hamilton's principle based on Euclidean geometry is provided. For other details, please refer to the relevant literatures [2, 7].

Consider the following functional:

$$I = \int_0^t L\left(t, q_1, q_2, \ldots, q_n, \dot{q}_1, \dot{q}_2, \ldots, \dot{q}_n\right) \mathrm{d}t \tag{2.26}$$

where "$\dot{*}$" represents the first derivative of the physical quantity "*" with respect to time t; $L = T\text{-}V\text{-}W$ is the Lagrange function, and W is the potential energy due to external forces. The self-variable function in the functional $q(t)$ corresponds to the generalized displacement. The true motion of the system causes the functional I to reach a stationary value:

$$\delta I = 0 \tag{2.27}$$

This is Hamilton's variational principle, and its necessary and sufficient condition for the functional I to reach a stationary value is that its first variation is zero, which is given by

$$\begin{aligned}\delta I &= \sum_{i=1}^{n}\int_0^t\left(\frac{\partial L}{\partial q_i}\delta q_i+\frac{\partial L}{\partial \dot{q}_i}\delta \dot{q}_i\right)\mathrm{d}t \\ &= \sum_{i=1}^{n}[\int_0^t\left(\frac{\partial L}{\partial q_i}-\frac{\mathrm{d}}{\mathrm{d}t}\frac{\partial L}{\partial \dot{q}_i}\right)\delta q_i\mathrm{d}t+\frac{\partial L}{\partial \dot{q}_i}\delta q_i\Big|_0^t]=0\end{aligned} \tag{2.28}$$

Simplifying, we get

$$\frac{\partial L}{\partial q_i}-\frac{\mathrm{d}}{\mathrm{d}t}\frac{\partial L}{\partial \dot{q}_i}=0 \tag{2.29}$$

$$\frac{\partial L}{\partial \dot{q}_i}\delta q_i\Big|_0^t=0 \tag{2.30}$$

The above equation is the Lagrange equation, also known as the dynamic equilibrium equation. Equation (2.30) is the condition of Hamilton's variational principle, which states that the generalized displacement variation at the initial and final moments of time is zero. The physical meaning is that the configuration of the mechanical system remains unchanged at the start and end times [8].

2.6 Typical Solution Methods

2.6.1 Rayleigh-Ritz Method

The Rayleigh-Ritz method, also known as the Ritz method, is an approximate method for finding unknown functions by applying the stationary value condition of a functional. First introduced by Rayleigh in 1877 and later proposed as an effective method by Ritz in 1908, this method has been applied in many problems in mechanics, physics, and quantum chemistry. It is also a widely used classical numerical method in applied mathematics and mechanical engineering, particularly for calculating the lower-order natural frequencies of structures. The Rayleigh-Ritz method is a direct variational method, based on the minimum potential energy principle. It approximates the exact solution to a problem by selecting a trial function, substituting it into the functional of the scientific problem, and then finding the stationary value of the functional. This process determines the undetermined parameters in the trial function, providing an approximate solution to the problem [9, 10].

First, select a trial function:

$$\tilde{\mathbf{u}} = \begin{Bmatrix} \tilde{\mathbf{u}}_1 \\ \tilde{\mathbf{u}}_2 \\ \tilde{\mathbf{u}}_3 \end{Bmatrix} = \begin{Bmatrix} \sum_{i=1}^{n} \alpha_i \varphi_{xi} \\ \sum_{i=1}^{n} \beta_i \varphi_{yi} \\ \sum_{i=1}^{n} \gamma_i \varphi_{zi} \end{Bmatrix} = \sum_{i=1}^{n} \boldsymbol{\varphi}_i \mathbf{a}_i = \mathbf{N}\mathbf{a} \tag{2.31}$$

where

$$\boldsymbol{\varphi}_i = \begin{bmatrix} \varphi_{xi} & 0 & 0 \\ 0 & \varphi_{yi} & 0 \\ 0 & 0 & \varphi_{zi} \end{bmatrix} \tag{2.32}$$

$$\mathbf{a}_i = \left[\alpha_i \ \beta_i \ \gamma_i\right]^{\mathrm{T}} \tag{2.33}$$

$$\mathbf{N} = \left[\varphi_1 \ \varphi_2 \ldots \varphi_n\right]^{\mathrm{T}} \tag{2.34}$$

$$\mathbf{a} = \left[a_1 \ a_2 \ldots a_n\right]^{\mathrm{T}} \tag{2.35}$$

where $\boldsymbol{a}$ consists of $3n$ undetermined parameters, and φ_{xi}, φ_{yi}, and φ_{zi} are continuous functions of the coordinates that satisfy all displacement boundary conditions. The total potential energy of the system can be expressed as

$$\Pi = \int_V \left[U_\varepsilon - \tilde{\mathbf{u}}^{\mathrm{T}} \boldsymbol{f} \right] \mathrm{d}V - \int_{S_\sigma} \tilde{\mathbf{u}}^{\mathrm{T}} \bar{\boldsymbol{f}} \mathrm{d}S \tag{2.36}$$

where U_ε is the elastic strain energy, and $\boldsymbol{f}$ is the body force per unit volume, which can be expressed as $\boldsymbol{f} = [f_x \ f_y \ f_z]^{\mathrm{T}}$, and $\bar{\boldsymbol{f}}$ is the surface force per unit area, expressed as $\bar{\boldsymbol{f}} = \left[\bar{f}_x \ \ \bar{f}_y \ \ \bar{f}_z\right]^{\mathrm{T}}$.

Using the minimum potential energy principle, we obtain

$$\frac{\partial \Pi}{\partial \boldsymbol{a}} = 0 \tag{2.37}$$

where

$$\frac{\partial \Pi}{\partial \alpha_i} = 0, \quad \frac{\partial \Pi}{\partial \beta_i} = 0, \quad \frac{\partial \Pi}{\partial \gamma_i} = 0 \tag{2.38}$$

Combining Eqs. (2.36) and (2.37), we derive

$$\begin{cases} \int_V \left(\dfrac{\partial U_\varepsilon}{\partial \alpha_i} - \varphi_{xi} f_x \right) \mathrm{d}V - \int_{S_\sigma} \varphi_{xi} \bar{f}_x \mathrm{d}S = 0 \\ \int_V \left(\dfrac{\partial U_\varepsilon}{\partial \beta_i} - \varphi_{yi} f_y \right) \mathrm{d}V - \int_{S_\sigma} \varphi_{yi} \bar{f}_y \mathrm{d}S = 0 \\ \int_V \left(\dfrac{\partial U_\varepsilon}{\partial \gamma_i} - \varphi_{zi} f_z \right) \mathrm{d}V - \int_{S_\sigma} \varphi_{zi} \bar{f}_z \mathrm{d}S = 0 \end{cases} \tag{2.39}$$

or

$$\int_V \left(\frac{\partial U_\varepsilon}{\partial \mathbf{a}} - \boldsymbol{N}^{\mathrm{T}} \boldsymbol{f} \right) \mathrm{d}V - \int_{S_o} \boldsymbol{N}^{\mathrm{T}} \bar{\boldsymbol{f}} \mathrm{d}S = 0 \tag{2.40}$$

This results in a system of equations in terms of the parameters a_i, with the number of equations equal to the number of undetermined parameters. Once the parameters a_i are determined, an approximate solution can be obtained.

The accuracy of the approximate solution using the Ritz method depends on the number of parameters in the trial solution and the appropriate choice of continuous functions. Generally, accuracy increases with the number of terms. However, even with fewer terms, a well-chosen continuous function can yield a displacement solution with good accuracy. However, the accuracy of the stress solution obtained with this method is typically lower.

2.6.2 Galerkin Method

The Rayleigh-Ritz method is based on global assumptions; however, constructing global trial functions is extremely challenging, particularly when solving problems with complex geometries. This is one of the main reasons why the Rayleigh-Ritz method is not widely applicable to engineering problems. To address these challenges, the weighted residual method was developed, which is also referred to as the weighted residual approach or weighted residual technique. The Galerkin method is a representative approach within the framework of the weighted residual method [11, 12].

In this method, the solution to the problem is expressed as a linear combination of trial functions, and the approximate solution is obtained by minimizing the weighted residuals. The core idea of the Galerkin method is to map the solution space of the original problem to a finite-dimensional approximation space and approximate the true solution by solving the weighted residual problem within this space. When applied to complex problems, the Galerkin method offers a high degree of flexibility and effectiveness. However, its accuracy and stability still depend on the appropriate selection of trial functions and proper numerical implementation.

The approximate solution for the displacement is assumed to have the following form:

$$\tilde{\mathbf{u}} = \begin{Bmatrix} \tilde{\mathbf{u}}_1 \\ \tilde{\mathbf{u}}_2 \\ \tilde{\mathbf{u}}_3 \end{Bmatrix} = \begin{Bmatrix} \sum_{i=1}^{n} \alpha_i \phi_{xi} \\ \sum_{i=1}^{n} \beta_i \phi_{yi} \\ \sum_{i=1}^{n} \gamma_i \phi_{zi} \end{Bmatrix} = \sum_{i=1}^{n} \phi_i a_i \tag{2.41}$$

where $\phi_i = [\phi_{xi} \ \phi_{yi} \ \phi_{zi}]^{\mathrm{T}}$ is a continuously differentiable function that satisfies both the displacement and stress boundary conditions, and $\mathbf{a}_i = [\alpha_i \ \beta_i \ \gamma_i]^{\mathrm{T}}$ is an undetermined constant. The stress derived from this approximate displacement may not necessarily satisfy the equilibrium equation and thus has a residual:

$$\mathbf{A}\left(\tilde{\mathbf{u}}\right) = \begin{Bmatrix} \dfrac{\partial \sigma_x}{\partial x} + \dfrac{\partial \tau_{yx}}{\partial y} + \dfrac{\partial \tau_{zx}}{\partial z} + f_x \\ \dfrac{\partial \sigma_y}{\partial y} + \dfrac{\partial \tau_{zy}}{\partial z} + \dfrac{\partial \tau_{xy}}{\partial x} + f_y \\ \dfrac{\partial \sigma_z}{\partial z} + \dfrac{\partial \tau_{xz}}{\partial x} + \dfrac{\partial \tau_{yz}}{\partial y} + f_z \end{Bmatrix} \tag{2.42}$$

Selecting a weighting function ϕ_i, we have

$$\int_V \phi_i^{\mathrm{T}} \mathbf{A}\left(\tilde{\mathbf{u}}\right) \mathrm{d}V = 0 \quad \left(i = 1,2,3\right) \tag{2.43}$$

Integrating each term by parts, we obtain the weak form:

$$\begin{aligned} & -\int_V [\sigma_x \frac{\partial \phi_{xi}}{\partial x} + \sigma_y \frac{\partial \phi_{yi}}{\partial y} + \sigma_z \frac{\partial \phi_{zi}}{\partial z} \\ & + \tau_{xy}\left(\frac{\partial \phi_{xi}}{\partial y} + \frac{\partial \phi_{yi}}{\partial x}\right) + \tau_{yz}\left(\frac{\partial \phi_{yi}}{\partial z} + \frac{\partial \phi_{zi}}{\partial y}\right) + \tau_{zx}\left(\frac{\partial \phi_{xi}}{\partial z} + \frac{\partial \phi_{zi}}{\partial x}\right)]\mathrm{d}V \\ & -\int_V \left(\phi_{xi} b_x + \phi_{yi} b_y + \phi_{zi} b_z\right)\mathrm{d}V + \int_S [\phi_{xi}\left(\sigma_x n_x + \tau_{xy} n_y + \tau_{xz} n_z\right) \\ & + \tau_{xy}\left(\frac{\partial \phi_{xi}}{\partial y} + \frac{\partial \phi_{yi}}{\partial x}\right) + \tau_{yz}\left(\frac{\partial \phi_{yi}}{\partial z} + \frac{\partial \phi_{zi}}{\partial y}\right) + \tau_{zx}\left(\frac{\partial \phi_{xi}}{\partial z} + \frac{\partial \phi_{zi}}{\partial x}\right)]\mathrm{d}S \quad = 0 \end{aligned} \tag{2.44}$$

Expressing the above in matrix and vector form yields

$$\int_V (L'\phi_i)^{\mathrm{T}} \sigma \mathrm{d}V - \int_V \phi_i^{\mathrm{T}} f \mathrm{d}V - \int_S \phi_i^{\mathrm{T}} \bar{f} \mathrm{d}S = 0 \tag{2.45}$$

where

$$L' = \begin{bmatrix} \frac{\partial}{\partial x} & 0 & 0 \\ 0 & \frac{\partial}{\partial y} & 0 \\ 0 & 0 & \frac{\partial}{\partial z} \\ \frac{\partial}{\partial y} & \frac{\partial}{\partial z} & 0 \\ 0 & \frac{\partial}{\partial z} & \frac{\partial}{\partial x} \\ \frac{\partial}{\partial y} & 0 & \frac{\partial}{\partial x} \end{bmatrix} \tag{2.46}$$

This forms a linear system of equations with $3n$ equations, and once the constants a_i are determined, an approximate solution can be obtained.

References

1. Xu ZL (2016) Elasticity. Higher Education Press, Beijing
2. Xing YF (2019) Principles and methods of computational solid mechanics. Beihang University Press, Beijing
3. Hu HC (1981) Variational principles of elastic mechanics and applications. Science Press, Beijing
4. Hu HC (1987) Variational methods. China Architecture & Building Press, Beijing, p 133
5. Qian WC (1980) Variational methods and finite elements. Science Press, Beijing, p 260
6. Washizu K (1968) Variational methods in elasticity and plasticity. Pergamon Press Limited, London
7. Newmark NM (1959) A method of computation for structural dynamics. J Eng Mech Div 85(3):67–94
8. Lanczos C (1957) The variational principles of mechanics. Oxford University Press, London
9. Ilanko S, Monterrubio L, Mochida Y (2014) The Rayleigh-Ritz method for structural analysis. John Wiley & Sons
10. Kumar Y (2018) The Rayleigh-Ritz method for linear dynamic, static and buckling behaviour of beams, shells and plates: a literature review. J Vib Control 24(7):1205–1227
11. Galerkin BG (1915) Rods and plates. Series occurring in various questions concerning the elastic equilibrium of rods and plates. Eng Bullet (Vestnik inzhenerov) 19:897–908
12. Chandrupatla TR, Belegundu AD (2021) Introduction to finite elements in engineering. Cambridge University Press, Cambridge

Chapter 3
Basic Mechanics of Composite Materials

Abstract This chapter focuses on the fundamental mechanics of continuous fiber-reinforced resin-matrix composites from a macro-mechanical perspective, covering lamina and laminate theories. It begins with the basics of anisotropic elasticity, including 15 unknowns and fundamental equations (equilibrium, geometric, and constitutive equations), noting the reduction in independent elastic constants for materials with different symmetry planes. For lamina stiffness analysis, it deduces the stress-strain relationship of orthotropic materials under plane stress and transforms elastic coefficients between principal and off-axis directions. Lamina strength analysis introduces criteria (maximum stress, maximum strain, Tsai-Hill, etc.), while mechanical testing methods cover tensile, compression, in-plane shear, and bending tests. Laminate theory analyzes laminate characteristics, representations, and classifications, derives stiffness matrices based on classical assumptions, outlines strength analysis steps for ultimate load calculation, and details nine layup design principles (balanced, symmetric, load-bearing, etc.), providing theoretical support for composite structure design.

The study of composite mechanics aims to investigate the deformation, stress, and failure behavior of composites under external forces. Based on the refinement level of the mechanical models, the mechanical analysis methods for composite materials [1–3] can be categorized into macro-mechanics and micro-mechanics. In macro-mechanics, the mechanical properties of composite materials are analyzed under the assumption that the material is homogeneous, where the contribution of the constituent materials is simplified into the average apparent properties of the composite. The corresponding material parameters are typically obtained through experiments [4–8]. This chapter focuses on fiber-reinforced composite materials consisting of continuous fibers and a resin matrix and introduces the stiffness and strength analysis of a single ply (referred to as lamina) and laminated composites (referred to as laminates) from a macro-mechanical perspective [9–11].

J.-B. Bai et al., *Folding Mechanics of Deployable Composite Structures*, Solid Mechanics and Its Applications 283,
https://doi.org/10.1007/978-3-031-99254-4_3

3.1 Basic Mechanics of Anisotropic Elasticity

The macroscopic elastic properties of the lamina are generally homogeneous and anisotropic. Some constituent materials inherently exhibit significant anisotropy, making the theory of anisotropic elasticity broadly applicable in composite mechanics. The primary difference between the fundamental equations of anisotropic and isotropic elasticity lies in the constitutive equations, where Hooke's law for anisotropic materials replaces the isotropic version. This section simply discusses the fundamental equations of anisotropic elasticity.

The fundamental equations of anisotropic elasticity include 15 unknowns:

Displacement components: u, v, w
Strain components: ε_x, ε_y, ε_z, γ_{yz}, γ_{xz}, γ_{xy}
Stress components: σ_x, σ_y, σ_z, τ_{yz}, τ_{xz}, τ_{xy}

The following basic equations are given; equilibrium equations can be expressed as

$$\begin{cases} \dfrac{\partial \sigma_x}{\partial x} + \dfrac{\partial \tau_{xy}}{\partial y} + \dfrac{\partial \tau_{xz}}{\partial z} + f_x = \rho_\mathrm{d} \dfrac{\partial^2 u}{\partial t^2} \\ \dfrac{\partial \tau_{xy}}{\partial x} + \dfrac{\partial \sigma_y}{\partial y} + \dfrac{\partial \tau_{yz}}{\partial z} + f_y = \rho_\mathrm{d} \dfrac{\partial^2 v}{\partial t^2} \\ \dfrac{\partial \tau_{xz}}{\partial x} + \dfrac{\partial \tau_{xy}}{\partial y} + \dfrac{\partial \sigma_z}{\partial z} + f_z = \rho_\mathrm{d} \dfrac{\partial^2 w}{\partial t^2} \end{cases} \tag{3.1}$$

where f_x, f_y, and f_z represent the body force per unit volume, ρ_d is density, and t is time.

The geometric equations are as follows:

$$\begin{cases} \varepsilon_x = \dfrac{\partial u}{\partial x}, \ \gamma_{yz} = \dfrac{\partial w}{\partial y} + \dfrac{\partial v}{\partial z} \\ \varepsilon_y = \dfrac{\partial v}{\partial y}, \ \gamma_{xz} = \dfrac{\partial u}{\partial z} + \dfrac{\partial w}{\partial x} \\ \varepsilon_z = \dfrac{\partial w}{\partial z}, \ \gamma_{xy} = \dfrac{\partial v}{\partial x} + \dfrac{\partial u}{\partial y} \end{cases} \tag{3.2}$$

According to deformation compatibility equations, the strain components satisfy the relationship

$$\begin{cases} \dfrac{\partial^2 \varepsilon_x}{\partial y^2} + \dfrac{\partial^2 \varepsilon_y}{\partial x^2} = \dfrac{\partial^2 \gamma_{xy}}{\partial x \partial y} \\ \dfrac{\partial^2 \varepsilon_y}{\partial z^2} + \dfrac{\partial^2 \varepsilon_z}{\partial y^2} = \dfrac{\partial^2 \gamma_{yz}}{\partial y \partial z} \\ \dfrac{\partial^2 \varepsilon_z}{\partial x^2} + \dfrac{\partial^2 \varepsilon_x}{\partial z^2} = \dfrac{\partial^2 \gamma_{xz}}{\partial z \partial x} \\ \dfrac{\partial}{\partial x}\left(\dfrac{\partial \gamma_{xz}}{\partial y} + \dfrac{\partial \gamma_{xy}}{\partial z} - \dfrac{\partial \gamma_{yz}}{\partial x} \right) = 2\dfrac{\partial^2 \varepsilon_x}{\partial y \partial z} \\ \dfrac{\partial}{\partial y}\left(\dfrac{\partial \gamma_{xy}}{\partial z} + \dfrac{\partial \gamma_{yz}}{\partial x} - \dfrac{\partial \gamma_{zx}}{\partial y} \right) = 2\dfrac{\partial^2 \varepsilon_y}{\partial z \partial x} \\ \dfrac{\partial}{\partial z}\left(\dfrac{\partial \gamma_{yz}}{\partial x} + \dfrac{\partial \gamma_{zx}}{\partial y} - \dfrac{\partial \gamma_{xy}}{\partial z} \right) = 2\dfrac{\partial^2 \varepsilon_z}{\partial x \partial y} \end{cases} \tag{3.3}$$

The boundary conditions for prescribed forces are

$$\begin{cases} \sigma_x l + \tau_{xy} m + \tau_{xz} n = \overline{X} \\ \tau_{xy} l + \sigma_y m + \tau_{yz} n = \overline{Y} \\ \tau_{xz} l + \tau_{yz} m + \sigma_z n = \overline{Z} \end{cases} \tag{3.4}$$

and for prescribed displacements

$$\begin{cases} u = \overline{u} \\ v = \overline{v} \\ w = \overline{w} \end{cases} \tag{3.5}$$

The stress-strain relationship (constitutive equations) for an anisotropic elastic body is

$$\begin{Bmatrix} \sigma_x \\ \sigma_y \\ \sigma_z \\ \tau_{yz} \\ \tau_{zx} \\ \tau_{xy} \end{Bmatrix} = \begin{bmatrix} C_{11} & C_{12} & C_{13} & C_{14} & C_{15} & C_{16} \\ C_{21} & C_{22} & C_{23} & C_{24} & C_{25} & C_{26} \\ C_{31} & C_{32} & C_{33} & C_{34} & C_{35} & C_{36} \\ C_{41} & C_{42} & C_{43} & C_{44} & C_{45} & C_{46} \\ C_{51} & C_{52} & C_{53} & C_{54} & C_{55} & C_{56} \\ C_{61} & C_{62} & C_{63} & C_{64} & C_{65} & C_{66} \end{bmatrix} \begin{Bmatrix} \varepsilon_x \\ \varepsilon_y \\ \varepsilon_{xy} \\ \varepsilon_{yz} \\ \varepsilon_{xz} \\ \varepsilon_{xy} \end{Bmatrix} \tag{3.6}$$

or

$$\begin{Bmatrix} \varepsilon_x \\ \varepsilon_y \\ \varepsilon_z \\ \gamma_{yz} \\ \gamma_{xz} \\ \gamma_{xy} \end{Bmatrix} = \begin{bmatrix} S_{11} & S_{12} & S_{13} & S_{14} & S_{15} & S_{16} \\ S_{21} & S_{22} & S_{23} & S_{24} & S_{25} & S_{26} \\ S_{31} & S_{32} & S_{33} & S_{34} & S_{35} & S_{36} \\ S_{41} & S_{42} & S_{43} & S_{44} & S_{45} & S_{46} \\ S_{51} & S_{52} & S_{53} & S_{54} & S_{55} & S_{56} \\ S_{61} & S_{62} & S_{63} & S_{64} & S_{65} & S_{66} \end{bmatrix} \begin{Bmatrix} \sigma_x \\ \sigma_y \\ \sigma_z \\ \sigma_{yz} \\ \sigma_{xz} \\ \sigma_{xy} \end{Bmatrix} \tag{3.7}$$

where [$\boldsymbol{C}$] and [$\boldsymbol{S}$] represent the stiffness matrix and compliance matrix, respectively, and both are symmetric matrices (e.g., $C_{ij} = C_{ji}$, $S_{ij} = S_{ji}$). Therefore, the general anisotropic Hooke's law includes 21 independent elastic constants. In specific cases, the number of independent constants is simplified: for materials with a single elastic symmetric plane, there are 13 constants; for orthotropic materials with two symmetric planes, there are 9; for transversely isotropic materials, there are 5; and for isotropic materials, the constitutive equations simplify further, with only 2 independent elastic constants.

These mechanical, geometric, physical, and boundary conditions form the 15 fundamental equations of anisotropic elasticity mechanics, which allow for the determination of 15 unknowns: the stress, strain, and displacement components. Compared to the fundamental equations of isotropic elasticity, the primary distinction lies in the stress-strain relationships.

3.2 Basic Mechanics of a Lamina

3.2.1 Stiffness Analysis

In most cases, a lamina is not used alone but as a basic unit in a composite laminated structure. At this point, the thickness along with the lamina (with the thickness direction set as the 3-axis direction, perpendicular to the plane of the ply) is generally much smaller compared to the dimensions in other in-plane directions (the 1-axis and 2-axis principal directions, 1-axis along the fiber direction, 2-axis perpendicular to the fibers of the ply plane). Therefore, it can be approximated as $\sigma_z = \tau_{yz} = \tau_{xz} = 0$, which defines the plane stress condition. For orthotropic materials, the strain-stress relationship under plane stress is

$$\begin{cases} \begin{bmatrix} \varepsilon_1 \\ \varepsilon_2 \\ \gamma_{12} \end{bmatrix} = \begin{bmatrix} S_{11} & S_{12} & 0 \\ S_{12} & S_{22} & 0 \\ 0 & 0 & S_{66} \end{bmatrix} \begin{bmatrix} \sigma_1 \\ \sigma_2 \\ \tau_{12} \end{bmatrix} = \mathbf{S}\sigma \\ \gamma_{31} = \gamma_{23} = 0 \\ \varepsilon_3 = S_{13}\sigma_1 + S_{23}\sigma_2 \end{cases} \tag{3.8}$$

where

$$\begin{cases} S_{11} = \dfrac{1}{E_1}, \quad S_{22} = \dfrac{1}{E_2}, \quad S_{66} = \dfrac{1}{G_{12}} \\ S_{12} = \dfrac{-\nu_{21}}{E_1} = \dfrac{-\nu_{12}}{E_2}, \quad S_{13} = \dfrac{-\nu_{31}}{E_1}, \quad S_{23} = \dfrac{-\nu_{32}}{E_2} \end{cases} \tag{3.9}$$

Writing Eq. (3.8) as a relationship expressing stress in terms of strain,

$$\begin{bmatrix} \sigma_1 \\ \sigma_2 \\ \tau_{12} \end{bmatrix} = \begin{bmatrix} Q_{11} & Q_{12} & 0 \\ Q_{12} & Q_{22} & 0 \\ 0 & 0 & Q_{66} \end{bmatrix} \begin{bmatrix} \varepsilon_1 \\ \varepsilon_2 \\ \gamma_{12} \end{bmatrix} \tag{3.10}$$

This is the stress-strain relationship for the principal direction of the lamina, where [***Q***] is the two-dimensional stiffness matrix; it can be obtained by inverting the two-dimensional compliance matrix [***S***]:

$$\begin{cases} Q_{11} = \dfrac{S_{22}}{S_{11}S_{22} - S_{12}^2}, \quad Q_{22} = \dfrac{S_{11}}{S_{11}S_{22} - S_{12}^2} \\ Q_{12} = \dfrac{-S_{12}}{S_{11}S_{22} - S_{12}^2}, \quad Q_{22} = \dfrac{1}{S_{66}} \end{cases} \tag{3.11}$$

where Q_{ij} is used instead of C_{ij} as the stiffness matrix coefficient; under plane stress, there is a practical difference between the two. That is, $Q_{ij} \neq C_{ij}$, which is generally reduced. Therefore, in some texts, Q_{ij} is referred to as components of the reduced stiffness matrix, while the compliance matrix remains represented by [***S***]. If all the S_{ij} coefficients are inverted overall, then $[\boldsymbol{C}] = [\boldsymbol{S}^{-}]$. The relationship between C_{ij} and the two-dimensional stiffness matrix [***Q***] for plane stress problems is as follows:

$$\begin{cases} Q_{11} = C_{11} - \dfrac{C_{12}^2}{C_{22}}, \quad Q_{12} = C_{12} - \dfrac{C_{12}C_{23}}{C_{22}} \\ Q_{22} = C_{22} - \dfrac{C_{23}^2}{C_{22}}, \quad Q_{66} = C_{66} \end{cases} \tag{3.12}$$

From this equation, it is evident that, except for C_{66}, $Q_{ij} < C_{ij}$ generally.

The components of Q_{ij} in terms of the reduced stiffness coefficients are given as

$$\begin{cases} Q_{11} = C_{11} - \dfrac{C_{12}^2}{C_{22}}, \quad Q_{12} = C_{12} - \dfrac{C_{12}C_{23}}{C_{22}} \\ Q_{22} = C_{22} - \dfrac{C_{23}^2}{C_{22}}, \quad Q_{66} = C_{66} \end{cases} \tag{3.13}$$

The above stress-strain relationship is based on the stiffness and stress analysis of the lamina subjected to external forces, where stress and strain are defined in the principal directions.

Due to $\frac{v_{12}}{E_2} = \frac{v_{21}}{E_1}$, for orthotropic materials under plane stress, there are four independent elastic constants, E_1, E_2, v_{21}, and G_{12}, and the stiffness and compliance coefficients each have four independent terms.

For isotropic materials under plane stress state, the strain-stress relationship is

$$\begin{bmatrix} \varepsilon_1 \\ \varepsilon_2 \\ \gamma_{12} \end{bmatrix} = \begin{bmatrix} S_{11} & S_{12} & 0 \\ S_{12} & S_{11} & 0 \\ 0 & 0 & 2(S_{11} - S_{12}) \end{bmatrix} \begin{bmatrix} \sigma_1 \\ \sigma_2 \\ \tau_{12} \end{bmatrix} \tag{3.14}$$

where $S_{11} = \frac{1}{E}, S_{12} = \frac{-v}{E}$ and $2(S_{11} - S_{12}) = \frac{1}{G} = \frac{2(1+v)}{E}$. Conversely, the stress-strain relationship is

$$\begin{bmatrix} \sigma_1 \\ \sigma_2 \\ \tau_{12} \end{bmatrix} = \begin{bmatrix} Q_{11} & Q_{12} & 0 \\ Q_{12} & Q_{22} & 0 \\ 0 & 0 & Q_{66} \end{bmatrix} \begin{bmatrix} \varepsilon_1 \\ \varepsilon_2 \\ \gamma_{12} \end{bmatrix} \tag{3.15}$$

where

$$Q_{11} = \frac{E}{1-v^2}, \quad Q_{12} = \frac{vE}{1-v^2}, \quad Q_{66} = G = \frac{E}{2(1+v)} \tag{3.16}$$

The above discussion focuses on the stress-strain relationship for orthotropic materials along the principal direction. However, the principal direction of the lamina is often not aligned with the global coordinate system of composite laminate. Therefore, in order to calculate the stiffness in a unified coordinate system, it is necessary to understand the relationship of elastic coefficients between the non-principal directions (also referred to as off-axis coordinates) and those in the principal direction. Here, a brief discussion of the stress transformation and strain transformation equations under plane stress is provided.

Consider a representative volume element in a composite laminate subjected to in-plane off-axis normal stresses σ_x and σ_y and shear stress τ_{xy}. Here, x and y denote two arbitrary coordinate directions, referred to as the off-axis directions, the x–y-axis collectively termed the off-axis coordinate system. As shown in Fig. 3.1a, the relationship between the principal coordinate system (1–2) and the off-axis coordinate system (x–y) is depicted. When a ply is incorporated in a laminate, its fibers will make some prescribed angle β, defined as the angle between the outward normal to the x-axis and the 1-axis principal direction. The angle β is also termed the orientation angle of the lamina. By convention, a counterclockwise direction from the off-axis x-axis to the principal 1-axis is considered positive, while a clockwise direction is considered negative. In the material mechanics, transformation

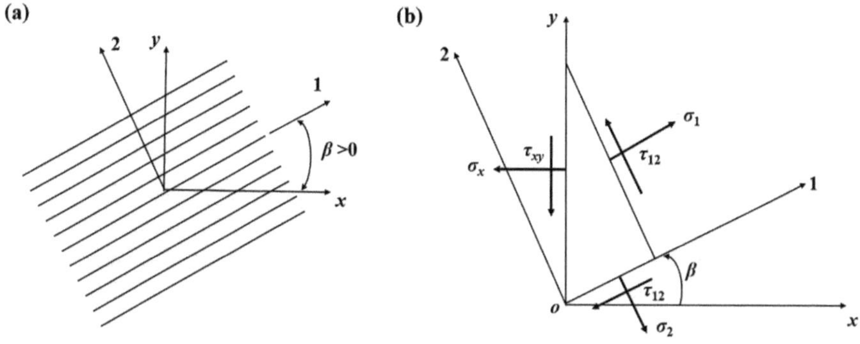

Fig. 3.1 Stress-strain relationships under plane stress condition: (**a**) Relationship between two coordinate systems. (**b**) Elementary force equilibrium

equations for expressing the stress components in the *x*-*y* coordinate system in terms of the 1–2 principal coordinate system are given as

$$\begin{bmatrix} \sigma_x \\ \sigma_y \\ \tau_{xy} \end{bmatrix} = \begin{bmatrix} \cos^2\beta & \sin^2\beta & -2\sin\beta\cos\beta \\ \sin^2\beta & \cos^2\beta & 2\sin\theta\cos\beta \\ \sin\beta\cos\beta & -\sin\beta\cos\beta & \cos^2\beta - \sin^2\beta \end{bmatrix} \begin{bmatrix} \sigma_1 \\ \sigma_2 \\ \tau_{12} \end{bmatrix} \tag{3.17}$$

Stress transformation is used to determine the relationship between the stress components of an elastic body in two different coordinate systems. The transformation from the off-axis to the principal coordinate system, specifically, determining the stress on an arbitrary inclined plane from the known stress components in a given orientation, can be derived using the section method from the mechanics of materials. Consider the stress state of a triangular element in a lamina, which is obtained by sectioning along an inclined plane, as shown in Fig. 3.1b. Based on the conditions of static equilibrium, the following relationships can be derived:

$$\sigma_x = \sigma_1\cos^2\beta + \sigma_2\sin^2\beta - 2\tau_{12}\sin\beta\cos\beta \tag{3.18}$$

Similarly, we get

$$\begin{aligned} \sigma_y &= \sigma_1\sin^2\beta + \sigma_2\cos^2\beta + 2\tau_{12}\sin\beta\cos\beta \\ \tau_{xy} &= \sigma_1\sin\beta\cos\beta - \sigma_2\sin\beta\cos\beta + \tau_{12}\left(\cos^2\beta - \sin^2\beta\right) \end{aligned} \tag{3.19}$$

which can be rewritten in matrix form as

$$\begin{bmatrix} \sigma_x \\ \sigma_y \\ \tau_{xy} \end{bmatrix} = \mathbf{T}^{-1} \begin{bmatrix} \sigma_1 \\ \sigma_2 \\ \tau_{12} \end{bmatrix} \tag{3.20}$$

The stress components in the x–y coordinate system can be represented in terms of the 1–2 system as follows:

$$\begin{bmatrix} \sigma_1 \\ \sigma_2 \\ \tau_{12} \end{bmatrix} = \mathbf{T} \begin{bmatrix} \sigma_x \\ \sigma_y \\ \tau_{xy} \end{bmatrix} \tag{3.21}$$

where $\boldsymbol{T}$ is called the coordinate transformation matrix, and $\boldsymbol{T}^1$ is its inverse matrix. Their expanded forms are

$$\mathbf{T} = \begin{bmatrix} \cos^2\beta & \sin^2\beta & 2\sin\beta\cos\beta \\ \sin^2\beta & \cos^2\beta & -2\sin\beta\cos\beta \\ -\sin\beta\cos\beta & \sin\beta\cos\beta & \cos^2\beta - \sin^2\beta \end{bmatrix}$$
$$\mathbf{T}^{-1} = \begin{bmatrix} \cos^2\beta & \sin^2\beta & -2\sin\beta\cos\beta \\ \sin^2\beta & \cos^2\beta & 2\sin\beta\cos\beta \\ \sin\beta\cos\beta & -\sin\beta\cos\beta & \cos^2\beta - \sin^2\beta \end{bmatrix} \tag{3.22}$$

Similarly, the strain transformation equations are

$$\begin{bmatrix} \varepsilon_1 \\ \varepsilon_2 \\ \gamma_{12} \end{bmatrix} = \left(\mathbf{T}^{-1}\right)^{\mathrm{T}} \begin{bmatrix} \varepsilon_x \\ \varepsilon_y \\ \gamma_{xy} \end{bmatrix}$$
$$\begin{bmatrix} \varepsilon_x \\ \varepsilon_y \\ \gamma_{xy} \end{bmatrix} = \mathbf{T}^{\mathrm{T}} \begin{bmatrix} \varepsilon_1 \\ \varepsilon_2 \\ \gamma_{12} \end{bmatrix} \tag{3.23}$$

Combining Eqs. (3.17) with (3.23), we derive the off-axis stress-strain relationship as

$$\begin{bmatrix} \sigma_x \\ \sigma_y \\ \tau_{xy} \end{bmatrix} = \bar{Q} \begin{bmatrix} \varepsilon_x \\ \varepsilon_y \\ \gamma_{xy} \end{bmatrix} = \begin{bmatrix} \bar{Q}_{11} & \bar{Q}_{12} & \bar{Q}_{16} \\ \bar{Q}_{12} & \bar{Q}_{22} & \bar{Q}_{26} \\ \bar{Q}_{16} & \bar{Q}_{26} & \bar{Q}_{66} \end{bmatrix} \begin{bmatrix} \varepsilon_x \\ \varepsilon_y \\ \gamma_{xy} \end{bmatrix} \tag{3.24}$$

where

$$
\begin{aligned}
\bar{Q}_{11} &= Q_{11}m^4 + 2\left(Q_{12} + 2Q_{66}\right)m^2n^2 + Q_{22}n^4 \\
\bar{Q}_{12} &= \left(Q_{11} + Q_{22} - 4Q_{66}\right)m^2n^2 + Q_{12}\left(m^4 + n^4\right) \\
\bar{Q}_{22} &= Q_{11}n^4 + 2\left(Q_{12} + 2Q_{66}\right)m^2n^2 + Q_{22}m^4 \\
\bar{Q}_{16} &= \left(Q_{11} - Q_{12} - 2Q_{66}\right)m^3n + \left(Q_{12} - Q_{22} + 2Q_{66}\right)mn^3 \\
\bar{Q}_{26} &= \left(Q_{11} - Q_{12} - 2Q_{66}\right)mn^3 + \left(Q_{12} - Q_{22} + 2Q_{66}\right)m^3n \\
\bar{Q}_{66} &= \left(Q_{11} + Q_{22} - 2Q_{12} - 2Q_{66}\right)m^2n^2 + Q_{66}\left(m^4 + n^4\right)
\end{aligned}
\tag{3.25}
$$

where $m =$ cos β and $n =$ sin β; the definition of β follows the previously defined sign convention. The matrix $\left[\bar{Q}\right]$ represents the transformation matrix of the two-dimensional stiffness matrix $[\boldsymbol{Q}]$ from the principal direction, also known as the off-axis modulus matrix. Since $Q_{ij} = Q_{ji}$, we have $\bar{Q}_{ij} = \bar{Q}_{ji}$, meaning that the off-axis modulus retains symmetry. Among the six listed coefficients, $\bar{Q}_{11}, \bar{Q}_{12}, \bar{Q}_{22}, \bar{Q}_{66}$ are even functions of β, while $\bar{Q}_{16}$ and $\bar{Q}_{26}$ are odd functions of β.

Similarly, the off-axis strain-stress relationship can be expressed as

$$
\begin{bmatrix} \varepsilon_x \\ \varepsilon_y \\ \gamma_{xy} \end{bmatrix} = \begin{bmatrix} \bar{S}_{11} & \bar{S}_{12} & \bar{S}_{16} \\ \bar{S}_{12} & \bar{S}_{22} & \bar{S}_{26} \\ \bar{S}_{16} & \bar{S}_{26} & \bar{S}_{66} \end{bmatrix} \begin{bmatrix} \sigma_x \\ \sigma_y \\ \tau_{xy} \end{bmatrix}
\tag{3.26}
$$

where

$$
\begin{aligned}
\bar{S}_{11} &= S_{11}m^4 + 2\left(S_{12} + 2S_{66}\right)m^2n^2 + S_{22}n^4 \\
\bar{S}_{12} &= \left(S_{11} + S_{22} - 4S_{66}\right)m^2n^2 + S_{12}\left(m^4 + n^4\right) \\
\bar{S}_{22} &= S_{11}n^4 + 2\left(S_{12} + 2S_{66}\right)m^2n^2 + S_{22}m^4 \\
\bar{S}_{16} &= \left(S_{11} - S_{12} - 2S_{66}\right)m^3n + \left(S_{12} - S_{22} + 2S_{66}\right)mn^3 \\
\bar{S}_{26} &= \left(S_{11} - S_{12} - 2S_{66}\right)mn^3 + \left(S_{12} - S_{22} + 2S_{66}\right)m^3n \\
\bar{S}_{66} &= \left(S_{11} + S_{22} - 2S_{12} - 2S_{66}\right)m^2n^2 + S_{66}\left(m^4 + n^4\right)
\end{aligned}
\tag{3.27}
$$

In this context, $[\boldsymbol{S}]$ becomes the off-axis compliance matrix, where its elements $\bar{S}_{ij}$ (i,j=1,2,6) are termed as the components of off-axis compliance matrix. Among them, $\bar{S}_{11}, \bar{S}_{12}, \bar{S}_{22}, \bar{S}_{66}$ are even functions of β, and $\bar{S}_{16}, \bar{S}_{26}$ are odd functions of β.

3.2.2 Strength Analysis

The strength of a material is a complex concept, referring to the material's ability to resist failure when subjected to external loads. It is influenced by various factors that contribute to material failure, depending not only on the inherent properties of the material but also on loading conditions and environmental factors specific to different materials. Most experimental measurements of material strength are based on uniaxial stress conditions; however, practical issues often involve plane or spatial stress states. Here, we discuss the strength of orthotropic lamina materials, focusing on in-plane strength indices and theories.

Assuming that the material is macroscopically homogeneous and disregarding certain microlevel failure mechanisms, isotropic materials have a single strength indicator: yield strength σ_s for plastic materials and ultimate strength σ_b for brittle materials. For orthotropic lamina materials in a plane stress state, there are five primary strength indicators: longitudinal tensile strength X_t, longitudinal compressive strength X_c, transverse tensile strength Y_t, transverse compressive strength Y_c, and in-plane shear strength S. The strength criteria use these basic lamina strengths to establish rules for determining whether the orthotropic lamina fails under various plane stress conditions. This section introduces several widely applied composite material failure criteria in plane stress conditions.

- Maximum stress criterion

  The maximum stress criterion posits that a composite material enters failure under complex stress states when one of its stress components reaches the corresponding strength. In other words, if the material does not fail, the principal direction stresses must be less than the basic strength in their respective directions, and the following set of inequalities must be satisfied:

$$\begin{cases} -X_c < \sigma_1 < X_t \\ -Y_c < \sigma_2 < Y_t \\ |\tau_{12}| < S \end{cases} \tag{3.28}$$

  Note that if any inequality in the above relation is not satisfied, the material fails according to the failure mechanism associated with X_t, X_c, Y_t, Y_c, or S. In this criterion, it does not account for possible interaction between the stress components; that is, the different failure modes are independent of each other. It is important to convert the stresses in the material to principal directions to assess failure accurately when applying this strength criterion.
- Maximum strain criterion

  The maximum strain strength criterion is similar to the maximum stress strength criterion, but here the strain is the limiting factor. This criterion predicts failure when any principal material axis strain component exceeds the corresponding ultimate strain. In order to avoid failure, the following inequality applies:

$$\begin{cases} -\varepsilon_{Xc} < \varepsilon_1 < \varepsilon_{Xt} \\ -\varepsilon_{Yc} < \varepsilon_2 < \varepsilon_{Yt} \\ |\gamma_{12}| < \gamma_S \end{cases} \tag{3.29}$$

  If any part of the equation is not satisfied, the material is considered to fail. ε_{Xt} and ε_{Xc} are the maximum tensile and compressive strains in the 1 direction, respectively, while ε_{Yt} and ε_{Yc} are the maximum tensile and compressive strains in the 2 direction, respectively. γ_S represents the maximum shear strain in the 1–2 plane. Similar to the maximum stress strength criterion, the maximum strain strength criterion consists of three independent inequalities. If any of the inequalities are violated, failure occurs. This criterion also considers the influence of another principal direction of the material, i.e., the Poisson coupling effect.

- Tsai-Hill criterion

  The Tsai-Hill criterion is an extension of the von Mises yield failure criterion for isotropic materials to orthotropic materials. In 1948, Hill proposed a yield criterion for anisotropic materials:

$$\begin{aligned}&(G+H)\sigma_1^2+(F+H)\sigma_2^2+(F+G)\sigma_3^2-2H\sigma_1\sigma_2\\&-2G\sigma_1\sigma_3-2F\sigma_2\sigma_3+2L\tau_{23}^2+2M\tau_{31}^2+2N\tau_{12}^2=1\end{aligned} \tag{3.30}$$

  where F, G, H, L, M, and N are failure strength parameters for the anisotropic material. If $L = M = N = 3F = 3G = 3H$ and $2F = 1/\sigma_s$, substituting these values into the equation gives

$$(\sigma_1-\sigma_2)^2+(\sigma_2-\sigma_3)^2+(\sigma_3-\sigma_1)^2+6\left(\tau_{12}^2+\tau_{23}^2+\tau_{31}^2\right)=2\sigma_s^2 \tag{3.31}$$

  where σ_s is the yield limit for isotropic material. The Tsai-Hill criterion considers that, using the basic failure strengths X, Y, and S to replace F, G, H, L, M, and N, the strength conditions of orthotropic lamina materials are represented by

$$\frac{\sigma_1^2}{X^2}-\frac{\sigma_1\sigma_2}{X^2}+\frac{\sigma_2^2}{Y^2}+\frac{\tau_{12}^2}{S^2}=1 \tag{3.32}$$

  This is the basic failure criterion expressed by the strength of composite laminates in terms of X, Y, and S and is known as the Tsai-Hill criterion. It integrates the interactions of the three principal stresses and their corresponding basic strengths on the lamina. It should be noted that the Tsai-Hill criterion is mainly applicable to composite materials where the tensile and compressive strengths in the principal directions are the same (i.e., $X_c = X_t = X$, $Y_c = Y_t = Y$). The Tsai-Hill criterion links the three principal stresses and their corresponding basic strengths in a single expression, considering their interdependence. If the equality in Eq. (3.32) is satisfied, failure occurs. For the material to function properly, the sum on the left-hand side of the equality must be <1. This criterion showed a good agreement between the analytical prediction and experimental values for E-glass/epoxy laminate at different rotational angles.

- Hoffman criterion

  The Tsai-Hill criterion does not account for the different tensile and compressive strengths of the laminates and their effects on material failure. Hoffman modified the Tsai-Hill criterion by adding linear stress terms, thus proposing the Hoffman criterion:

$$\frac{\sigma_1^2}{X_t X_c}-\frac{\sigma_1\sigma_2}{X_t X_c}+\frac{\sigma_2^2}{Y_t Y_c}+\frac{X_c-X_t}{X_t X_c}\sigma_1+\frac{Y_c-Y_t}{Y_t Y_c}\sigma_2+\frac{\tau_{12}^2}{S^2}=1 \tag{3.33}$$

The linear terms for σ_1 and σ_2 reflect the influence of unequal tensile and compressive strengths on material failure. When the tensile and compressive strengths are equal (i.e., $X_t = X_c$, $Y_t = Y_c$), the equation simplifies to the Tsai-Hill criterion.

- Tsai-Wu criterion

In fiber-reinforced composites, the tensile and compressive strengths along the material's principal directions are generally not equal, especially the transverse tensile and compressive strengths, which can differ by several times. To address this, Tsai and Wu proposed a tensor polynomial failure theory, where they assumed that the failure surface in stress space takes the following form:

$$F_i\sigma_i + F_{ij}\sigma_i\sigma_j = 1 \quad (i,j = 1,2,\cdots,6) \tag{3.34}$$

For orthotropic lamina under in-plane stress state, Eq. (3.34) simplifies to

$$\begin{aligned}&F_1\sigma_1+F_2\sigma_2+F_6\sigma_6+F_{11}\sigma_1^2+F_{22}\sigma_2^2+F_{66}\sigma_6^2\\&+2F_{16}\sigma_1\sigma_6+2F_{26}\sigma_2\sigma_6+2F_{12}\sigma_1\sigma_2=1\end{aligned} \tag{3.35}$$

All strength parameters except F_{12} can be obtained from unidirectional tensile tests along the principal axes. The equation becomes

$$\begin{cases}F_1X_t+F_{11}X_t^2=1, & (\sigma_1\neq 0, \sigma_1>0)\\F_1X_c+F_{11}X_c^2=1, & (\sigma_1\neq 0, \sigma_1<0)\\F_2Y_t+F_{22}Y_t^2=1, & (\sigma_2\neq 0, \sigma_2>0)\\F_2Y_c+F_{22}Y_c^2=1, & (\sigma_2\neq 0, \sigma_2<0)\\F_{66}S^2=1, & (\tau_{12}\neq 0)\end{cases} \tag{3.36}$$

From this system of equations, we can solve for

$$\begin{cases}F_1=\dfrac{1}{X_t}-\dfrac{1}{X_c}, & F_{11}=\dfrac{1}{X_tX_c}\\F_2=\dfrac{1}{Y_t}-\dfrac{1}{Y_c}, & F_{22}=\dfrac{1}{Y_tY_c}\\F_{66}=\dfrac{1}{S^2}\end{cases} \tag{3.37}$$

The remaining unknown tensor coefficient F_{12} is the coefficient of the term $\sigma_1\sigma_2$, which reflects the interaction of biaxial normal stresses. At this point, Eq. (3.35) becomes

$$F_1\sigma_1+F_2\sigma_2+F_{11}\sigma_1^2+F_{22}\sigma_2^2+F_{66}\sigma_6^2+2F_{12}\sigma_1\sigma_2=1 \tag{3.38}$$

We can use a biaxial tensile test, where $\sigma_1 = \sigma_2 = \sigma_m$ and all other stress components are zero, and substitute into the above equation to get

$$F_{12} = \frac{1}{2\sigma_m^2}\left[1-\left(\frac{1}{X_t}-\frac{1}{X_c}+\frac{1}{Y_t}-\frac{1}{Y_c}\right)\sigma_m-\left(\frac{1}{X_tX_c}+\frac{1}{Y_tY_c}\right)\sigma_m^2\right] \tag{3.39}$$

where F_{12} depends on X_t, X_c, Y_t, Y_c, and the biaxial tensile failure stress σ_m.

3.2.3 Mechanical Testing Methods

For the orthotropic lamina with identical tensile and compressive properties, its stiffness characteristics are elastic modulus E_1 in the 1 direction and E_2 in the 2 direction. The principal Poisson's ratio $\nu_{21} = \frac{-\varepsilon_2}{\varepsilon_1}$, and the secondary Poisson's ratio $\nu_{12} = \frac{-\varepsilon_1}{\varepsilon_2}$. The shear modulus G_{12} is in the 1–2 plane. Since $\frac{\nu_{21}}{E_1} = \frac{\nu_{12}}{E_2}$, there are only four independent parameters in this system. The strength characteristics include the longitudinal strength X, transverse strength Y, and shear strength S. However, for an orthotropic lamina with differing tensile and compressive properties, the elastic constants E_1 and E_2 are divided into two values each: E_{1t}, E_{1c} and E_{2t}, E_{2c}, while the strength has one value for each direction: X_t, X_c, Y_t, Y_c, and S, where the subscript "t" represents tensile and "c" represents compressive properties.

The above basic stiffness and strength characteristics can be determined through experimental tests. This section mainly introduces the experimental methods for determining the basic mechanical properties of the orthotropic lamina.

- Tensile test

Samples with different fiber orientations are prepared according to national standards (dimensions of the tensile samples are referenced in [6]), as shown in Fig. 3.2a. Aluminum or fiberglass sheets are used as reinforcement at both ends of the sample, with a thickness of 1–2 mm. The reinforcement is bonded with adhesive to prevent it from detaching during the test. The tensile tests for longitudinal and transverse directions are shown in Fig. 3.2b, c, respectively.

(1) Longitudinal tensile test

A standard sample is used, and tensile force is gradually increased. The corresponding longitudinal strain ε_1 and transverse strain ε_2 are measured using an extensometer or resistance strain gauge until the sample fails. The final failure load is recorded. The longitudinal modulus E_{1t}, longitudinal tensile strength X_t, and Poisson's ratio ν_{21} are calculated using the following formulas: $E_{1t} = \frac{P_1}{\varepsilon_1 b_s t_s}$, $X_t = \frac{P_{L1}}{b_s t_s}$, and $\nu_{21} = -\frac{\varepsilon_2}{\varepsilon_1}$, where b_s is the sample width, t_s is the sample thickness, p_t is the load in the 1 direction, and P_{L1} is the limit load in the 1 direction. ε_1 and ε_2 represent the strains in the 1 and 2

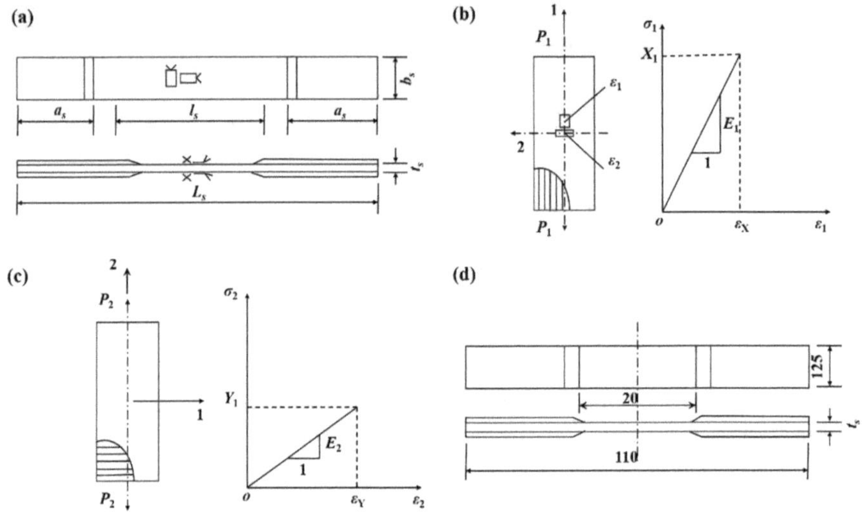

Fig. 3.2 Tensile and compression tests: (**a**) Schematic of tensile sample. (**b**) 0° tensile test. (**c**) 90° tensile test. (**d**) Schematic of compression sample

directions, respectively. If a strain gauge is used to measure the above strains, the transverse effect of the strain gauge must be considered for correction.

(2) Transverse tensile test

The testing procedure is the same as described above. The transverse modulus E_{2t}, transverse tensile strength X_t, and Poisson's ratio ν_{12} can be calculated with the following formulas: $E_{2t} = \dfrac{P_2}{\varepsilon_2 b_s t_s}$, $\nu_{12} = -\dfrac{\varepsilon_1}{\varepsilon_2}$, $Y_t = \dfrac{P_{L2}}{b_s t_s}$, where P_2 is the load in the 2 direction, and P_{L2} is the limit load in the 2 direction.

- Compression test

The compression test is primarily used to measure the longitudinal compressive modulus E_{1c}, the transverse compressive modulus E_{2c}, the transverse compressive strength X_c, the longitudinal compressive strength Y_c, and the Poisson's ratio ν_{21}, Due to the potential for eccentric loading, sample instability, and end damage, the technical challenges of this test are difficult to address fully. Therefore, short gage lengths are used for the samples, as shown in Fig. 3.2d.

- In-plane shear test

The in-plane shear test is used to determine the shear modulus G_{12} and shear strength S. Consider a unidirectional fiber-reinforced laminated plate subjected to a unidirectional tensile test at a 45° angle to the 1 direction, as shown in Fig. 3.3. The shear modulus can be calculated using the following formula:

$$\begin{aligned}
\frac{1}{E_x} &= \frac{1}{E_1}\cos^4\beta + \left(\frac{1}{G_{12}} - \frac{2\nu_{21}}{E_1}\right)\sin^2\beta\cos^2\beta + \frac{1}{E_2}\sin^4\beta \\
\frac{1}{E_{45}} &= \frac{1}{4}\left[\frac{1}{E_1} + \left(\frac{1}{G_{12}} - \frac{2\nu_{21}}{E_1}\right) + \frac{1}{E_2}\right] \\
G_{12} &= \frac{1}{\dfrac{4}{E_{45}} - \dfrac{1}{E_1} - \dfrac{1}{E_2} + \dfrac{2\nu_{21}}{E_1}}
\end{aligned} \tag{3.40}$$

Under the action of P_{Lx}, shear failure occurs in the 45° sample. The shear strength S can be given by the following equation:

$$S = \frac{P_{Lx}}{2b_s t_s} \tag{3.41}$$

- Bending test

The bending modulus and bending strength of unidirectional fiber-reinforced composites are determined using a simply supported beam with three-point loading. The dimensions of the beam sample and the loading diagram are shown in Fig. 3.4. The span l_s is calculated based on the span-to-thickness ratio l_s/t_s. The selection of the span-to-thickness ratio should ensure that failure occurs in the outermost fibers under bending moment; a recommended value is 16 or 32. Standard bending test sample dimensions are thickness t_s = 2.0 mm ± 0.2 mm, width b_s = 12.5 mm ± 0.5 mm, and span-to-thickness ratio l_s/t_s is 16 ± 1 for glass fiber composites and 32 ± 1 for carbon fiber composites.

When determining bending strength, the loading speed v is calculated using the following equation:

$$v = \frac{l_s^2}{6t_s}\dot{\varepsilon} \tag{3.42}$$

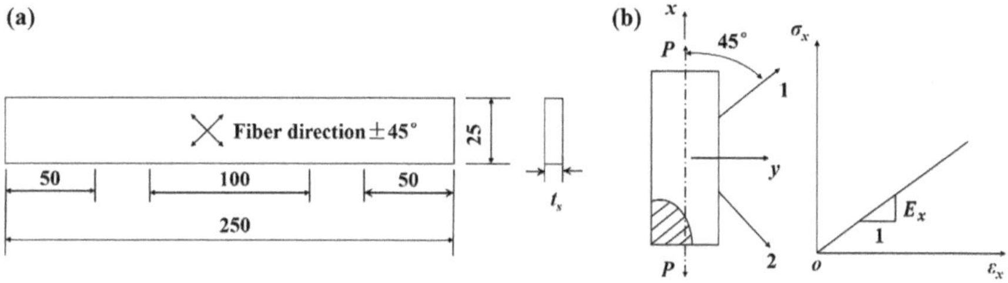

Fig. 3.3 In-plane shear test: (**a**) ±45° symmetric tensile sample. (**b**) Unidirectional off-axis tensile test

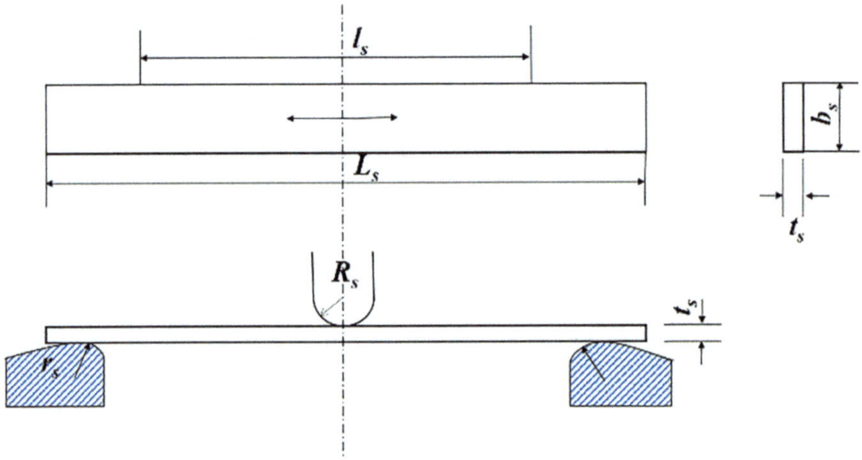

Fig. 3.4 Schematics of bending test and sample

where $\dot{\varepsilon}$ is the strain rate in the outermost fibers at the midpoint of the span, typically taken as 1%/min. $l_s/t_s = 16$ can be taken as $v = t_s/2$/min, and $l_s/t_s = 32$ can be taken as $v = 2t_s$/min. For general experiments, v can range from 5 to 10 mm/min. For measuring bending modulus and the load-deflection curve, v is typically set at 1–2 mm/min or a manual speed.

For three-point loading, the bending strength is

$$\sigma_Y = \frac{3P_{\max} l_s}{2b_s t_s^2} \tag{3.43}$$

In addition to three-point loading, four-point loading can also be used, where the loading points are located at $l_s/4$:

$$\sigma_Y = \frac{3P_{\max} l_s}{4bt^2} \tag{3.44}$$

When the ratio of the midpoint deflection f to the span l_s exceeds 10% during bending failure, to improve accuracy, the bending strength is calculated as

$$\sigma_Y = \frac{3P_{\max} l_s}{2b_s t_s^2}\left[1+4\left(\frac{f}{l_s}\right)^2\right] \tag{3.45}$$

The bending modulus is expressed as

$$E_Y = \frac{\Delta P l_s^3}{4b_s t_s^3 \Delta f} \tag{3.46}$$

where ΔP is the load increment corresponding to the linear segment of the load versus displacement curve, and Δf is the displacement increment at the midpoint of the span corresponding to ΔP.

Considering the effect of shear deformation in the composite material, a correction term is added in Eq. (3.46), as follows:

$$E_{\mathrm{Y}} = \frac{\Delta P l_s^{\ 3}}{4 b_s t_s^{\ 3} \Delta f}\left(1 + S_{\mathrm{m}}\right) \tag{3.47}$$

where S_{m} is the shear deformation correction factor; it can be defined as

$$S_{\mathrm{m}} = \frac{3 t_s^{\ 2}}{2 l_s^{\ 2}} \frac{E_{\mathrm{Y}}}{G_{13}} \tag{3.48}$$

where G_{13} is the shear modulus of the composite material in the longitudinal plane along the thickness direction.

3.3 Basic Mechanics of Composite Laminates

3.3.1 Introduction

In the process of product design for composite material, considering the diversity of loading characteristics and processing requirements for different products, the unidirectional lamina alone generally cannot meet the structural demands effectively. Taking advantage of the designability of the mechanical properties of composite materials, multiple laminae or plies are typically bonded together in a specific orientation to form an integrated structural unit, known as the composite laminated structure. Laminates not only meet the stiffness and strength requirements of structures under various loading conditions but also further achieve structural optimization. They are among the most widely used structural forms in composite structures. The performance of laminates is related to the properties of the ply materials, ply orientations, and ply-stacking sequence; altering the combinations considerably enhances the design flexibility inherent in composite structures, and a composite laminated structure with excellent performance can be achieved.

The performance of a lamina depends on its material and material's principal direction. By laying up the principal directions of materials of each single ply at different angles and in different sequences, laminates with various performance characteristics can be obtained. In this way, laminates with different mechanical properties can be designed to meet engineering requirements, without changing the material of the lamina. This is a characteristic that the lamina does not have, which is why laminated structures are commonly used in engineering.

Compared to the lamina, laminates have the following characteristics:

(1) For fiber-reinforced lamina, which is usually orthotropic, the principal directions of the materials of each ply are typically arranged at different angles, meaning that they do not necessarily have a fixed principal direction. The natural axes of the structure are generally chosen as the coordinate system.
(2) The mechanical properties of the laminate depend not only on the mechanical properties and thickness of the individual single ply but also on the fiber orientation, ply-stacking sequence, and number of plies.
(3) Laminates typically exhibit coupling effects, which means that in-plane tensile (compressive) and shear loads cause bending and twisting deformations, while bending and torsional loads induce tensile (compressive) and shear deformations.
(4) A lamina fails entirely when it is loaded to failure, but a laminate, made up of multiple single plies, may continue to bear load even when one or several plies fail, so the analysis of strength is more complex than for the lamina.
(5) The bonding process of the laminates requires heating and curing. After cooling, the thermal expansion behavior of the individual plies is inconsistent, leading to residual thermal stresses in the final state. Therefore, this factor must be considered in strength calculations.
(6) The stiffness of the individual plies in a laminate is not uniform, and to satisfy interlayer deformation compatibility, interlayer stresses exist.
(7) The analysis of the stiffness and strength of laminates is more complex than for the lamina. Typically, macro-mechanical analysis methods are used, treating each lamina as a homogeneous, anisotropic thin plate, and then the individual laminas are combined to form the laminate, and the stiffness and strength are analyzed accordingly.

The typical representation of laminates is as follows: the fiber direction of each ply may not necessarily align with the direction of the global coordinate axes of the laminate. The angle between the fiber direction and the coordinate axis is represented as the angle β. To indicate the ply-stacking sequence, the following notation can be used (as shown in Fig. 3.5): For a laminate made of four single-layer materials, as shown in Fig. 3.5, the sequence can be written as $[\alpha/0°/90°/\text{-}\alpha]$. Similarly, a laminated plate with a sequence of [60°/-60°/0°/0°/-60°/60°] can be represented as $[\pm 60°/0°]_s$, where "$_s$" indicates symmetry, and the "±" sign indicates that the two plies of positive and negative angles are alternated. A sequence such as [45°/90°/0°/0°/90°/45°] can be expressed as $[45°/90°/0°]_s$ and so on. For laminates composed of single layers with different thicknesses, the thickness of each layer needs to be specified. For example, $[0°h/90°2h/45°3h]_s$ indicates that the first, second, and third plies have thicknesses of h, 2 h, and 3 h, respectively.

According to the arrangement of individual plies relative to the midplane, laminates can be classified into three main types: symmetric laminates, unsymmetric laminates, and sandwich laminates. If the material, thickness, and ply orientation of each ply are symmetric with respect to the midplane of the laminate, it is referred to as a symmetric laminate, which is widely used in engineering structures. For

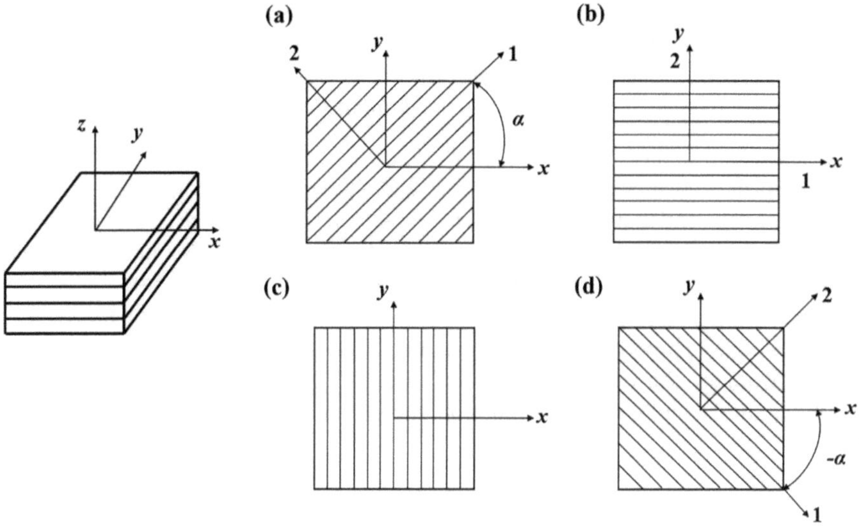

Fig. 3.5 Representation method of laminates: (**a**) $\beta = \alpha$. (**b**) $\beta = 0°$. (**c**) $\beta = 90°$. (**d**) $\beta = -\alpha$

symmetric laminates, the following relationships are satisfied: $\beta(z) = \beta(-z)$ and $Q_{ij}(z) = Q_{ij}(-z)$. A symmetric laminate with only two mutually perpendicular ply orientations is called an orthotropic symmetric laminate, such as $[0°/90°/0°]_S$. For unsymmetric laminates, the fiber orientation of each ply is not symmetric with respect to the midplane. An unsymmetric laminate satisfying the relationship $\beta(z) = -\beta(-z)$ is referred to as an antisymmetric laminate. Unsymmetric laminates that do not satisfy this relationship are generally referred to as general laminates. A sandwich laminate consists of two thin face sheets made of high-strength, high-modulus materials and a thick, low-density core sandwiched between them. This structure significantly enhances the bending stiffness of the laminated plate, makes efficient use of material strength, and improves the compressive stability of the laminate structure.

3.3.2 *Stiffness Analysis*

The derivation of the equations for classical laminate theory (CLT) is based on the following assumptions:

(1) The adhesive layer between the individual plies of the laminate is very thin and perfectly bonded, with no slip occurring at ply interfaces.
(2) Although the laminate is composed of multiple single layers, its total thickness still satisfies the thin plate assumption; that is, the ratio of thickness t to span L_s is $\left(\frac{1}{50} \sim \frac{1}{100}\right) < \frac{t}{L_s} < \left(\frac{1}{8} \sim \frac{1}{10}\right)$, and the entire laminate is of uniform thickness.

(3) The bending deformation of the laminate occurs within a small deflection range, and the straight lines perpendicular to the middle surface remain straight and perpendicular to the midplane after deformation, with the length of these lines unchanged. This is known as the straight-line assumption.

Consider a coordinate system x, y, z, where the plane xoy at $z = 0$ is defined as the middle surface. Displacements in the x, y, and z directions are denoted by u, v, and w, respectively, and they are small compared with the plate thickness, also expressed as only functions of x and y. u^0, v^0, w^0 are the displacements of the middle surface, with w^0 representing the deflection of the plate (as shown in Fig. 3.6a).

Since strain and displacement continuity within the laminate are independent of the directional properties of each ply, the strain-displacement equation can be derived from elasticity theory, as shown in Eq. (3.2).

Based on the straight-line assumption, the geometric equation can be written as

$$\begin{aligned} \varepsilon_z &= \frac{\partial w}{\partial z} = 0 \\ \gamma_{yz} &= \frac{\partial w}{\partial y} + \frac{\partial v}{\partial z} = 0 \\ \gamma_{xz} &= \frac{\partial u}{\partial z} + \frac{\partial w}{\partial x} = 0 \end{aligned} \tag{3.49}$$

By integrating the above equation, we find that

$$\begin{aligned} w(x,y,z) &= w^0 \\ u(x,y,z) &= u^0 - z\frac{\partial w}{\partial x} \\ v(x,y,z) &= v^0 - z\frac{\partial w}{\partial y} \end{aligned} \tag{3.50}$$

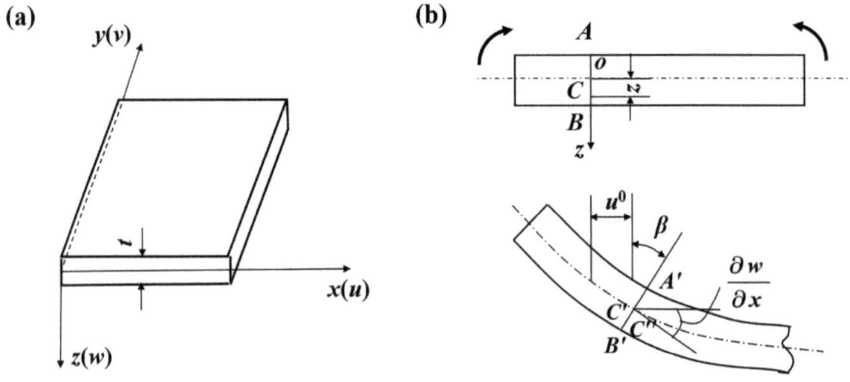

Fig. 3.6 Geometric model of the laminate: (**a**) Coordinate system. (**b**) Relationship of straight-line segment before and after deformation

Thus, the strain-displacement relations for the in-plane strains can be expressed as

$$\begin{aligned}
\varepsilon_x &= \frac{\partial u}{\partial x} = \frac{\partial u^0}{\partial x} + z\left(-\frac{\partial^2 w}{\partial x^2}\right) = \varepsilon_x^0 + z\kappa_x \\
\varepsilon_y &= \frac{\partial v}{\partial y} = \frac{\partial v^0}{\partial x} + z\left(-\frac{\partial^2 w}{\partial y^2}\right) = \varepsilon_y^0 + z\kappa_y \\
\gamma_{xy} &= \frac{\partial u}{\partial x} + \frac{\partial u}{\partial y} = \frac{\partial v^0}{\partial x} + \frac{\partial u^0}{\partial y} + z\left(-2\frac{\partial^2 w}{\partial x \partial y}\right) = \gamma_{xy}^0 + z\kappa_{xy}
\end{aligned} \tag{3.51}$$

where $\varepsilon_x^0, \varepsilon_y^0, \gamma_{xy}^0$ are the in-plane strains at the middle surface. Now, define

$$\kappa_x = \frac{-\partial^2 w}{\partial x^2}, \quad \kappa_y = -\frac{\partial^2 w}{\partial y^2}, \quad \kappa_{xy} = -2\frac{\partial^2 w}{\partial x \partial y} \tag{3.52}$$

where κ_x and κ_y are the bending curvatures associated with the middle surface, and κ_{xy} represents a twisting curvature associated with the middle surface, which lies in xy-plane before deformation.

By combining the above equations, Eq. (3.51) can be written as

$$\begin{bmatrix} \varepsilon_x \\ \varepsilon_y \\ \gamma_{xy} \end{bmatrix} = \begin{bmatrix} \varepsilon_x^0 \\ \varepsilon_y^0 \\ \gamma_{xy}^0 \end{bmatrix} + z \begin{bmatrix} \kappa_x \\ \kappa_y \\ \kappa_{xy} \end{bmatrix} \tag{3.53}$$

From this, we can see that the strain $\{\boldsymbol{\varepsilon}\}$ at any position z within the laminate can be expressed in terms of the in-plane strain $\{\boldsymbol{\varepsilon}^0\}$ and bending strain $\{\boldsymbol{\kappa}\}$ at the corresponding point on the middle surface and that the strain within the laminate varies linearly with thickness.

Internal force components in the laminate: As shown in Fig. 3.7, let N_x, N_y, and N_{xy} represent the internal forces (tensile, compressive, or shear) per unit width (or length) on the cross section of laminate, and M_x, M_y, and M_{xy} are the internal moments (bending or torsional moments) per unit width on the cross section of the laminate. These can be obtained by integrating the stresses on each layer of the laminate through its thickness, and the relationship is given by

$$\begin{bmatrix} N_x \\ N_y \\ N_{xy} \end{bmatrix} = \int_{\frac{-t}{2}}^{\frac{t}{2}} \begin{bmatrix} \sigma_x \\ \sigma_y \\ \tau_{xy} \end{bmatrix} \mathrm{d}z, \quad \begin{bmatrix} M_x \\ M_y \\ M_{xy} \end{bmatrix} = \int_{\frac{-t}{2}}^{\frac{t}{2}} \begin{bmatrix} \sigma_x \\ \sigma_y \\ \tau_{xy} \end{bmatrix} z\mathrm{d}z \tag{3.54}$$

where t is the laminate thickness.

The classical laminate theory only considers the plane stress state and does not account for interlaminar stresses between individual layers. Since the Q_{ij} matrices of the laminate are different, the stresses in the laminate are discontinuously

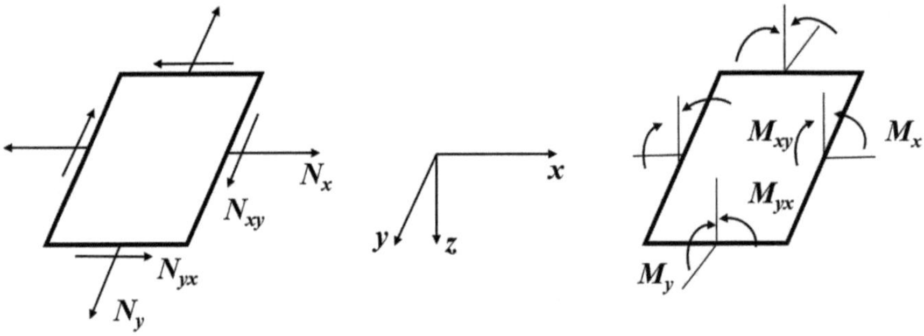

Fig. 3.7 The stress resultants for laminated plate

distributed and can only be integrated layer by layer. For example, using the z-coordinates of each layer as shown in Fig. 3.8, the above equations can be rewritten in the following form:

$$\begin{bmatrix} N_x \\ N_y \\ N_{xy} \end{bmatrix} = \sum_{k=1}^{n} \int_{z_{k-1}}^{z_k} \begin{bmatrix} \sigma_x \\ \sigma_y \\ \tau_{xy} \end{bmatrix}_k \mathrm{d}z, \quad \begin{bmatrix} M_x \\ M_y \\ M_{xy} \end{bmatrix} = \sum_{k=1}^{n} \int_{z_{k-1}}^{z_k} \begin{bmatrix} \sigma_x \\ \sigma_y \\ \tau_{xy} \end{bmatrix}_k z\mathrm{d}z \tag{3.55}$$

It is simplified as

$$\begin{aligned} \{N\} &= \sum_{k=1}^{n} \int_{z_{k-1}}^{z_k} \{\sigma^k\} \mathrm{d}z \\ \{M\} &= \sum_{k=1}^{n} \int_{z_{k-1}}^{z_k} \{\sigma^k\} z\mathrm{d}z \end{aligned} \tag{3.56}$$

where $\boldsymbol{\sigma}^k$ is the stress in the kth lamina, z_{k-1} is the distance from middle surface to inner surface of the kth lamina, and z_k is the corresponding distance from middle surface to outer surface of the kth lamina, as shown in Fig. 3.8.

Substituting the lamina stress-strain relationship $\sigma_k = \overline{Q}_k \varepsilon_k$ into Eq. (3.55), we find that

$$\begin{aligned} \begin{bmatrix} N_x \\ N_y \\ N_{xy} \end{bmatrix} &= \sum_{k=1}^{n} \begin{bmatrix} \bar{Q}_{11} & \bar{Q}_{12} & \bar{Q}_{16} \\ \bar{Q}_{12} & \bar{Q}_{22} & \bar{Q}_{26} \\ \bar{Q}_{16} & \bar{Q}_{26} & \bar{Q}_{66} \end{bmatrix}_k \times \left\{ \int_{z_{k-1}}^{z_k} \begin{bmatrix} \varepsilon_x^0 \\ \varepsilon_y^0 \\ \gamma_{xy}^0 \end{bmatrix} \mathrm{d}z + \int_{z_{k-1}}^{z_k} \begin{bmatrix} \kappa_x \\ \kappa_y \\ \kappa_{xy} \end{bmatrix} z\mathrm{d}z \right\} \\ \begin{bmatrix} M_x \\ M_y \\ M_{xy} \end{bmatrix} &= \sum_{k=1}^{n} \begin{bmatrix} \bar{Q}_{11} & \bar{Q}_{12} & \bar{Q}_{16} \\ \bar{Q}_{12} & \bar{Q}_{22} & \bar{Q}_{26} \\ \bar{Q}_{16} & \bar{Q}_{26} & \bar{Q}_{66} \end{bmatrix}_k \times \left\{ \int_{z_{k-1}}^{z_k} \begin{bmatrix} \varepsilon_x^0 \\ \varepsilon_y^0 \\ \gamma_{xy}^0 \end{bmatrix} z\,\mathrm{d}z + \int_{z_{k-1}}^{z_k} \begin{bmatrix} \kappa_x \\ \kappa_y \\ \kappa_{xy} \end{bmatrix} z^2\mathrm{d}z \right\} \end{aligned} \tag{3.57}$$

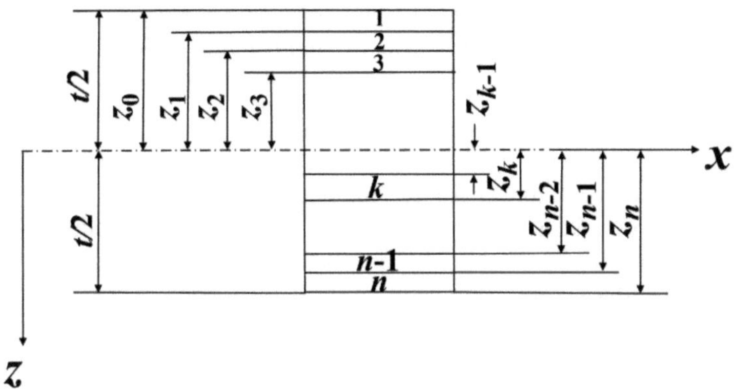

Fig. 3.8 Coordinate system for the stiffness matrix calculation

Therefore, the internal force-strain relationship in the laminate can be written in matrix form as

$$\begin{bmatrix} N_x \\ N_y \\ N_{xy} \\ M_x \\ M_y \\ M_{xy} \end{bmatrix} = \begin{bmatrix} A_{11} & A_{12} & A_{16} & B_{11} & B_{12} & B_{16} \\ A_{12} & A_{22} & A_{26} & B_{12} & B_{22} & B_{26} \\ A_{16} & A_{26} & A_{66} & B_{16} & B_{26} & B_{66} \\ B_{11} & B_{12} & B_{16} & D_{11} & D_{12} & D_{16} \\ B_{12} & B_{22} & B_{26} & D_{12} & D_{22} & D_{26} \\ B_{16} & B_{26} & B_{66} & D_{16} & D_{26} & D_{66} \end{bmatrix} \begin{bmatrix} \varepsilon_x^0 \\ \varepsilon_y^0 \\ \gamma_{xy}^0 \\ \kappa_x \\ \kappa_y \\ \kappa_{xy} \end{bmatrix} \tag{3.58}$$

or in partitioned form as

$$\begin{pmatrix} \mathbf{N} \\ \mathbf{M} \end{pmatrix} = \begin{bmatrix} \mathbf{A} & \mathbf{B} \\ \mathbf{B} & \mathbf{D} \end{bmatrix} \begin{bmatrix} \varepsilon^0 \\ \kappa \end{bmatrix} \tag{3.59}$$

where $\boldsymbol{A} = [A_{ij}]$, $\boldsymbol{B} = [B_{ij}]$, $\boldsymbol{D} = [D_{ij}]$, and $\boldsymbol{A}$ is the laminate extensional stiffness matrix that relates in-plane forces to midplane strains. $\boldsymbol{B}$ is the coupling stiffness matrix that relates in-plane forces with curvatures and the moments with in-plane strains. $\boldsymbol{D}$ is the bending stiffness matrix that relates bending moments with curvatures.

By comparing with Eq. (3.57), these matrix elements are given by

$$\begin{aligned} A_{ij} &= \sum_{k=1}^{n} \int_{z_{k-1}}^{z_k} \left(\bar{Q}_{ij}\right)^k \mathrm{d}z = \sum_{k=1}^{n} \left(\bar{Q}_{ij}\right)^k \left(z_k - z_{k-1}\right) \\ B_{ij} &= \sum_{k=1}^{n} \int_{z_{k-1}}^{z_k} \left(\bar{Q}_{ij}\right)^k z\mathrm{d}z = \frac{1}{2} \sum_{k=1}^{n} \left(\bar{Q}_{ij}\right)^k \left(z_k^2 - z_{k-1}^2\right), \quad (i,j = 1,2,6) \\ D_{ij} &= \sum_{k=1}^{n} \int_{z_{k-1}}^{z_k} \left(\bar{Q}_{ij}\right)^k z^2\mathrm{d}z = \frac{1}{3} \sum_{k=1}^{n} \left(\bar{Q}_{ij}\right)^k \left(z_k^3 - z_{k-1}^3\right) \end{aligned} \tag{3.60}$$

where the stiffness coefficients in the equation include in-plane stiffness coefficients A_{ij}, coupling stiffness coefficients B_{ij}, and bending stiffness coefficients D_{ij}. The physical meanings of each stiffness coefficient are as follows:

A_{11}, A_{12}, A_{22}: Stiffness coefficients for tension (compression) and midplane tensile (compressive) strains
A_{66}: Stiffness coefficient for shear force and midplane shear strain
A_{16}, A_{26}: Coupling stiffness coefficients between shear and tensile strains
B_{11}, B_{12}, B_{22}: Coupling stiffness coefficients between tension and bending
B_{66}: Coupling stiffness coefficient between shear and torsion.
B_{16}, B_{26}: Coupling stiffness coefficients between tension and torsion or shear and bending
D_{11}, D_{12}, D_{22}: Stiffness coefficients for bending moments and curvature
D_{66}: Stiffness coefficient for torsion and twist
D_{16}, D_{26}: Coupling stiffness coefficients between torsion and bending

Based on the descriptions of A_{ij}, B_{ij}, and D_{ij}, the $\boldsymbol{A}$, $\boldsymbol{B}$, and $\boldsymbol{D}$ matrices are all symmetric matrices. The 6×6 total matrix formed by these matrices is also symmetric.

The inverse relationship of Eq. (3.59) is

$$\begin{bmatrix} \varepsilon^0 \\ \kappa \end{bmatrix} = \begin{bmatrix} \boldsymbol{A}' & \boldsymbol{B}' \\ \boldsymbol{H}' & \boldsymbol{D}' \end{bmatrix} \begin{bmatrix} \mathbf{N} \\ \mathbf{M} \end{bmatrix} \tag{3.61}$$

The 6×6 symmetric matrix on the right-hand side is called the flexibility matrix. By using block matrix inversion, we can obtain

$$\begin{cases} \boldsymbol{A}' = \boldsymbol{A}^* - \boldsymbol{B}^* \boldsymbol{D}^{*-1} \boldsymbol{H} \\ \boldsymbol{B}' = \boldsymbol{B}^* \boldsymbol{D}^{*-1} \\ \boldsymbol{H}' = -\boldsymbol{D}^{*-1} \boldsymbol{H}^* \\ \boldsymbol{D}' = \boldsymbol{D}^{*-1} \end{cases} \tag{3.62}$$

where

$$\begin{cases} \boldsymbol{A}^* = \boldsymbol{A}^{-1} \\ \boldsymbol{B}^* = -\boldsymbol{A}^{-1} \boldsymbol{B} \\ \boldsymbol{H}^* = \boldsymbol{B} \boldsymbol{A}^{-1} \\ \boldsymbol{D}^* = -\boldsymbol{B} \boldsymbol{A}^{-1} \boldsymbol{B} + \boldsymbol{D} \end{cases} \tag{3.63}$$

3.3.3 Strength Analysis

Like stiffness, a laminate is composed of multiple laminas, which are the individual plies. Therefore, the strength of the laminate is primarily predicted based on the strength of its individual plies. The foundation of this method is to calculate the

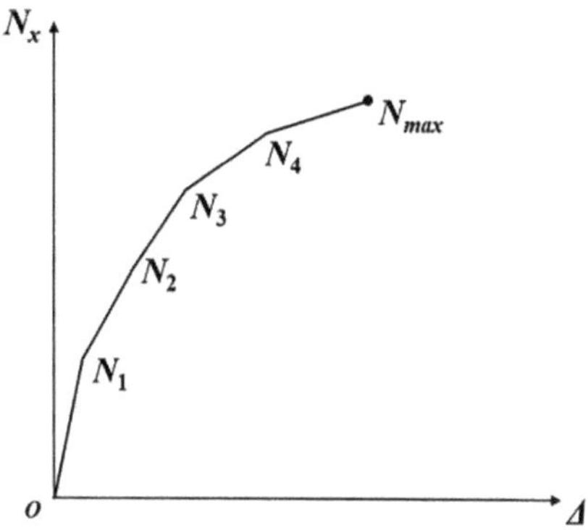

Fig. 3.9 Load-deformation curve of laminates

stress state of each ply. Due to the anisotropy and heterogeneity of composite materials, the failure modes are complex. For composite laminates, the failure of one ply does not necessarily correspond to the failure of the entire laminate. Although the failure of one or several plies may lead to a reduction in laminate stiffness, the laminate may still withstand higher loads and continue loading until the entire laminate fails. The external load at this point is defined as the laminate's ultimate load. The main objective of laminate strength analysis is to determine its ultimate load.

Figure 3.9 shows the load-strain curve of a loaded laminate. In this figure, N_1, N_2, N_3, N_4, …, $N_{\max}$ represent the loads at which the individual plies fail sequentially. N_1 is the "yield" point of the laminate, and $N_{\max}$ is the ultimate load of the laminate. The most significant feature of the laminate failure process is that individual plies may fail sequentially. Therefore, the primary aim of laminate strength analysis is to determine the maximum load that the laminate can withstand and to design the laminate with the required characteristics to support a given load. The corresponding calculation process involves determining the stress in each ply under the given external load and applying the appropriate strength criterion to assess the laminate's strength, as shown in Fig. 3.10. However, determining the ultimate load of the laminate is relatively complex, and the general steps for calculation are as follows:

(1) First, set the proportionality between the external loads; that is, apply the loads according to their component ratios.
(2) Based on the performance of the lamina, calculate the stiffness of the laminate.
(3) Determine the relationship between the stress in each ply in the material's principal direction and the external load.
(4) Apply the stresses of each ply to the relevant strength criteria and compare, determining which ply fails first.
(5) Remove the failed ply from the laminate, while maintaining the geometry of the remaining plies, and recalculate the laminate stiffness.
(6) Repeat the above process, calculating the stress in each layer; then, use strength criteria for comparison, check for failure in other plies, and recalculate stiffness. This is repeated until the laminate can no longer withstand the increasing load. When all the plies fail, the load at this point is the ultimate load of the laminate.

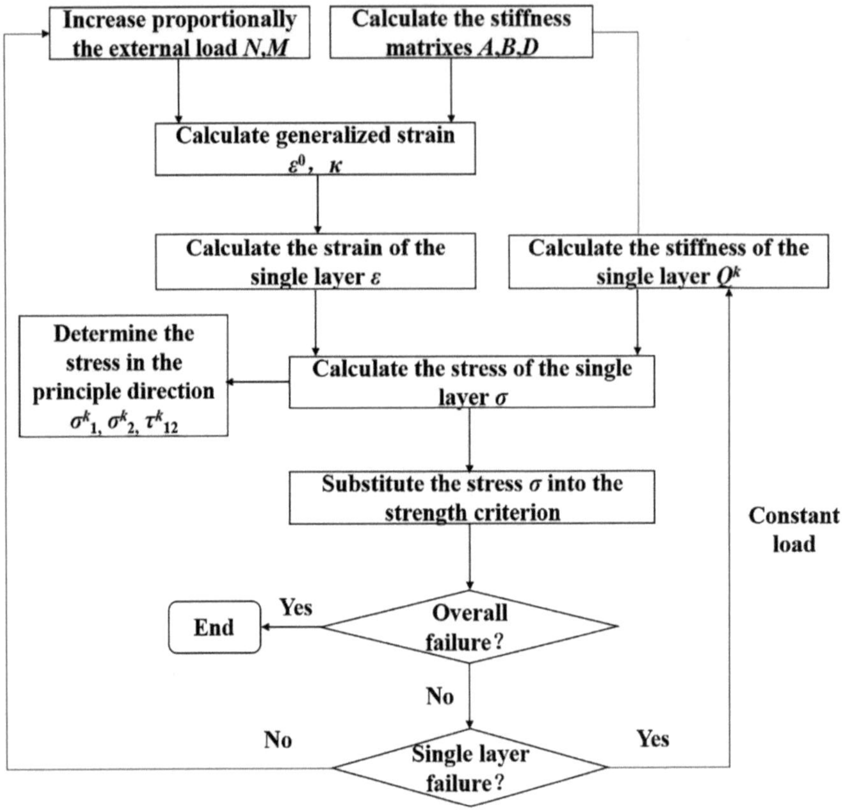

Fig. 3.10 Strength calculation process for laminates

3.3.4 Layup Design Principles

As time progresses, carbon composites are increasingly prevalent in aerospace, automotive, high-speed rail, sports equipment, and mechanical manufacturing industries. During the manufacturing process of products made from carbon fiber composites, significant differences in both processing and performance compared to traditional materials result in substantial design differences. Layup design is a key component of carbon fiber composite design; the principles are as follows:

(1) Balanced and symmetric principle

Except for special cases, composite structures are mostly designed as balanced and symmetric laminates to avoid warping deformation after curing caused by tension-shear or tension-bending coupling. If asymmetric or unbalanced laminates are required, the process-induced deformation constraints should be considered. Asymmetric or unbalanced layers are placed closer to the middle surface so that it reduces laminate process-induced deformations.

(2) Layup orientation principle

The number of ply orientations should be minimized under loading conditions to simplify design and construction work. Typically, four ply orientations are used: 0°, 90°, and ±45°. If a quasi-isotropic laminate is required, a laminate configuration can be used. For structures manufactured using filament winding techniques, winding angles are not restricted by the above orientations.

(3) Load-bearing principle

The fiber axis of the ply should align with the tensile or compressive direction of the internal force to maximize the use of the fiber's high-performance direction. Specifically, for uniaxial tension or compression loads, the fiber direction is aligned; for biaxial tension or compression loads, fibers are arranged orthogonally at 0° and 90°; for shear loads, fibers are paired at ±45°; and for combined tension (or compression) and shear loads, fibers are arranged in multi-directions (0°, 90°, +45°, −45°). The 90° direction fibers improve transverse strength and adjust the Poisson's ratio.

(4) Stacking sequencing principle

The ply orientation should be evenly distributed along the laminate thickness to avoid clustering of layers with the same ply angle. When this is unavoidable, it should generally not exceed four layers to minimize the crack and edge delamination between two different oriented layers. For laminates containing ±45° layers, 0° layers, and 90° layers, it is recommended to insert 0° or 90° layers between +45° and −45° layers, and +45° or −45° layers between 0° and 90° layers. Avoid grouping 90° layers together to reduce interlaminar stresses. For exposed laminates, placing fabric or ±45° layers on the surface improves maintainability and may enhance compressive and impact resistance. Additionally, ply sequencing has a significant impact on laminate stability and load-bearing capacity, which must be considered.

(5) Minimum ply ratio principle

To ensure that the composite material's matrix is not loaded along any direction, the minimum ply ratio in any direction for a laminate composed of 0°, 90°, and ±45° plies should range from 6% to 10%.

(6) Impact load area design principle

For laminates subject to concentrated in-plane impact loads, local reinforcement should be applied. Sufficient fibers should be laid in the impact load direction to resist the local impact. Additionally, a certain number of plies at ±45° to the load direction should be included to diffuse concentrated loads. Furthermore, local strengthening measures should be taken to ensure adequate strength. For structures susceptible to out-of-plane impact, the surface layers of the laminate should have fibers distributed in all directions, with the angles between adjacent layers as small as possible to reduce interlaminar delamination due to matrix loading. For areas that still do not meet impact resistance requirements, hybrid composites, such as a mixture of aramid or glass fibers with carbon fibers, should be used.

(7) Joint area design principle

The proportion of plies at ±45° to the fastener load direction should be no less than 40%, and the proportion of plies aligned with the fastener load direction should exceed 25%. This ensures sufficient shear and compression strength in the joint area as well as aids in load diffusion and reducing stress concentration around holes.

(8) Variable thickness design principle

In regions with varying thickness, the number of plies should increase or decrease in a stepwise manner, gradually transitioning, as sudden changes in thickness can lead to stress concentrations. Each step should have a similar width and more than 60°, with a height not exceeding 1/10 of its width. A continuous cover layer should be applied on the surface to prevent delamination at the step edges.

(9) Opening area ply design principle

In areas with openings in the structure, the angle between adjacent plies should be <60° to reduce interlaminar stresses. Openings should ideally be circular, as circular holes have less edge stress concentration. If rectangular holes must be used, the corners should have a larger radius. Additionally, the number of fibers cut during the creation of the opening should be minimized.

References

1. Shen GL, Hu GK, Liu B (2006) Mechanics of composite materials. Tsinghua University Press, Beijing
2. Chen JQ (2006) Introduction to the mechanics of composite materials. Science Press, Beijing
3. Wang YX (2012) Composite materials mechanics and structural design. East China University of Science and Technology Press, Shanghai
4. Li SL, Wang XY (1993) Fundamentals of composite material structural design. Wuhan University of Technology Press, Wuhan
5. Cai WL (1989) Composite material design. Translated by Liu FL, et al. Science Press, Beijing
6. Liu XY (2007) New developments in composite materials. Chemical Industry Press, Beijing
7. Yi XS (2007) Research and development in advanced composite materials technology. National Defense Industry Press, Beijing
8. Liu XL, Wang BQ (1984) Fundamentals of composite materials mechanics. China Building Industry Press, Beijing
9. Christensen RM (2005) Mechanics of composite materials. Courier Corporation, Massachusetts
10. Jones RM (2018) Mechanics of composite materials. CRC Press, Boca Raton
11. Kaw AK (2005) Mechanics of composite materials. CRC Press, Boca Raton

Chapter 4
Compressive Folding of Composite Cylindrical Shells

Abstract This chapter investigates the axial compressive folding behavior of deployable composite cylindrical shells, establishing a theoretical model based on the Euler-Bernoulli beam model to predict load-displacement curves, shape functions, and stress levels using geometric equations combined with mathematical methods (Maclaurin series, Chebyshev polynomials). The compressive process is divided into three stages: linear elastic pre-buckling, buckling with rapid load drop, and post-buckling. Equilibrium equations, non-dimensionalization, and the Galerkin method solve for load-bearing capacity under large deflections. Experiments using T300/epoxy composite samples validate the model, showing good agreement between theoretical predictions and tests. Parametric studies reveal that increasing arc length and thickness enhances compressive load, while increasing length decreases it.

Based on the classical Euler-Bernoulli beam model, geometric and physical equations were established to describe the axial compression folding deformation of deployable composite cylindrical shells. By integrating various mathematical methods, an analytical model was developed to predict the axial compression load versus displacement curve and shape function of the shells. Additionally, analytical expressions for stress levels were derived based on the minimum complementary energy principle and classical laminate theory. Through the preparation of structural samples, axial compressive folding tests were carried out to validate the effectiveness of the analytical model. Utilizing this model, the influence of different geometric parameters on the axial compression folding behavior was analyzed.

J.-B. Bai et al., *Folding Mechanics of Deployable Composite Structures*, Solid Mechanics and Its Applications 283,
https://doi.org/10.1007/978-3-031-99254-4_4

4.1 Analytical Model

4.1.1 Geometric Model

The geometric configuration of the deployable composite cylindrical shell in its deploying state is determined by its length L, thickness t, cross-sectional radius R, and cross-sectional angle θ, as shown in Fig. 4.1. The deployable composite cylindrical shell with a small cross-sectional angle can achieve axial compression folding under fixed boundary conditions at both ends through buckling and post-buckling. It can serve as a fundamental component for deployable composite hinges in space deployable structures. The cross-sectional angle of such a deployable composite cylindrical shell is typically <20°, since larger angles may result in failure or local buckling during the compression folding process, rather than global buckling. The

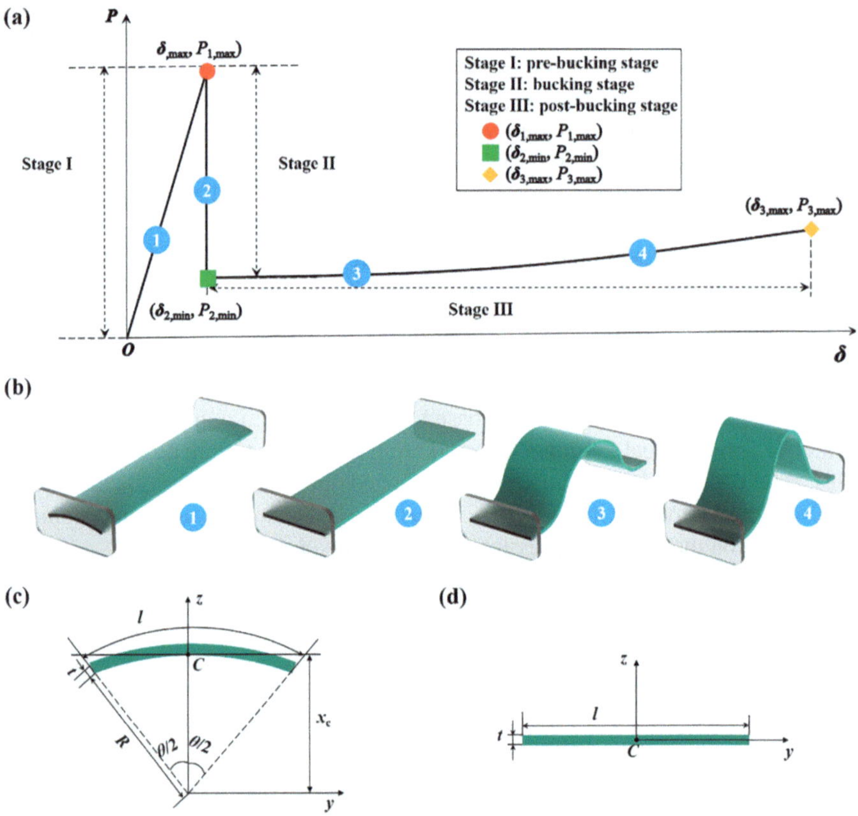

Fig. 4.1 Schematic of the axial compression folding process of the deployable composite cylindrical shell: (**a**) Compressive load versus displacement curve. (**b**) Geometric configuration. (**c**) Cross-sectional geometric configuration of pre-buckling. (**d**) Cross-sectional geometric configuration of post-buckling

compressive process can be divided into three stages, as shown in Fig. 4.1a. The first stage is the pre-buckling stage (i.e., Stage I), which is the linear elastic deformation process that exhibits a high stiffness characteristic. The second stage is the buckling stage (i.e., Stage II), marking a critical phase where the cross section undergoes a significant transformation, and the compressive load applied on the shell decreases rapidly. The main reason is that the cross section of the shell changes from circular arc shape to rectangular shape after the buckling. Therefore, the inertia moment of each shell drops rapidly, and then the external compressive load decreases significantly. The third stage is the post-buckling process (i.e., Stage III), in which the external compressive load exceeds the critical buckling load of the shell with a rectangular cross section; the shell does not lose its load-carrying capacity entirely. Instead, it enters a post-buckling state, maintaining some residual load-carrying capability.

To establish the analytical model for the axial compression folding performance of the deployable composite cylindrical shell [1, 2], the following assumptions are made:

(1) In Stage III, the neutral axis length of the deployable composite cylindrical shell remains constant, and the cross section fully transforms into a rectangle.
(2) The axial compression folding deformation of the deployable composite cylindrical shell is idealized as a quasi-static elastic folding process, as shown in Fig. 4.1b.

As shown in Fig. 4.1c, the Cartesian coordinate system is established with the center of the cross section of the deployable composite cylindrical shell as the coordinate origin. The static moments of the cross section about the z- and y-axes can be expressed as

$$\begin{cases} S_z = \int_A y\mathrm{d}A \\ S_y = \int_A z\mathrm{d}A \end{cases} \tag{4.1}$$

where A is the cross-section area of the deployable composite cylindrical shell. The cross section of the cylindrical shell is symmetric about the z-axis; therefore, the centroid coordinates can be expressed as

$$\begin{cases} z_C = \dfrac{S_y}{A} \\ y_C = 0 \end{cases} \tag{4.2}$$

Based on the transformation relationship between the Cartesian and polar coordinate systems, the expression of any minute element on the cross section in the polar coordinate system can be written as

$$\begin{cases} z = l'\cos\gamma \\ y = l'\sin\gamma \end{cases}, \quad l' \in \left[R - \frac{t}{2}, R + \frac{t}{2}\right] \tag{4.3}$$

where γ is the angle between the z-axis and the line connecting any point on the cross section of the deployable composite cylindrical shell to the coordinate start, and l' is the distance from any minute element to the origin of the polar coordinate system.

Therefore, the minute element area of the cross section of the deployable composite cylindrical shell can be expressed in polar coordinates as

$$\mathrm{d}A = l'\mathrm{d}l'\mathrm{d}\gamma, \quad \gamma \in \left[-\frac{\theta}{2}, \frac{\theta}{2}\right] \tag{4.4}$$

Using the parallel axis theorem, the moments of inertia of the cross section of the deployable composite cylindrical shell about the z-axis and y-axis can be expressed as, respectively,

$$\begin{cases} I_z = \int_A (y - y_C)^2 \, \mathrm{d}A \\ I_y = \int_A (z - z_C)^2 \, \mathrm{d}A \end{cases} \tag{4.5}$$

Substituting Eqs. (4.1)–(4.4) into (4.5) and performing the integration, the moments of inertia of the cross section of the deployable composite cylindrical shell about the z-axis and y-axis can be deduced as follows:

$$\begin{cases} I_z = \dfrac{\theta - \sin\theta}{8}\left(4tR^3 + t^3R\right) \\ I_y = \dfrac{\theta + \sin\theta}{2}\left(tR^3 + \dfrac{t^3R}{4}\right) + t\theta R\left[\dfrac{2}{\theta}\sin\dfrac{\theta}{2}\left(R + \dfrac{t^2}{12R}\right)\right]^2 \\ \quad -t\sin\dfrac{\theta}{2}\left(4R^2 + \dfrac{t^2}{3}\right)\left[\dfrac{2}{\theta}\sin\dfrac{\theta}{2}\left(R + \dfrac{t^2}{12R}\right)\right] \end{cases} \tag{4.6}$$

The arc length of the deployable composite cylindrical shell can be expressed as

$$l = \theta R \tag{4.7}$$

The inertia moment and area of the rectangular cross section of the deployable composite cylindrical shell (shown in Fig. 4.1d) can be expressed as

$$I_r = \frac{lt^3}{12} \tag{4.8}$$

$$A = lt \tag{4.9}$$

4.1.2 Compressive Load Versus Displacement Curve

In Stage I of the compressive process of the deployable composite cylindrical shell, the critical buckling load of the shell can be obtained using the classical Euler equation, as follows:

$$P_{1,\max} = \frac{\pi^2 E_x I_y}{(\mu L)^2} \tag{4.10}$$

where E_x and μ are the axial elastic modulus and effective length factor of the deployable composite cylindrical shell, respectively. The axial elastic modulus can be calculated using classical laminate theory, and the detailed formula can be found in Chap. 3. In this chapter, the boundary conditions of the cylindrical shell are fixed at both ends, so $\mu = 0.5$.

Before reaching the critical buckling load in Stage I, the deployable composite cylindrical shell undergoes elastic deformation. The deformation at the point of buckling can be expressed as

$$\delta_{1,\max} = \frac{P_{1,\max} L}{E_x A} \tag{4.11}$$

Substituting Eq. (4.10) and $\mu = 0.5$ into (4.11), we get

$$\delta_{1,\max} = \frac{4\pi^2 I_y}{AL} \tag{4.12}$$

Therefore, the analytical formula of the load versus displacement curve in Stage I can be shown as

$$P_1(\delta_1) = \frac{E_x A}{L}\delta_1, \quad (0 \le \delta_1 \le \delta_{1,\max}) \tag{4.13}$$

In Stage II, the external load on the deployable composite cylindrical shell decreases rapidly. This is primarily due to the fact that after buckling, the cross section of the shell transitions from a curved shape to a rectangular shape. As a result, the moment of inertia of the cross section decreases sharply, leading to a rapid decrease in the external compressive load. During Stage II, the critical instability load of the shell with a rectangular cross section can also be obtained using the classical Euler equation, as follows:

$$P_{2,\min} = \frac{\pi^2 E_x I_r}{(\mu L)^2} = \frac{4\pi^2 E_x I_r}{L^2} \tag{4.14}$$

Figure 4.2a shows the initial state of the deployable composite cylindrical shell with a rectangular cross section in Stage III. In Stage III, when the external compressive load exceeds the critical instability load of the shell with a rectangular cross section, the shell does not completely lose its load-bearing capacity. Instead, it enters the post-buckling stage, where it can continue to carry load. In order to determine the load-bearing capacity of the shell with a rectangular cross section based on geometric nonlinear theory, it is assumed that large deflections occur after buckling, as shown in Fig. 4.2b.

Figure 4.2c shows the force analysis of the minute element in Stage III; the equilibrium equation is given by [3]

$$M(S) = M(S) + \mathrm{d}M(S) + P_3 \sin\alpha \mathrm{d}S \tag{4.15}$$

Simplifying Eq. (4.15) gives

$$\mathrm{d}M(S) + P_3 \sin\alpha \mathrm{d}S = 0 \tag{4.16}$$

According to the classical Euler-Bernoulli beam theory, it is known that

$$M(S) = E_x I_r \frac{\mathrm{d}\alpha}{\mathrm{d}S} \tag{4.17}$$

Substituting Eqs. (4.17) into (4.16) and simplifying yield

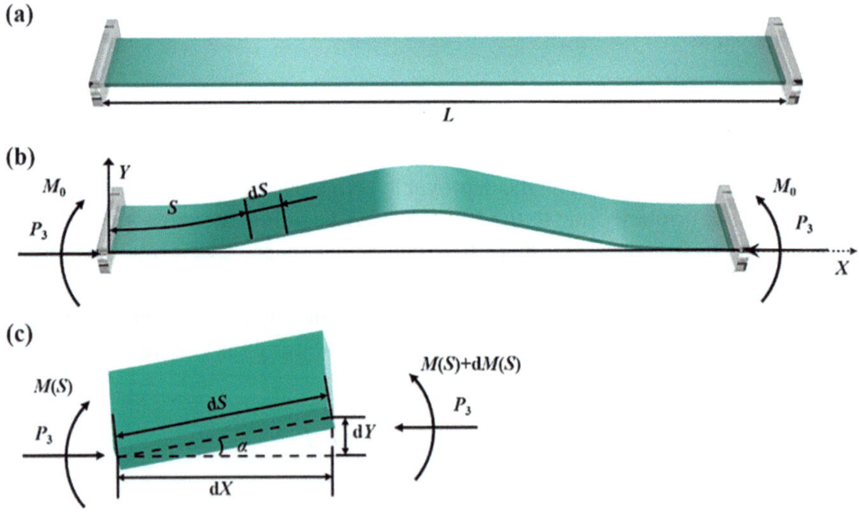

Fig. 4.2 Schematic of the composite shell with a rectangular cross section in Stage III: (**a**) Initial state. (**b**) Force analysis. (**c**) Force analysis of the minute element

$$\frac{\mathrm{d}^2\alpha}{\mathrm{d}S^2}+\frac{P_3}{E_x I_r}\sin\alpha=0 \tag{4.18}$$

The geometric relationship of the minute element gives

$$\frac{\mathrm{d}Y}{\mathrm{d}S}=\sin\alpha \tag{4.19}$$

$$\frac{\mathrm{d}X}{\mathrm{d}S}=\cos\alpha \tag{4.20}$$

where α is the angle between the tangent direction of any point on the neutral axis of the rectangular shell and the horizontal axis, P_3 is the force acting in the X direction on any section of the shell with a rectangular cross section, and the (X, Y) constitutes the Cartesian coordinate describing the deformation of the rectangular shell.

By introducing the dimensionless parameters $P_{3,1}$, s, y, and x, the quantities P_3, S, Y, and X can be non-dimensionalized, yielding

$$\begin{cases} P_{3,1}=\dfrac{P_3L^2}{E_x I_r} \\ s=\dfrac{S}{L} \\ y=\dfrac{Y}{L} \\ x=\dfrac{X}{L} \end{cases} \tag{4.21}$$

Substituting Eq. (4.21) into Eqs. (4.18) to (4.20), the nondimensional form can be deduced as

$$\frac{\mathrm{d}^2\alpha}{\mathrm{d}s^2}+P_{3,1}\sin\alpha=0 \tag{4.22}$$

$$\frac{\mathrm{d}y}{\mathrm{d}s}=\sin\alpha \tag{4.23}$$

$$\frac{\mathrm{d}x}{\mathrm{d}s}=\cos\alpha \tag{4.24}$$

where $P_{3,1}$, s, y, and x are the nondimensional compressive load, arc length of the rectangular shell, transverse displacement, and axial displacement, respectively.

During the post-buckling process of the deployable composite cylindrical shell with a rectangular cross section, the post-buckling configuration of the shell with fixed ends is symmetric, with the maximum deflection occurring at $s = 1/2$, $\alpha_{\max}$ is the maximum angle of rotation, and the maximum angle of the deformed rectangular cross section of the composite shell occurs at $s = 1/4$. Therefore, the dimensionless boundary conditions and geometric conditions are

$$\begin{cases} \alpha\big|_{s=0} = 0 \\ \alpha\big|_{s=1/2} = 0 \\ \alpha\big|_{s=1/4} = \alpha_{\max} \\ x\big|_{s=0} = 0 \\ y\big|_{s=0} = 0 \\ y\big|_{s=1} = 0 \end{cases} \tag{4.25}$$

Introducing a new independent variable τ, it is defined as

$$\tau = 2\pi s - \frac{\pi}{2}, \tau \in \left[-\frac{\pi}{2}, \frac{3\pi}{2}\right] \tag{4.26}$$

Substituting Eq. (4.26) into (4.22) allows Eq. (4.22) to be rewritten as

$$\frac{\mathrm{d}^2\alpha}{\mathrm{d}\tau^2} + P_{3,2}\sin\alpha = 0 \tag{4.27}$$

where

$$P_{3,2} = \frac{P_{3,1}}{4\pi^2} \tag{4.28}$$

The dimensionless boundary conditions can also be reformulated as

$$\begin{cases} \alpha\Big|_{\tau=-\frac{\pi}{2}} = 0 \\ \alpha\Big|_{\tau=\frac{3\pi}{2}} = 0 \\ \alpha\big|_{\tau=0} = \alpha_{\max} \end{cases} \tag{4.29}$$

Equation (4.27) contains trigonometric functions, which makes it difficult to solve it. In this chapter, to eliminate the influence of strong nonlinearity caused by trigonometric functions, a new variable $e = \alpha/\alpha_{\max}$ is introduced, so that the Maclaurin series expansion and orthogonal Chebyshev polynomials are applied to replace sin and cos in equations. Thus, the complex nonlinear problem can be written as

$$\frac{\mathrm{d}y}{\mathrm{d}s} = \sin\alpha = H_1 u + H_2 u^3 \tag{4.30}$$

$$\frac{\mathrm{d}x}{\mathrm{d}s} = \cos\alpha = J_0 + J_1 u^2 + J_2 u^4 \tag{4.31}$$

where H_1, H_2, J_0, J_1, and J_2 are intermediate variables related to $\alpha_{\max}$, as detailed below:

$$\begin{aligned} H_1 &= \alpha_{\max} - \frac{\alpha_{\max}^5}{384} + \frac{\alpha_{\max}^7}{11520} - \frac{\alpha_{\max}^9}{737280} \\ H_2 &= -\frac{\alpha_{\max}^3}{6} + \frac{\alpha_{\max}^5}{96} - \frac{\alpha_{\max}^7}{3840} + \frac{\alpha_{\max}^9}{276480} \\ J_0 &= 1 - \frac{\alpha_{\max}^6}{23040} + \frac{\alpha_{\max}^8}{737280} - \frac{\alpha_{\max}^{10}}{51609600} \\ J_1 &= -\frac{\alpha_{\max}^2}{2} + \frac{\alpha_{\max}^6}{1280} - \frac{\alpha_{\max}^8}{46080} + \frac{\alpha_{\max}^{10}}{3440640} \\ J_2 &= \frac{\alpha_{\max}^4}{24} - \frac{\alpha_{\max}^6}{480} + \frac{\alpha_{\max}^8}{23040} - \frac{\alpha_{\max}^{10}}{1935360} \end{aligned} \tag{4.32}$$

Substituting Eq. (4.30) into (4.27) yields

$$\frac{\mathrm{d}^2 e}{\mathrm{d}\tau^2} + \frac{P_{3,2}\left(H_1 e + H_2 e^3\right)}{\alpha_{\max}} = 0 \tag{4.33}$$

According to the harmonic balance method [4], the initial solution that satisfies the boundary conditions can be assumed in the following form:

$$e(\tau) = \cos\tau \tag{4.34}$$

Substituting Eq. (4.34) into (4.33), and applying the Galerkin method [5], yields

$$\lambda_1 P_{3,2} + \lambda_2 = 0 \tag{4.35}$$

where

$$\begin{cases} \lambda_1 = \left(H_1 + \dfrac{3}{4} H_2\right) \dfrac{\pi}{\alpha_{\max}} \\ \lambda_2 = -4\pi^3 \end{cases} \tag{4.36}$$

Substituting Eq. (4.36) into (4.35) gives the dimensionless compressive load for the deployable composite cylindrical shell with a rectangular cross section, as follows:

$$P_{3,2} = \frac{16\pi^2 \alpha_{\max}}{4H_1 + 3H_2} \tag{4.37}$$

Substituting Eq. (4.26) and $e = \alpha/\alpha_{\max}$ into Eq. (4.34) yields the expression for the post-buckling rotation angle of the deployable composite cylindrical shell with a rectangular cross section and clamped ends, as follows:

$$\alpha(s) = \alpha_{\max}\cos\left(2\pi s - \frac{\pi}{2}\right) = \alpha_{\max}\sin 2\pi s, s \in [0,1] \tag{4.38}$$

Substituting Eq. (4.38) and $e = \alpha/\alpha_{\max}$ into Eq. (4.31) yields

$$\frac{\mathrm{d}x}{\mathrm{d}s} = J_0 + J_1\sin^2 2\pi s + J_2\sin^4 2\pi s \tag{4.39}$$

Integrating Eq. (4.39) gives the expression for x as a function of s:

$$\begin{aligned} x &= J_0 s + \frac{J_1}{2}\left(s - \frac{\sin 4\pi s}{4\pi}\right) \\ &+ J_2\left[-\frac{1}{8\pi}\sin^3 2\pi s\cos 2\pi s + \frac{3}{8}\left(s - \frac{\sin 4\pi s}{4\pi}\right)\right] + C_1 \end{aligned} \tag{4.40}$$

Substituting the boundary conditions into Eq. (4.40) allows the determination of the constant C_1, as follows:

$$C_1 = 0 \tag{4.41}$$

Therefore, Eq. (4.40) can be rewritten as

$$\begin{aligned} x &= J_0 s + \frac{J_1}{2}\left(s - \frac{\sin 4\pi s}{4\pi}\right) \\ &+ J_2\left[-\frac{1}{8\pi}\sin^3 2\pi s\cos 2\pi s + \frac{3}{8}\left(s - \frac{\sin 4\pi s}{4\pi}\right)\right] \end{aligned} \tag{4.42}$$

When $s = 1$, it corresponds to the position of the free end during the post-buckling process. Substituting $s = 1$ into Eq. (4.42) yields

$$x = J_0 + \frac{J_1}{2} + \frac{3}{8}J_2 \tag{4.43}$$

The dimensionless displacement of the free end is given by

$$\Delta = 1 - x = 1 - J_0 - \frac{J_1}{2} - \frac{3}{8}J_2 \tag{4.44}$$

As shown in Fig. 4.3a, the dimensionless load versus displacement curve of the deployable composite cylindrical shell with a rectangular cross section during the compressive process can be expressed by the parametric equation

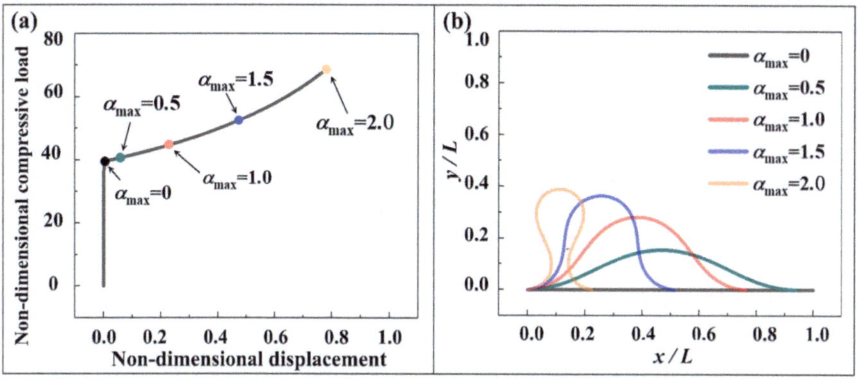

Fig. 4.3 Dimensionless axial compression behavior of the deployable composite cylindrical shell with a rectangular cross section in Stage III: (**a**) Compressive load versus displacement curve. (**b**) Post-buckling shape

$$\begin{cases} P_{3,2} = \dfrac{16\pi^2\alpha_{max}}{4H_1 + 3H_2} \\ \Delta = 1 - x = 1 - J_0 - \dfrac{J_1}{2} - \dfrac{3}{8}J_2 \end{cases} \tag{4.45}$$

Substituting Eqs. (4.21) and (4.28) into (4.45) yields the load versus displacement curve, as follows:

$$\begin{cases} P_3 & = \dfrac{16\pi^2 E_x I_r \alpha_{max}}{(4H_1 + 3H_2)L^2} \\ \delta_3 & = \left(1 - J_0 - \dfrac{J_1}{2} - \dfrac{3}{8}J_2\right)L \end{cases} \tag{4.46}$$

4.1.3 Shape Functions

Next, we calculate the dimensionless shape functions for the post-buckling behavior of the deployable composite cylindrical shell with a rectangular cross section. By substituting Eq. (4.37) and $u = \alpha/\alpha_{max}$ into Eq. (4.30), we obtain

$$\frac{dy}{ds} = H_1 \sin 2\pi s + H_2 \sin^3 2\pi s \tag{4.47}$$

By integrating Eq. (4.47), the expression for y as a function of s is obtained:

$$y = -\frac{H_1 \cos 2\pi s}{2\pi} - \frac{H_2}{2\pi}\left(\cos 2\pi s - \frac{\cos^3 2\pi s}{3}\right) + C_2 \tag{4.48}$$

By substituting the boundary condition $y|_{s=0} = 0$ into Eq. (4.48), the constant C_2 can be determined, as follows:

$$C_2 = \frac{H_1}{2\pi} + \frac{H_2}{3\pi} \tag{4.49}$$

Substituting Eq. (4.49) into (4.48) yields the expression for y as a function of s, as follows:

$$y = -\frac{H_1 \cos 2\pi s}{2\pi} - \frac{H_2}{2\pi}\left(\cos 2\pi s - \frac{\cos^3 2\pi s}{3}\right) + \frac{H_1}{2\pi} + \frac{H_2}{3\pi} \tag{4.50}$$

As shown in Fig. 4.3b, the dimensionless shape function curve of the deployable composite cylindrical shell with a rectangular cross section can be expressed as

$$\begin{cases} x = J_0 s + \dfrac{J_1}{2}\left(s - \dfrac{\sin 4\pi s}{4\pi}\right) \\ \quad + J_2\left[-\dfrac{1}{8\pi}\sin^3 2\pi s \cos 2\pi s + \dfrac{3}{8}\left(s - \dfrac{\sin 4\pi s}{4\pi}\right)\right] \\ y = -\dfrac{H_1 \cos 2\pi s}{2\pi} \\ \quad - \dfrac{H_2}{2\pi}\left(\cos 2\pi s - \dfrac{\cos^3 2\pi s}{3}\right) + \dfrac{H_1}{2\pi} + \dfrac{H_2}{3\pi} \end{cases}, s \in [0,1] \tag{4.51}$$

The shape function curve during the compressive process of the deployable composite cylindrical shell with a rectangular cross section can be expressed by the parametric equations as follows:

$$\begin{cases} X = J_0 S + \dfrac{J_1}{2}\left(S - \dfrac{\sin 4\pi S}{4\pi}\right) \\ \quad + J_2\left[-\dfrac{1}{8\pi}\sin^3 2\pi S \cos 2\pi S + \dfrac{3}{8}\left(S - \dfrac{\sin 4\pi S}{4\pi}\right)\right] \\ Y = -\dfrac{H_1 \cos 2\pi S}{2\pi} \\ \quad - \dfrac{H_2}{2\pi}\left(\cos 2\pi S - \dfrac{\cos^3 2\pi S}{3}\right) + \dfrac{H_1}{2\pi} + \dfrac{H_2}{3\pi} \end{cases}, S \in [0,L] \tag{4.52}$$

4.1.4 Stress Analysis

If the design is inappropriate, the deployable composite cylindrical shell may fail during the folding deformation process. Under the action of axial force and bending moment, large stresses are generated in the rectangular thin shell. In Stage III, the axial force and bending moment on any cross section can be expressed as

$$\begin{cases} F(S) = P_3 \cos\alpha \\ M(S) = M_0 - P_3 Y \end{cases} \tag{4.53}$$

Substituting Eq. (4.38) into (4.53), it is possible to show as

$$\begin{cases} F(S) & = P_3 \cos(\alpha_{\max} \sin 2\pi S) \\ M(S) & = M_0 - P_3 Y \end{cases} \tag{4.54}$$

In Eq. (4.54), there exists one undetermined bending moment M_0. The relationship between M_0 and P_3 can be derived using the minimum complementary energy principle. In Stage III, the total complementary energy can be expressed as

$$\Pi = \frac{1}{E_x I_r} \int_0^L M^2(S) \mathrm{d}S + \frac{1}{E_x A} \int_0^L F^2(S) \mathrm{d}S \tag{4.55}$$

By minimizing the total complementary potential energy of the deployable composite cylindrical shell, one has

$$\frac{d\Pi}{dM_0} = 0 \tag{4.56}$$

Thus, the variable of M_0 is formulated based on P_3, as follows:

$$M_0 = P_3 \int_0^L Y \mathrm{d}S \tag{4.57}$$

Substituting Eq. (4.52) into (4.57), the relationship between M_0 and P_3 can be rewritten as

$$M_0 = \left(\frac{H_1}{2\pi} + \frac{H_2}{3\pi} \right) P_3 L \tag{4.58}$$

For the deployable composite cylindrical shell, selecting a symmetric laminate (i.e., $\boldsymbol{B} = \mathbf{0}$) and simplifying Eq. (3.61) according to classical laminate theory, we obtain

$$\begin{bmatrix} \varepsilon^0 \\ \kappa \end{bmatrix} = \begin{bmatrix} \mathbf{A}^{-1}\mathbf{N} \\ \boldsymbol{D}^{-1}\mathbf{M} \end{bmatrix} \tag{4.59}$$

Equation (4.59) can also be expressed as

$$\begin{bmatrix} \varepsilon_x \\ \varepsilon_y \\ \gamma_{xy} \end{bmatrix} = \begin{bmatrix} A_{11} & A_{12} & A_{16} \\ A_{12} & A_{22} & A_{26} \\ A_{16} & A_{26} & A_{66} \end{bmatrix}^{-1} \begin{bmatrix} N_x \\ N_y \\ N_{xy} \end{bmatrix} \tag{4.60}$$

$$\begin{bmatrix} \kappa_x \\ \kappa_y \\ \kappa_{xy} \end{bmatrix} = \begin{bmatrix} D_{11} & D_{12} & D_{16} \\ D_{12} & D_{22} & D_{26} \\ D_{16} & D_{26} & D_{66} \end{bmatrix}^{-1} \begin{bmatrix} M_x \\ M_y \\ M_{xy} \end{bmatrix} \tag{4.61}$$

The stress-strain relationship of the kth ply ($k = 1, 2, 3, \ldots, n$) in the laminate is

$$\begin{bmatrix} \sigma_x \\ \sigma_y \\ \tau_{xy} \end{bmatrix}_k = \begin{bmatrix} \bar{Q}_{11} & \bar{Q}_{12} & \bar{Q}_{16} \\ \bar{Q}_{12} & \bar{Q}_{22} & \bar{Q}_{26} \\ \bar{Q}_{16} & \bar{Q}_{26} & \bar{Q}_{66} \end{bmatrix}_k \left\{ \begin{bmatrix} \varepsilon_x \\ \varepsilon_y \\ \gamma_{xy} \end{bmatrix} + z \begin{bmatrix} \kappa_x \\ \kappa_y \\ \kappa_{xy} \end{bmatrix} \right\} \tag{4.62}$$

Substituting Eqs. (4.60) and (4.61) into (4.62), it is possible to have

$$\begin{bmatrix} \sigma_x \\ \sigma_y \\ \tau_{xy} \end{bmatrix}_k = \begin{bmatrix} \bar{Q}_{11} & \bar{Q}_{12} & \bar{Q}_{16} \\ \bar{Q}_{12} & \bar{Q}_{22} & \bar{Q}_{26} \\ \bar{Q}_{16} & \bar{Q}_{26} & \bar{Q}_{66} \end{bmatrix}_k \left\{ \begin{matrix} \begin{bmatrix} A_{11} & A_{12} & A_{16} \\ A_{12} & A_{22} & A_{26} \\ A_{16} & A_{26} & A_{66} \end{bmatrix}^{-1} \begin{bmatrix} N_x \\ N_y \\ N_{xy} \end{bmatrix} \\ + z \begin{bmatrix} D_{11} & D_{12} & D_{16} \\ D_{12} & D_{22} & D_{26} \\ D_{16} & D_{26} & D_{66} \end{bmatrix}^{-1} \begin{bmatrix} M_x \\ M_y \\ M_{xy} \end{bmatrix} \end{matrix} \right\} \tag{4.63}$$

According to the force analysis of the deployable composite cylindrical shell, the rectangular shell only bears the compressive load and bending moment in the x direction, and this leads to

$$N_x = \frac{P_3}{l}, N_y = 0, N_{xy} = 0 \tag{4.64}$$

$$M_x = \frac{M_0}{l}, M_y = 0, M_{xy} = 0 \tag{4.65}$$

Substituting Eq. (4.45) into (4.64), it is possible to deduce as

$$N_x = \frac{16\pi^2 E_x I_r \alpha_{\max}}{(4H_1 + 3H_2) l L^2}, N_y = 0, N_{xy} = 0 \tag{4.66}$$

Substituting Eq. (4.57) into (4.65), it is possible to determine as

$$M_x = \frac{16\pi^2 E_x I_r \alpha_{\max}}{(4H_1 + 3H_2)lL}\left(\frac{H_1}{2\pi} + \frac{H_2}{3\pi}\right), M_y = 0, M_{xy} = 0 \tag{4.67}$$

According to the coordinate transformation equation of stress components of the kth ply in the laminate, the maximum principal stresses of the kth ply in the laminate can be expressed as

$$\begin{bmatrix} \sigma_1 \\ \sigma_2 \\ \tau_{12} \end{bmatrix}_k = \begin{bmatrix} \cos^2\beta & \sin^2\beta & 2\sin\beta\cos\beta \\ \sin^2\beta & \cos^2\beta & -2\sin\beta\cos\beta \\ -\sin\beta\cos\beta & \sin\beta\cos\beta & \cos^2\beta - \sin^2\beta \end{bmatrix}_k \begin{bmatrix} \sigma_x \\ \sigma_y \\ \tau_{xy} \end{bmatrix}^k \tag{4.68}$$

Substituting Eqs. (4.63), (4.66), and (4.67) into (4.68), the principal stress components of the kth ply in the laminate can be obtained, denoted as σ_1^k, σ_2^k, and τ_{12}^k. Once the stress components in each ply are determined, the stress level of the deployable composite cylindrical shell during the folding process can be analyzed using failure criterion analysis. When the failure index reaches or exceeds 1, the shell will fail; otherwise, no failure occurs. In this chapter, both the Tsai-Hill criterion and the maximum stress criterion are used to analyze the stress levels of the deployable composite cylindrical shell.

The Tsai-Hill failure index $I_{f,1}$ is computed by the following equations [6]:

$$I_{f,1}^2 = \frac{(\sigma_1^k)^2}{X_1^2} - \frac{\sigma_1^k \sigma_2^k}{X_2^2} + \frac{(\sigma_2^k)^2}{Y^2} + \frac{(\tau_{12}^k)^2}{S_{12}^2} \tag{4.69}$$

where

$$X_1 = \begin{cases} X_t, \text{if } \sigma_1^k > 0 \\ X_c, \text{if } \sigma_1^k < 0 \end{cases} \tag{4.70}$$

$$X_2 = \begin{cases} X_t, \text{if } \sigma_2^k > 0 \\ X_c, \text{if } \sigma_2^k < 0 \end{cases} \tag{4.71}$$

$$Y = \begin{cases} Y_t, \text{if } \sigma_2^k > 0 \\ Y_c, \text{if } \sigma_2^k < 0 \end{cases} \tag{4.72}$$

The maximum stress failure index $I_{f,2}$ can be calculated by using the following formula [6]:

$$I_{f,2}^2 = \max\left\{\left|\frac{\sigma_1^k}{X_1}\right|, \left|\frac{\sigma_2^k}{Y}\right|, \frac{\tau_{12}^k}{S_{12}}\right\} \tag{4.73}$$

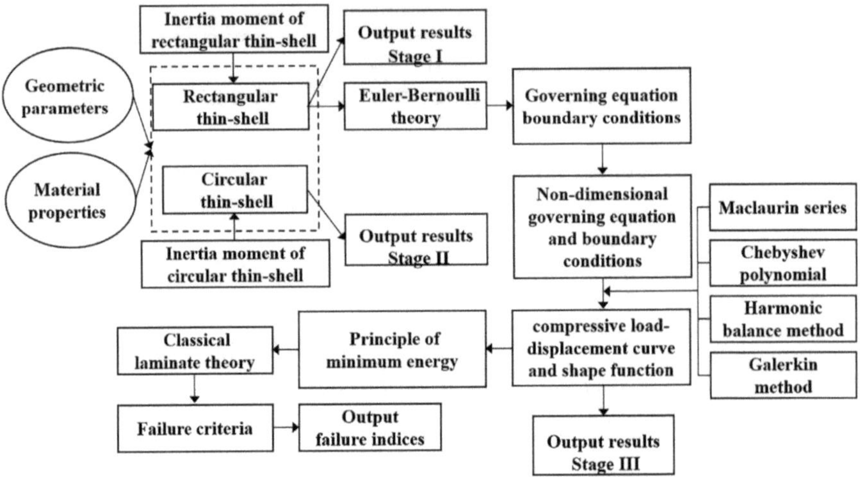

Fig. 4.4 Flowchart for predicting the folding behavior of the deployable composite cylindrical shell

Figure 4.4 shows the flowchart of the solution procedure for predicting the compressive folding behavior of the deployable composite cylindrical shell. Firstly, according to the mechanical properties of constituent materials, the geometric parameters, and the classical Euler equation, the load versus displacement curves of Stage I and Stage II of the cylindrical shell can be obtained, respectively. Based on the Euler-Bernoulli beam model, the governing equations and boundary conditions of Stage III are established. Then, the nondimensional forms of the governing equations and boundary conditions can be determined. By combining Maclaurin series expansion, orthogonal Chebyshev polynomials, Galerkin method, and harmonic balance method, the load versus displacement curve and shape function of Stage III of the shell are deduced. Finally, the minimum complementary energy principle and classical laminate theory are applied to obtain the maximum principal stresses of each ply, and the compressive strength of the cylindrical shell is obtained through the failure criteria.

4.2 Validation Analysis

4.2.1 Experiment Design

The deployable composite cylindrical shell samples were fabricated using T300/epoxy composite ply. Following ASTM standards, the samples were prepared, and their basic mechanical properties were tested. To meet the testing accuracy requirements of the MTS-100KN testing machine, a sample containing eight deployable composite cylindrical shells was designed, as shown in Fig. 4.5. Table 4.1 lists the

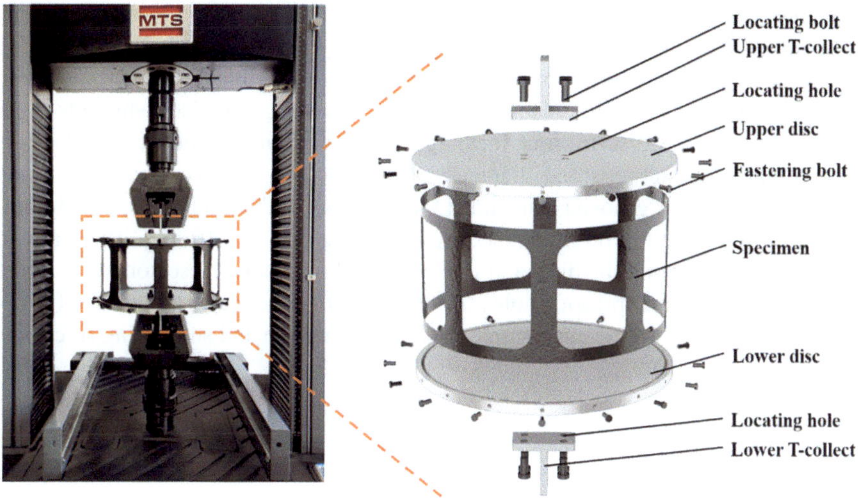

Fig. 4.5 Schematics of compression fixture

Table 4.1 Properties and specifications of T300/epoxy composite ply and samples

Mechanical properties and specifications	Values
E_1 (GPa)	114.38
E_2 (GPa)	7.06
G_{12} (GPa)	2.96
G_{13} (GPa)	2.96
G_{23} (GPa)	2.50
ν_{12}	0.29
X_t (MPa)	1825.73
Y_t (MPa)	74.62
X_c (MPa)	1020.50
Y_c (MPa)	124.90
S_{12} (MPa)	88
t_0 (mm)	0.04
R (mm)	150
l (mm)	30
L (mm)	125
Ply scheme	$[45°/90°/-45°/0°]_S$

basic mechanical properties and specifications of the composite materials and the samples.

As shown in Fig. 4.5, the compression fixture mainly consists of the upper and lower bolts, with component types (including T-type collect, fastening bolts, locating bolts, and disc), installation methods, and sizes being completely the same. The upper and lower fixtures are tightly fixed to the structure through fastening bolts, providing sufficient space for compression deformation. One end of the T-type collect serves as the clamping end for the testing machine, while the other end is fixed to the disc by four locating bolts. The diameter of locating bolts is 9 mm, and the thickness and diameter of the disc are, respectively, 12 mm and 320 mm. A circular groove is machined into the disc, with an inner and outer diameter of 300 mm and 310 mm, respectively, and the depth of 10 mm. There are 16 threaded holes around the disc, and the fastening bolts make close contact with the sample through these threaded holes. The function of the circular groove and fastening bolts is to fix the disc and sample together by friction and supporting forces.

4.2.2 Axial Compression Experiments

Under room-temperature conditions, the compression test was conducted on the sample using the MTS-100KN testing machine. Firstly, the weight of a sample and the compression fixture was balanced before the compression experiments were performed. Secondly, the upper fixture, the sample, and the lower fixture were then sequentially placed from top to bottom and aligned coaxially. The bottom surfaces of the disc groove on the upper and lower fixtures need to be in complete contact with the free ends of the sample. After that, the fastening bolts around the disc were tightened through threaded holes to compress the sample. Finally, compression experiments were conducted at a loading rate of 2 mm/min, while a camera recorded the deformation process under compressive load, and compressive displacement and compressive load were recorded using the MTS-100KN testing machine, as shown in Fig. 4.6a.

Five compression experiments were carried out, with a maximum compression displacement of 60 mm selected, as shown in Fig. 4.6b. After each compression experiment, the sample recovered to its initial configuration without any material damage observed. During the compression experiments, the sample underwent multiple loadings and unloadings within a short time. The recovery rate to the initial configuration was assessed by measuring the height of the sample, and the recovery rate was found to be 100%, indicating that the sample effectively achieved the folding and deploying functions. From Fig. 4.6b, it is evident that during the compressive deformation process, the load versus displacement curves of the sample are consistent, with clear three-stage nonlinear characteristics as described in Sect. 4.1.1.

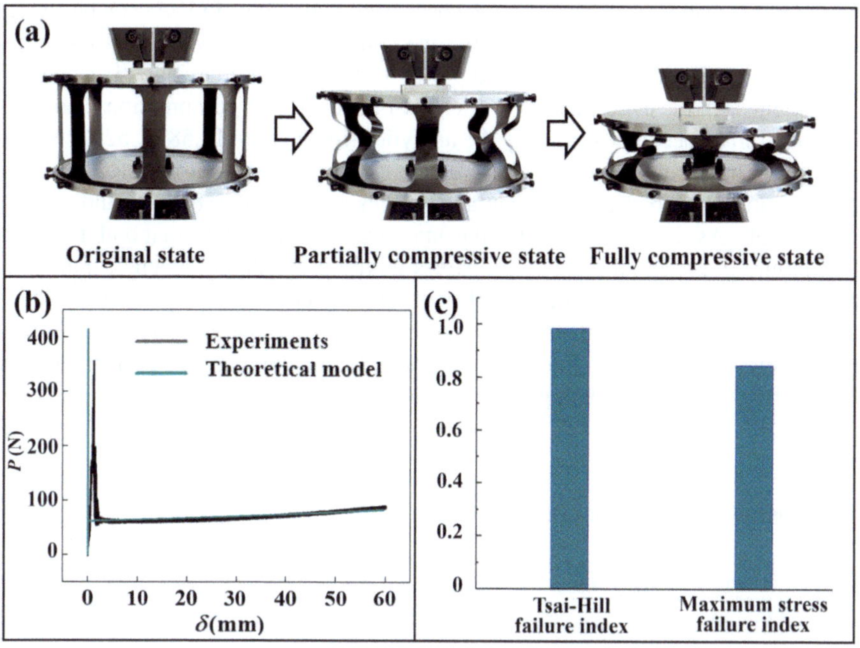

Fig. 4.6 Compressive folding behavior of the sample: (**a**) Deformation process. (**b**) The load versus displacement curves determined by experiment and analytical model. (**c**) Failure indices at 60 mm displacement

Table 4.2 Comparison of analytical model predictions against experiments

		$P_{1,max}$ (N)	$P_{2,min}$ (N)	$P_{3,max}$ (N)
Experiment 1		325.89	57.86	86.80
Experiment 2		353.96	54.93	89.11
Experiment 3		335.96	57.83	86.00
Experiment 2		326.57	58.61	85.52
Experiment 5		355.65	58.57	88.91
Experimental average value		339.61	57.56	87.26
Analytical model	Predictions	413.58	62.41	83.21
	Relative error	21.78%	8.43%	4.64%

4.2.3 Analytical Model Validation

Based on the analytical model of the compressive behavior of the deployable composite cylindrical shell (i.e., compressive load versus displacement curve, shape function, and compressive strength) established in Sect. 4.1, a MATLAB program was written to calculate the compressive folding performance of the shell. To validate the effectiveness of the analytical model proposed, the predicted results were compared with experimental data, as listed in Table 4.2, and the corresponding compressive load versus displacement curve is plotted in Fig. 4.6b. As can be clearly observed from Table 4.2 and Fig. 4.6b, the compressive load versus displacement curve predicted by the analytical model with the three-stage characteristic is in good agreement with the experimental averages. Specifically, the compression loads predicted at three special positions by the analytical model are 413.58 N, 62.41 N, and 83.21 N. The relative errors of the predicted final compression load for the three stages compared to the experimental averages are 21.78%, 8.43%, and 4.64%, respectively. As shown in Fig. 4.6c, the analytical model predicts that both the Tsai-Hill failure index and the maximum stress failure index do not reach 1, indicating that the sample does not undergo failure during the compressive process, which is consistent with the experimental results. In conclusion, the analytical model proposed demonstrates good predictive accuracy for the compressive performance of the cylindrical shell, validating the effectiveness by experiments.

4.3 Results and Discussion

4.3.1 Error Analysis

The analytical model shows a relatively large discrepancy with the experimental results during Stage I, while the relative error is smaller in Stage III. In Stage I of the compression load, the relative error between the analytical prediction and the experimental results reaches 21.78%, which is mainly attributed to the unavoidable defects during the preparation process. During the manufacturing process, factors such as process control and material uniformity inevitably lead to the presence of small defects in the sample. These defects significantly affect the mechanical properties and buckling behavior of the sample, resulting in a large error in the analytical model's prediction for the final compression load during Stage I. However, as the compression process progresses, the impact of these defects on the load gradually diminishes. Therefore, in the latter two stages of buckling, the error becomes relatively smaller, with relative errors of 8.43% and 4.64%, respectively. This indicates that, despite the presence of defects, the analytical model is still relatively accurate in describing the overall trend of the buckling process.

4.3.2 Nonlinear Mechanical Response Mechanism

The analytical model's predictions show good agreement with the experimental results. In Stage I, the deployable composite cylindrical shell does not undergo buckling, and its stiffness is relatively high. This is primarily because the deformation is entirely driven by the axial force, with no bending moment acting on the cylindrical shell. In Stage II, the external compressive load of the shell decreases sharply. This is mainly due to the fact that after buckling, the cross section changes from a curved shape to a rectangular one, causing a rapid decrease in the moment of inertia of the cross section, and consequently, a sharp reduction in the load-bearing capacity. In Stage III, the cylindrical shell with a rectangular cross section does not lose its load-bearing capacity but enters the post-buckling stage, where it can still bear load. Compared to Stage I, the stiffness in Stage III decreases significantly. This is primarily because, during large deformations, the bending strain in the shell with a rectangular cross section becomes dominant over the axial strain.

4.4 Parametric Study

Analytical model analysis indicates that the geometric parameters significantly affect the compressive performance of the deployable composite cylindrical shell. This section uses the analytical model to investigate the influence of the arc length, axial length, and thickness on the folding behavior of the cylindrical shell. The material properties and other geometric parameters remain consistent with the samples described in Sect. 4.2.2. To align with the aforementioned experiment, the compressive load versus displacement curves are all based on the computational results of the samples. The specific variable values are as follows:

- The arc length l is varied as 25 mm, 27.5 mm, 30 mm, 32.5 mm, and 35 mm.
- The axial length L is varied as 105 mm, 115 mm, 125 mm, 135 mm, and 145 mm.
- The thickness t is varied as 0.24 mm, 0.28 mm, 0.32 mm, 0.36 mm, and 0.40 mm.

From Fig. 4.7a, b, and Table 4.3, it can be seen that as the arc length increases, the compressive load at the three special positions also increases, though to varying degrees. The maximum compressive load in Stage I increases significantly because the cross-sectional angle of the corresponding cylindrical shell increases as the arc length increases, leading to a significant increase in the moment of inertia of the cylindrical shells. In contrast, the increase in the minimum compressive load in Stage II and the maximum compressive load in the third stage is relatively small, as the moment of inertia of the rectangular cross-sectional shell is only influenced by the arc length and not by the cross-sectional angle. Additionally, as the arc length increases, the Tsai-Hill failure index-displacement curves do not change with the increase of arc length. When the compressive displacement reaches 60 mm, the Tsai-Hill failure index for all five curves does not exceed 1, indicating that all five types of cylindrical shells can be compressed to 60 mm without failure.

From Fig. 4.7c, d, and Table 4.3, it can be observed that as the axial length increases, the compressive loads at all three stages gradually decreases. Moreover, the Tsai-Hill failure index-displacement curves differ with increase in the axial length. When the compression displacement reaches 60 mm, the Tsai-Hill failure index for the shells with axial lengths of 105 mm and 115 mm exceeds 1, indicating that these two types of cylindrical shells will fail before reaching a compression of 60 mm. In contrast, the Tsai-Hill failure index for cylindrical shells with axial

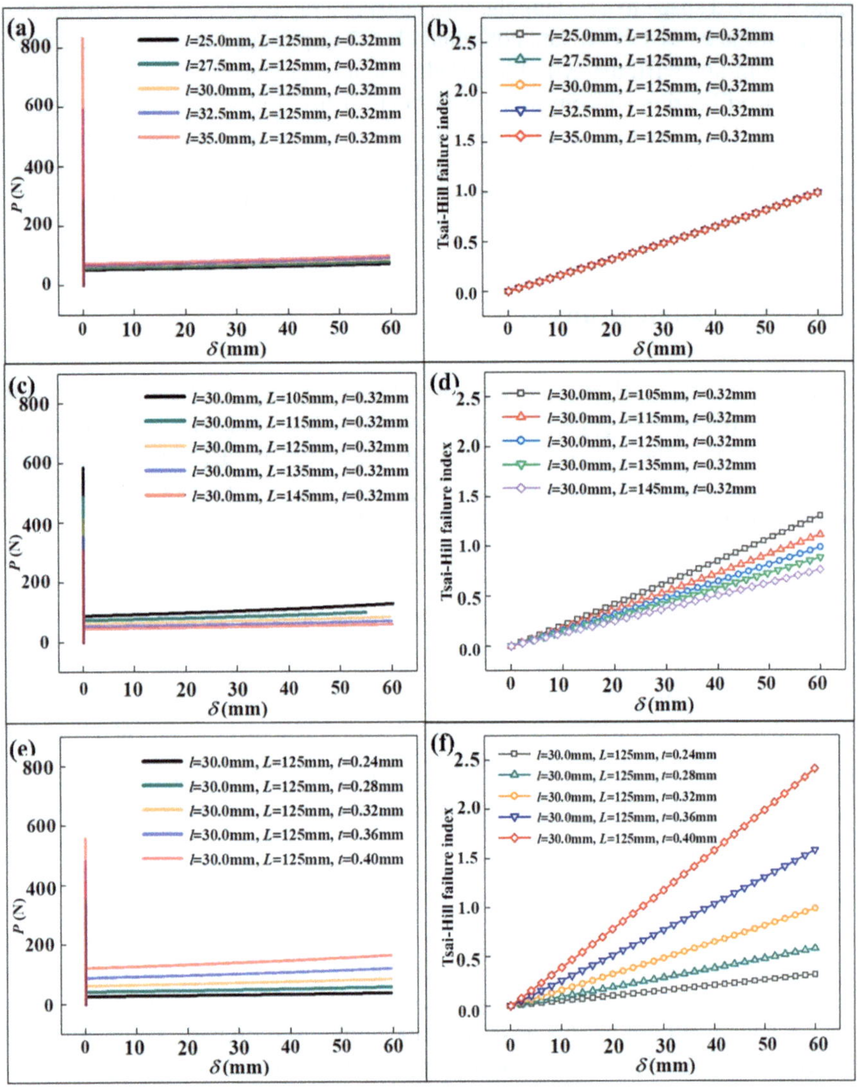

Fig. 4.7 Effect of geometric parameters of the deployable composite cylindrical shells on the folding behavior: (**a**) Compressive load versus displacement curves with different arc lengths. (**b**) Tsai-Hill failure index-displacement curves with different arc lengths. (**c**) Compressive load versus displacement curves with different axial lengths. (**d**) Tsai-Hill failure index versus displacement curves with different axial length. (**e**) Compressive load versus displacement curves with different thicknesses. (**f**) Tsai-Hill failure index versus displacement curves with different thicknesses

lengths of 125 mm, 135 mm, and 145 mm does not exceed 1, showing that these three types can be compressed by 60 mm of compression without failure.

From Fig. 4.7e, f, and Table 4.3, it can be seen that as the thickness increases, the compressive loads at the three special positions also increase. In particular, the

Table 4.3 Effect of geometric parameters of the deployable composite cylindrical shells on the compressive behavior

	l (mm)	L (mm)	t (mm)	$P_{1,max}$ (N)	$P_{2,min}$ (N)	$P_{3,max}$ (N)	Tsai-Hill failure index
Groups of different l	25.0	125	0.32	192.30	52.01	69.34	0.986
	27.5	125	0.32	283.99	57.21	76.27	0.986
	30.0	125	0.32	413.58	62.41	83.21	0.986
	32.5	125	0.32	592.35	67.61	90.14	0.986
	35	125	0.32	833.57	72.81	97.07	0.986
Groups of different L	30.0	105	0.32	586.15	88.45	127.05	1.303
	30.0	115	0.32	488.64	73.74	98.31	1.111
	30.0	125	0.32	413.58	62.41	83.21	0.986
	30.0	135	0.32	354.58	53.51	69.65	0.881
	30.0	145	0.32	307.36	46.38	29.13	0.759
Groups of different t	30.0	125	0.24	290.44	26.33	35.10	0.312
	30.0	125	0.28	349.54	41.81	55.74	0.578
	30.0	125	0.32	413.58	62.41	83.21	0.986
	30.0	125	0.36	483.27	88.86	118.47	1.580
	30.0	125	0.40	559.31	121.90	162.51	2.408

maximum compressive load in Stage I is not sensitive to the shell thickness, while the minimum compressive load in Stage II and the maximum compression load in Stage III increase significantly. Moreover, as the thickness increases, the Tsai-Hill failure index-displacement curves differ. When the compressive displacement reaches 60 mm, the Tsai-Hill failure indexes for cylindrical shells with thicknesses of 0.36 mm and 0.40 mm exceed 1, indicating that these two configurations will fail before reaching 60 mm of compression. However, the Tsai-Hill failure index for the cylindrical shells with thicknesses of 0.24 mm, 0.28 mm, and 0.32 mm remains below 1, meaning these three configurations can be compressed up to 60 mm without failure.

References

1. Liu TW, Bai JB (2022) Folding behaviour of a deployable composite cabin for space habitats-part 1: experimental and numerical investigation. Comp Struct 302:116244
2. Liu TW, Bai JB (2022) Folding behaviour of a deployable composite cabin for space habitats-Part 2: analytical investigation. Comp Struct 297:115929
3. Yu Y, Chen L, Sun W et al (2018) An analytical investigation on large post-buckling behaviour of a drilling shaft modeled as a rotating beam with various boundary conditions. Int J Mech Sci 148:486–495
4. Woiwode L, Balaji NN, Tubita F et al (2020) Comparison of two algorithms for harmonic balance and path continuation. Mech Syst Signal Process 136:106503
5. Fletcher CAJ (1984) Computational Galerkin methods. Springer-Verlag, Berlin
6. Shen GL, Hu GK (2006) Mechanics of composite materials. Tsinghua University Press, Beijing, Beijing

Chapter 5
Bending-Folding of Composite Tape Springs

Abstract This chapter focuses on the bending folding behavior of composite tape springs, developing a theoretical model based on classical laminate theory and the minimum energy principle with the Newton-Raphson method to predict geometric configurations, folding moments, and longitudinal curvatures, considering tensile and bending strain energy contributions. Bending folding is classified into equal-sense (with bending-torsion coupling) and opposite-sense (highly nonlinear) modes, with the latter showing nonlinear moment-angle curves that peak, drop, and stabilize. The minimum energy principle deduces the propagation moment, dependent only on cross-sectional angle and bending stiffness. Experiments on metallic and composite tape springs validate the model, showing higher accuracy than Wuest, Yee, and Yao models. Strain energy analysis shows tensile energy dominance initially, shifting to bending energy when longitudinal curvature exceeds 0.0013 mm^{-1}. Parametric studies find cross-sectional radius affects peak moment, while angle and thickness influence both peak and propagation moments.

Geometric equations describing the bending-folding behavior of composite tape springs are constructed in this chapter. Utilizing the Newton-Raphson method and the minimum potential energy principle, an analytical model was established to predict the bending-folding behavior of composite tape springs, encompassing key indicators such as geometric configuration, folding moment, and longitudinal curvature. This model comprehensively considers the contributions of tensile and bending strain energy to the total strain energy. Through comparisons with three classical models and validation using experimental data, the proposed model in this chapter demonstrates the highest prediction accuracy.

J.-B. Bai et al., *Folding Mechanics of Deployable Composite Structures*, Solid Mechanics and Its Applications 283,
https://doi.org/10.1007/978-3-031-99254-4_5

5.1 Analytical Model

5.1.1 Geometric Model

Applying a bending moment at both ends of the composite tape spring can achieve bending-folding functionality, making it a fundamental element for deployable hinges in space deployable structures. The cross-sectional angle of the composite tape spring is typically <180°, as shown in Fig. 5.1a. There are two types of bending-folding modes for composite tape spring: when the directions of the longitudinal and transverse curvatures are the same, it is referred to as "equal-sense folding," as shown in Fig. 5.1b. When the directions of the longitudinal and transverse curvatures are opposite, it is referred to as "opposite-sense folding," as shown in Fig. 5.1c.

The equal-sense folding process of the composite tape spring is difficult to predict theoretically, as it involves both bending and torsion modes simultaneously during the deformation process [1, 2]. This section develops a new analytical model to predict the opposite-sense folding behavior [3–6]. The relationship between the bending angle and the folding moment during the opposite-sense folding process exhibits high nonlinearity, as shown in Fig. 5.2a.

When the bending angle is small, the folding moment increases nearly linearly with the increase in the bending angle. As the bending angle increases, the cross

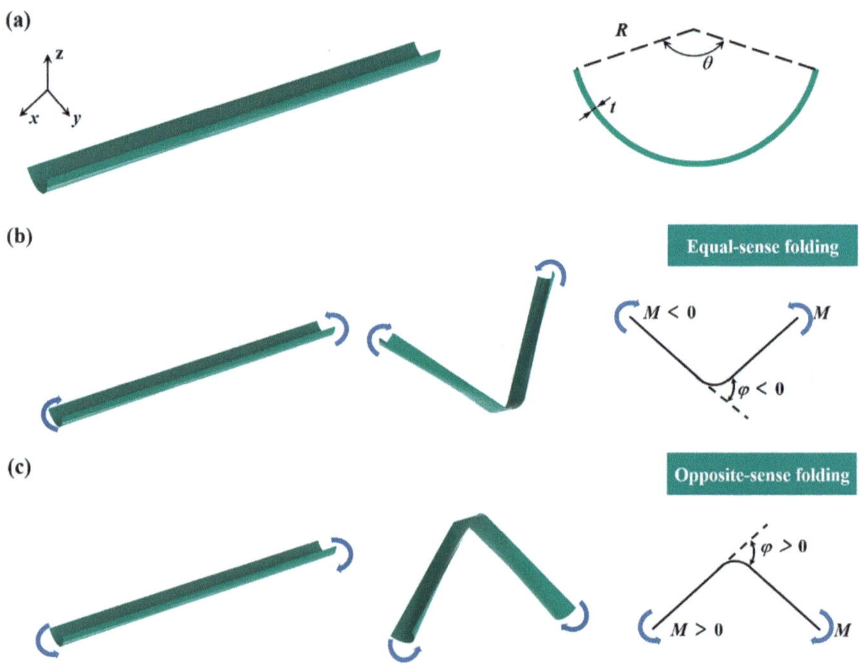

Fig. 5.1 Schematic folding behavior of the composite tape spring: (**a**) Deployed state. (**b**) Equal-sense folding. (**c**) Opposite-sense folding

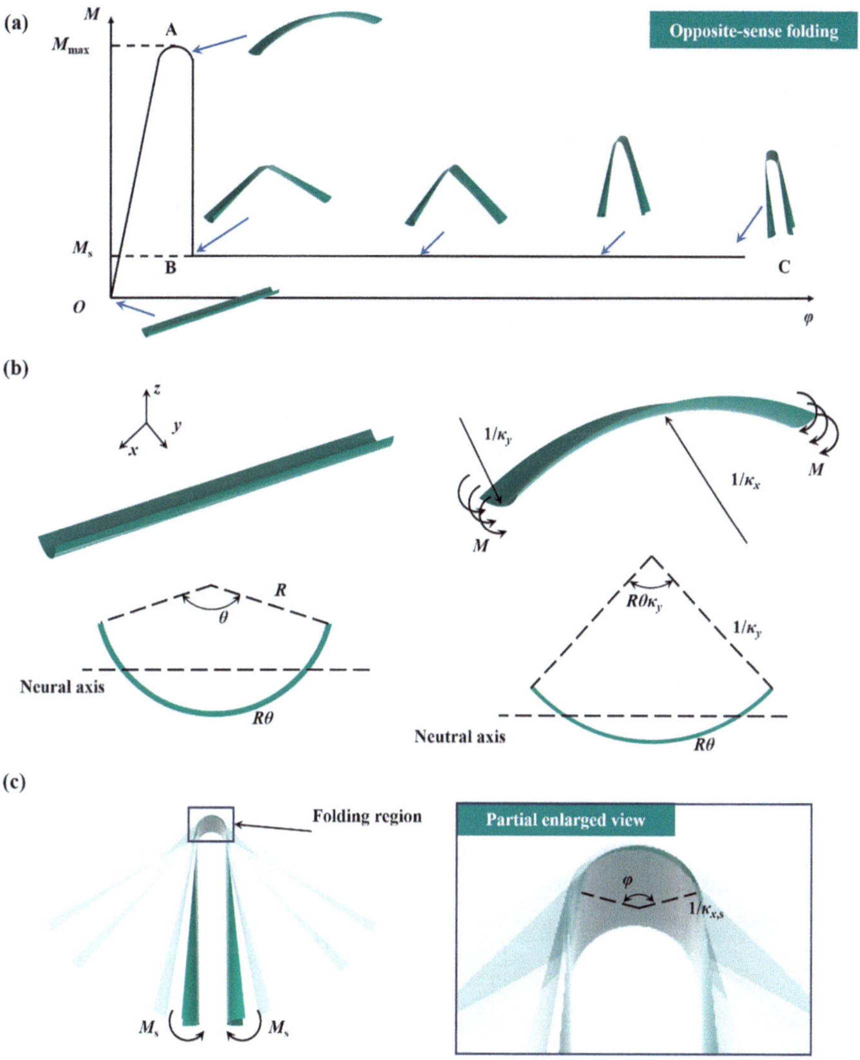

Fig. 5.2 Geometric configuration and force analysis of the folding behavior of the composite tape spring: (**a**) Folding moment versus longitudinal curvature curve. (**b**) Overall bending process. (**c**) Stable bending process

section of the composite tape spring begins to flatten. When the folding moment reaches point A (i.e., the peak moment), the composite tape spring experiences a sudden inversion. At this point, the folding moment sharply decreases from point A to point B, as shown in Fig. 5.2a, b. Subsequently, as the bending angle continues to increase (within the BC range), the folding region in the middle of the composite tape spring begins to increase, and the folding moment essentially stabilizes. During this phase, the longitudinal curvature of the folding region remains unchanged, i.e., $\kappa_{x,\,s}$ is constant, as shown in Fig. 5.2a, c.

5.1.1.1 Folding Moment Versus Longitudinal Curvature Curve

To characterize the geometric equation of the composite tape spring during the folding deformation process, the following basic assumptions are made:

(1) The membrane strain in the y-direction ε_y, twisting strain γ_{xy}, and twisting curvature κ_{xy} in the folding deformation process are neglected [7].
(2) Prior to reaching the folded stable state, the cross section of the composite tape spring remains an arc, but the transverse curvature continuously changes.
(3) When the composite tape spring is in the folded stable state, the transverse curvature is zero and the longitudinal curvature of the folding region remains constant.
(4) The boundary effect occurring at the edge is ignored when the composite tape spring is completely flat.

The **ABD** matrix can be utilized to express the elastic deformation of the composite tape spring during the folding process. Typically, isotropic materials or symmetric laminates (i.e., $\boldsymbol{B} = \boldsymbol{0}$) are chosen as the fabrication materials for the composite tape spring. Based on the basic assumption (1), it can be obtained as

$$\begin{bmatrix} M_x \\ M_y \end{bmatrix} = \begin{bmatrix} D_{11} & D_{12} \\ D_{12} & D_{22} \end{bmatrix} \begin{bmatrix} \Delta\kappa_x \\ \Delta\kappa_y \end{bmatrix} \tag{5.1}$$

The bending strain energy per unit length of the composite tape spring during the folding process can be expressed as [8]

$$u_{\mathrm{B}} = \frac{R\theta}{2} \begin{bmatrix} \Delta\kappa_x & \Delta\kappa_y \end{bmatrix} \begin{bmatrix} M_x \\ M_y \end{bmatrix} \tag{5.2}$$

where

$$\begin{cases} \Delta\kappa_x = \kappa_x \\ \Delta\kappa_y = \kappa_y - \dfrac{1}{R} \end{cases} \tag{5.3}$$

Substituting Eqs. (5.1) and (5.3) into (5.2) yields

$$u_{\mathrm{B}} = \frac{R\theta}{2} \left[D_{11}\kappa_x^2 - 2D_{12}\kappa_x \left(\kappa_y - \frac{1}{R} \right) + D_{22}\kappa_y^2 \right] \tag{5.4}$$

Based on the basic assumption (2), the cross section of the composite tape spring remains an arc during the folding process (shown in Fig. 5.2b). Due to the nonzero

Gaussian curvature during the bending process, the composite tape spring shell exhibits stretching strain energy. The stiffness matrix $\boldsymbol{A}$ in Eq. (3.59), which describes the tensile and compressive deformation of the material, can be simplified to $[A_{11}]$, and the corresponding calculation formula can be simplified as

$$N_x = A_{11}\varepsilon_x \tag{5.5}$$

The stretching strain energy per unit length of the deployable composite cylindrical shell during the folding process can be expressed as [8].

$$u_{\mathrm{S}} = \frac{A_{11}\kappa_x^2}{4\kappa_y^2}\left[R\theta + \frac{\sin(R\theta\kappa_y)}{\kappa_y} - \frac{8\sin^2(R\theta\kappa_y/2)}{R\theta\kappa_y^2}\right] \tag{5.6}$$

Thus, the total strain energy per unit length of the composite tape spring during the folding process can be deduced as

$$\begin{aligned} u(\kappa_x,\kappa_y) &= \frac{R\theta}{2}\left[D_{11}\kappa_x^2 - 2D_{12}\kappa_x\left(\kappa_y - \frac{1}{R}\right) + D_{22}\left(\kappa_y - \frac{1}{R}\right)^2\right] \\ &+ \frac{A_{11}\kappa_x^2}{4\kappa_y^2}\left[R\theta + \frac{\sin(R\theta\kappa_y)}{\kappa_y} - \frac{8\sin^2(R\theta\kappa_y/2)}{R\theta\kappa_y^2}\right] \end{aligned} \tag{5.7}$$

Based on the minimum potential energy principle, the value of κ_y is chosen to minimize the total energy per unit length of the composite tape spring for a given value of κ_x. Therefore, the curvature that minimizes the strain energy $u(\kappa_x, \kappa_y)$ for the given value of κ_x is equal to κ_y:

$$\begin{cases} \dfrac{\partial u}{\partial \kappa_y} = 0 \\ \dfrac{\partial^2 u}{\partial \kappa_y^2} \geq 0 \end{cases} \tag{5.8}$$

Substituting Eq. (5.7) into (5.8), the quantitative relationship between κ_x and κ_y can be derived as follows:

$$D_{12}R\theta\kappa_x + D_{22}R\theta\left(\kappa_y - \frac{1}{R}\right) + \frac{A_{11}\kappa_x^2(K_1 + K_2)}{2} = 0 \tag{5.9}$$

where K_1 and K_2 are transformation variables, as follows:

$$K_1 = -\frac{R\theta}{\kappa_y^3} + \frac{R\theta\kappa_y\cos(R\theta\kappa_y) - 3\sin(R\theta\kappa_y)}{2\kappa_y^4} \tag{5.10}$$

$$K_2 = \frac{4R\theta\kappa_y \sin\left(R\theta\kappa_y/2\right)\cos\left(R\theta\kappa_y/2\right) - 16\sin^2\left(R\theta\kappa_y/2\right)}{R\theta\kappa_y^5} \tag{5.11}$$

Clearly, Eq. (5.9) is an implicit function that can be numerically solved using the Newton-Raphson method to obtain the quantitative relationship between κ_x and κ_y.

The potential energy due to the bending of per unit length of the composite tape spring caused by the folding moment at both ends can be expressed as

$$W = M\varphi \tag{5.12}$$

where M represents the folding moment at both ends of per unit length of the composite tape spring, and φ is the bending angle due to uniform deformation of per unit length of the composite tape spring, which can be determined as

$$\varphi = \kappa_x \tag{5.13}$$

Therefore, the total potential energy of per unit length of the composite tape spring during the folding process can be obtained as

$$\Pi\left(\kappa_x, \kappa_y\right) = u\left(\kappa_x, \kappa_y\right) - W \tag{5.14}$$

By applying the minimum potential energy principle, it can be obtained that

$$\frac{\partial \Pi\left(\kappa_x, \kappa_y\right)}{\partial \kappa_x} = 0 \tag{5.15}$$

Substituting Eqs. (5.12)–(5.14) into (5.15) yields

$$M\left(\kappa_x, \kappa_y\right) = \frac{\partial u\left(\kappa_x, \kappa_y\right)}{\partial \kappa_x} \tag{5.16}$$

Substituting Eq. (5.7) into (5.16) and simplifying yield

$$\begin{aligned} M\left(\kappa_x, \kappa_y\right) &= D_{11}R\theta\kappa_x + D_{12}R\theta\left(\kappa_y - \frac{1}{R}\right) \\ &+ \frac{A_{11}\kappa_x}{2\kappa_y^2}\left[R\theta + \frac{\sin\left(R\theta\kappa_y\right)}{\kappa_y} - \frac{8\sin^2\left(R\theta\kappa_y/2\right)}{R\theta\kappa_y^2}\right] \end{aligned} \tag{5.17}$$

Substituting the quantitative relationship between κ_x and κ_y determined by Eq. (5.9) into (5.17) and using the Newton-Raphson method yield the quantitative relationship between the longitudinal curvature and folding moment. By maximizing the folding moment in Eq. (5.17), the peak moment $M_{\max}$ during the folding process of the composite tape spring can be obtained.

5.1.2 Propagation Moment

According to the basic assumption (3), when the composite tape spring is in a folded stable state, the transverse curvature is zero, and the longitudinal curvature $\kappa_{x,s}$ in the folding region remains constant. Therefore, there is no stretching strain energy in the folding region, only bending strain energy. The bending strain energy can be derived as

$$U\left(\kappa_{x,s},\varphi\right)=\frac{R\theta\varphi}{2}\left(D_{11}\kappa_{x,s}+\frac{2D_{12}}{R}+\frac{D_{22}}{\kappa_{x,s}R^2}\right) \tag{5.18}$$

Similarly, according to the minimum potential energy principle, the value of $\kappa_{x,s}$ is chosen to minimize the energy per unit length of the cross section for a given value of φ. By applying the minimum potential energy principle to Eq. (5.18), the following can be obtained:

$$\frac{\partial U\left(\kappa_{x,s},\varphi\right)}{\partial \kappa_{x,s}}=\frac{R\theta\varphi}{2}\left(-D_{11}\kappa_{x,s}^2+\frac{D_{22}}{R^2}\right)=0 \tag{5.19}$$

Simplifying Eq. (5.19) yields

$$\kappa_{x,s}=\frac{1}{R}\sqrt{\frac{D_{22}}{D_{11}}} \tag{5.20}$$

Substituting Eq. (5.20) into (5.18) yields [9].

$$U(\varphi)=\theta\varphi\left(\sqrt{D_{11}D_{22}}+D_{22}\right) \tag{5.21}$$

Similar to Eq. (5.16), the derivative of U yields

$$M_s=\frac{\mathrm{d}U}{\mathrm{d}\varphi}=\theta\left(\sqrt{D_{11}D_{22}}+D_{12}\right) \tag{5.22}$$

Equation (5.20) provides the analytical expression for the propagation moment, showing that the propagation moment depends only on the cross-sectional central angle and the bending stiffness matrix of the laminate. Figure 5.3a, b presents a clear flowchart of the analytical modeling process of the folding behavior for the composite tape spring discussed in this section. Additionally, unlike the use of the minimum energy method to obtain the propagation moment, Seffen and Pellegrino [2] applied the Maxwell construction method to predict propagation moment and demonstrated high predictive accuracy. The method is based on the assumption that there is a strain energy function in the balloon material, so the strain energy is path independent in a particular configuration. This method solves for the propagation moment by equalizing areas.

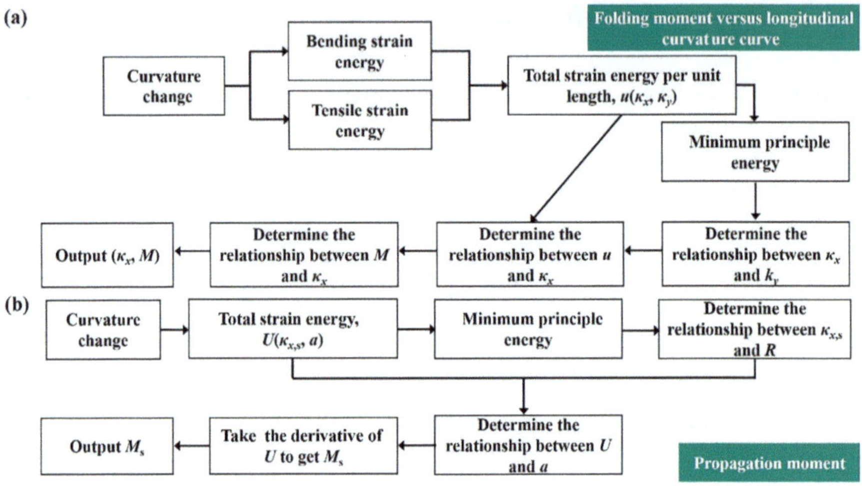

Fig. 5.3 Flowchart for predicting the folding behavior of the composite tape spring: (**a**) Folding moment versus longitudinal curvature curve. (**b**) Propagation moment

5.2 Validation Analysis

Based on the proposed model, as well as the classical Wuest model [1] and Yee model [9], and Yao model [7], a computational program was developed using MATLAB software to calculate the folding behavior of the composite tape spring. The four models were used to predict the folding performance of the composite tape spring made of two different materials and sizes, namely, the metal cylindrical tape spring [2] and composite tape spring [10]. The proposed model was validated and demonstrated the highest prediction accuracy.

5.2.1 Metallic Tape Springs

The folding performance of the metallic tape spring is analyzed; the corresponding specifications and properties are listed in Table. 5.1. Figure 5.4a presents the folding moment versus bending angle curve for the metallic tape spring as reported in literature [2]; the curve exhibits obvious nonlinearity, with the folding moment initially increasing with the bending angle, then decreasing, and finally stabilizing. Figure 5.4b shows the folding moment versus longitudinal curvature curves predicted by the proposed model, along with those predicted by the Wuest and Yao models. As seen in Fig. 5.4b, the predictions of all three models exhibit nonlinear characteristics similar to the experiment. However, compared to the predictions from the Wuest and Yao models, the proposed model provides results that are more consistent with the experimental data. The peak moment predicted by the Wuest model is lower, which is consistent with conclusions from the literature [2, 9].

Table 5.1 Specifications and properties of the metallic tape spring

Mechanical properties and specifications	Values
E (GPa)	131
ν	0.3
L (mm)	200
t (mm)	0.3
R (mm)	13.3
θ (rad)	1.85

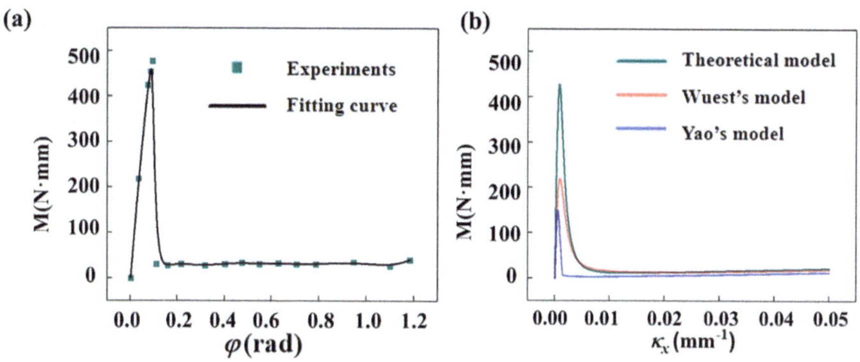

Fig. 5.4 Folding processes of the metallic tape spring: (**a**) Folding moment versus bending angle curve determined by the experiments. (**b**) Folding moment versus longitudinal curvature curves predicted by the three models

Table 5.2 Peak moment and propagation moment of the metallic tape spring

Metallic tape springs		Folding moment (N·mm)	
		M_{max}	M_s
Experiments [2]		476.41	31.44
Proposed model	Value	428.70	28.85
	Error	10.01%	8.24%
Wuest model	Value	220.64	N.A.
	Error	54.73%	N.A.
Yao model	Value	151.74	N.A.
	Error	68.15%	N.A.

Note: N.A. represents not available data in the literature

Table 5.2 compares the peak and propagation moments of the metallic tape spring as determined by the three models and experimental data. For the peak moment, the error of the proposed model is only 10.01%, which is much lower than the 54.73% error of the Wuest model and 68.15% error of the Yao model. For the propagation moment, the error of the proposed model is 10.38%, while both the Wuest and Yao models fail to predict a constant propagation moment. The proposed model demonstrates higher accuracy in predicting the folding moment versus bending angle curve for the metallic tape spring, especially in predicting the peak moment and propagation moment, with errors significantly smaller than those of the Wuest and Yao models, thus proving the model's effectiveness in predicting nonlinear bending and folding performance.

5.2.2 Composite Tape Springs

Next, the folding performance of the composite tape spring is analyzed using the aforementioned three models. Table 5.3 lists the basic mechanical properties and specifications of the composite tape spring. Yee et al. [9] extended the Wuest model from isotropic materials to orthotropic materials, using the Yee model instead of the Wuest model to predict the folding performance of the composite tape spring. Figure 5.5a presents the folding moment versus bending angle curve for the composite tape spring as reported in literature [10]. Figure 5.5b shows the folding moment versus longitudinal curvature curves determined by experiments and three models. The predictions from all three models exhibit nonlinear characteristics

Table 5.3 Specifications and properties of the composite tape spring [10]

Mechanical properties and specifications	Values
E_1 (GPa)	50
E_2 (GPa)	50
G_{12} (GPa)	6.2
ν_{12}	0.3
L (mm)	180
t (mm)	0.3
R (mm)	20
θ (rad)	π/2
Ply scheme	[0°/90°/0°/90°]

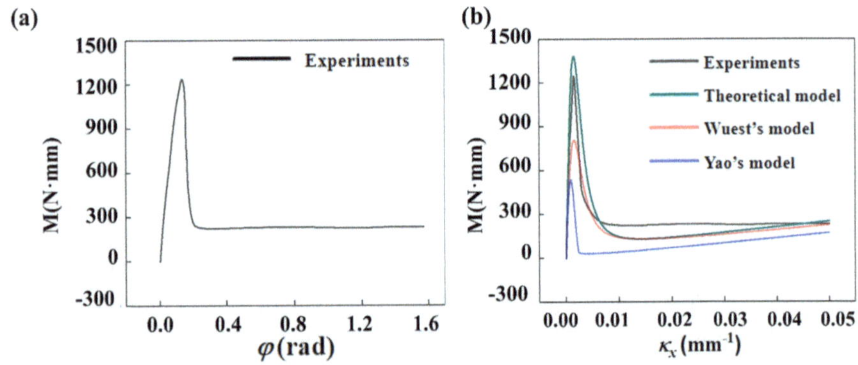

Fig. 5.5 Folding process of the composite tape spring: (**a**) Folding moment versus bending angle curve determined by the experiments. (**b**) Folding moment versus longitudinal curvature curves predicted by the experiments and three models

Table 5.4 Peak moment and propagation moment of the composite tape spring

Composite tape spring		Folding moment (N·mm)	
		M_{max}	M_s
Experiment [10]		1237.6	230.98
Proposed model	Value	1381.89	252.45
	Error	11.66%	9.30%
Yee model	Value	809.78	N.A.
	Error	36.56%	N.A.
Yao model	Value	542.06	N.A.
	Error	57.54%	N.A.

similar to the experimental data, and the proposed model provides results that are more consistent with the experiments. Table 5.4 compares the peak and propagation moments of the composite tape spring as determined by the three models and experimental data. For the peak moment, the error of the proposed model is only 11.66%, much lower than the 36.56% error of the Yee model and the 57.54% error of the Yao model. For the propagation moment, both the proposed model and the Yee model have a prediction error of 9.30%, while the Yao model is unable to predict a constant propagation moment.

5.3 Results and Discussion

5.3.1 Error Analysis

The proposed model, along with the three classic comparative models, is based on a core assumption: the composite tape spring is infinitely long. However, this assumption neglects the significant influence of boundary conditions applied at the ends of the composite tape spring on its folding performance, thereby limiting the applicability of these models. In practical experiments and finite element analyses, the composite tape springs of finite length are used, with rigid constraints applied at both ends to generate the required folding moments. Clearly, both the length of the composite tape spring and the boundary conditions at its ends are key factors determining its folding performance. Therefore, if the boundary conditions applied at the ends are ignored, the predictions of the analytical model will be underestimated and will fail to accurately reflect the actual physical behavior. To enhance the predictive accuracy and practicality of the model, future research should investigate in depth the impact of boundary conditions on the folding behavior of the composite tape springs and make necessary adjustments to improve and refine the model.

Additionally, the treatment method for the cross section is a crucial factor in determining the predictive accuracy of analytical models. In the Wuest and Yee models, the folding moment at both ends of the longitudinal curvature is obtained by calculating the moment of the entire cross section of the composite tape spring about the z-axis, with the integration range set from $R\theta/2$ to $-R\theta/2$. However, Seffen and Pellegrino [2] specifically noted that this approach is only valid when the transverse curvature is zero. In contrast, the proposed model adopts an approach for cross-sectional treatment inspired by the work of Iqbal et al. [8], assuming that the cross section remains arc shaped during folding while gradually reducing its curvature. This treatment method avoids the limitations present in the Wuest and Yee models, thereby improving the predictive accuracy of the model.

5.3.2 Comparison Between Proposed Model and Yao Model

The study indicates that accurately predicting the folding performance of the composite tape spring establishes a quantitative relationship between transverse and longitudinal curvature during the folding process. The Yao model simplifies this quantitative relationship between transverse and longitudinal curvature. Figure 5.6 shows the relationship between transverse and longitudinal curvature during the folding process for two types of tape springs, as predicted by the proposed model and Yao model.

The proposed model reveals a pronounced nonlinear relationship between the transverse and longitudinal curvatures: As longitudinal curvature increases, transverse curvature rapidly decreases and approaches zero once the longitudinal curvature reaches 0.01 mm^{-1}. In contrast, the Yao model predicts a linear relationship, where transverse curvature declines more gradually and remains significant even when longitudinal curvature reaches 0.05 mm^{-1}. Existing studies have shown that when the longitudinal curvature of the composite tape spring reaches a certain threshold, the transverse curvature is nearly zero and can be disregarded [2]. Therefore, compared to the Yao model, the proposed model provides a more accurate prediction of the quantitative relationship between transverse and longitudinal curvatures during the bending and folding process.

5.3.3 Relationship Between Strain Energy and Longitudinal Curvature

Figure 5.7 presents the relationship between three types of strain energy (stretching strain energy, bending strain energy, and total strain energy) and longitudinal curvature for metallic and composite tape springs. The total strain energy versus longitudinal curvature curve exhibits a distinct two-stage nonlinear behavior, where the energy increases rapidly during the initial folding phase. Once the longitudinal curvature exceeds 0.005 mm^{-1}, the rate of increase gradually approaches linearity. The bending strain energy versus longitudinal curvature curve shows a similar trend to the total strain energy, also featuring a two-stage nonlinear relationship. After the longitudinal curvature surpasses 0.01 mm^{-1}, the two curves nearly coincide. In

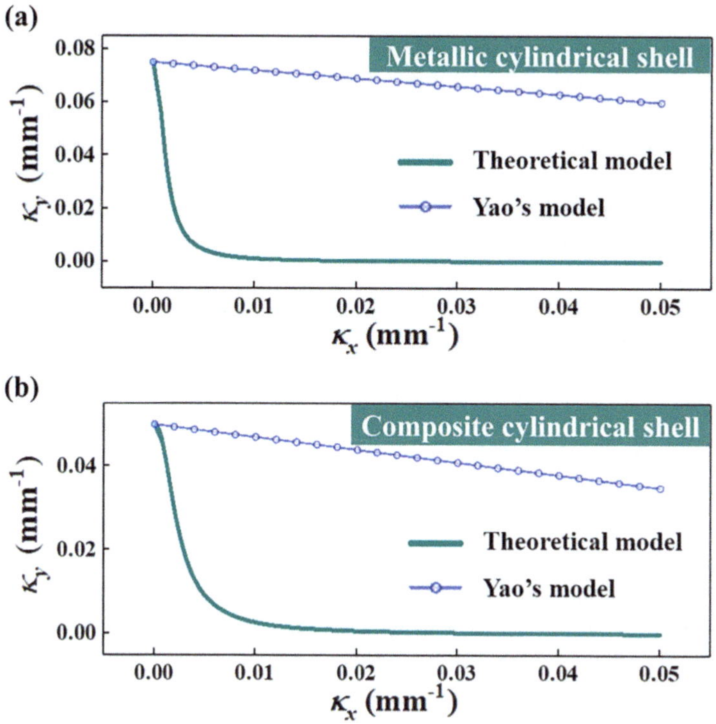

Fig. 5.6 Relationship between transverse curvature and longitudinal curvature predicted by the proposed model and Yao model: (**a**) Metallic tape springs. (**b**) Composite tape springs

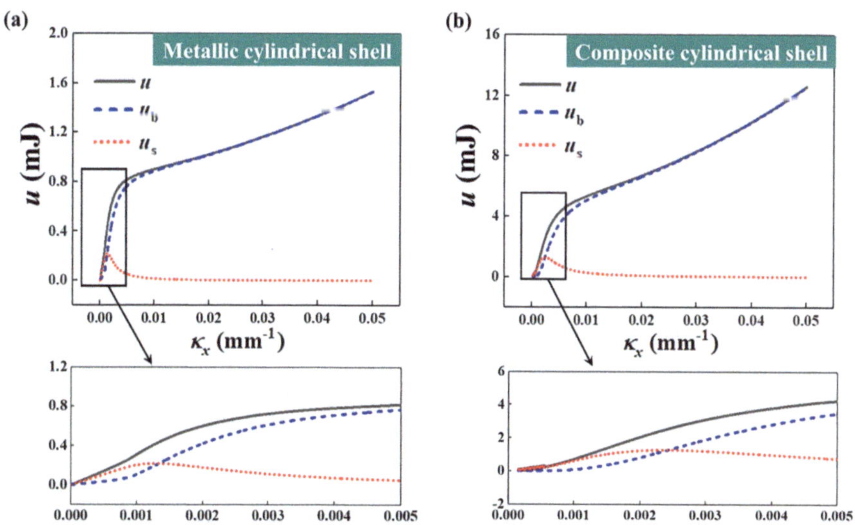

Fig. 5.7 Relationship between strain energy and longitudinal curvature: (**a**) Metallic tape springs. (**b**) Composite tape springs

contrast, the stretching strain energy versus longitudinal curvature curve behaves differently: it initially increases and then decreases, approaching zero once the longitudinal curvature exceeds 0.01 mm^{-1}. Additionally, it is clear that stretching strain energy dominates during the initial bending and folding phase. For metallic tape springs, bending strain energy begins to dominate once the longitudinal curvature exceeds 0.0013 mm^{-1}, whereas for composite tape springs, bending strain energy takes over at a longitudinal curvature beyond 0.0025 mm^{-1}. Based on this analysis, it is evident that as longitudinal curvature increases, the strain energy behavior differs significantly between metallic and composite tape springs, and stretching strain energy predominates at the initial stage, while bending strain energy gradually becomes the dominant component.

5.4 Parametric Study

In this section, the proposed model is used to investigate the effects of the cross-sectional radius R, the cross-sectional angle θ, and the thickness t on the folding behavior of the composite tape spring. The material properties and other geometric parameters remain consistent with the aforementioned composite tape spring in Sect. 5.2. The specific variable values are as follows:

- The cross-sectional radius is varied as 15 mm, 17.5 mm, 20 mm, 22.5 mm, and 25 mm.
- The cross-sectional angle is varied as 70°, 80°, 90°, 100°, and 110°.
- The thickness is varied as 0.2 mm, 0.25 mm, 0.3 mm, 0.35 mm, and 0.4 mm.

According to Fig. 5.8a, b and Table 5.5, it is evident that as the cross-sectional radius increases, the peak moment also gradually increases, while the propagation moment remains unchanged. This is because the propagation moment in Eq. (5.22) is not affected by the cross-sectional radius but is only related to the cross-sectional angle and the bending stiffness matrix. As the cross-sectional angle and thickness increase, both the peak and propagation moments gradually increase. As shown in Fig. 5.8c, the three geometric parameters also have a significant impact on the relationship between the transverse curvature and longitudinal curvature. For all deployable composite cylindrical shells, the transverse curvature eventually stabilizes at a small value. Increasing the cross-sectional radius and angle, along with reducing the thickness, helps to stabilize the transverse curvature more quickly. From Fig. 5.8d, it is observed that the longitudinal curvature reaches 0.05 mm^{-1}; there is no significant difference in the strain energy stored during the folding process for the composite tape springs with different cross-sectional radii. This is because, although the increase in cross-sectional radius leads to a longer arc length, it also reduces the cross-sectional curvature, thereby decreasing the strain energy that can be stored in per unit area. Moreover, as the cross-sectional angle and

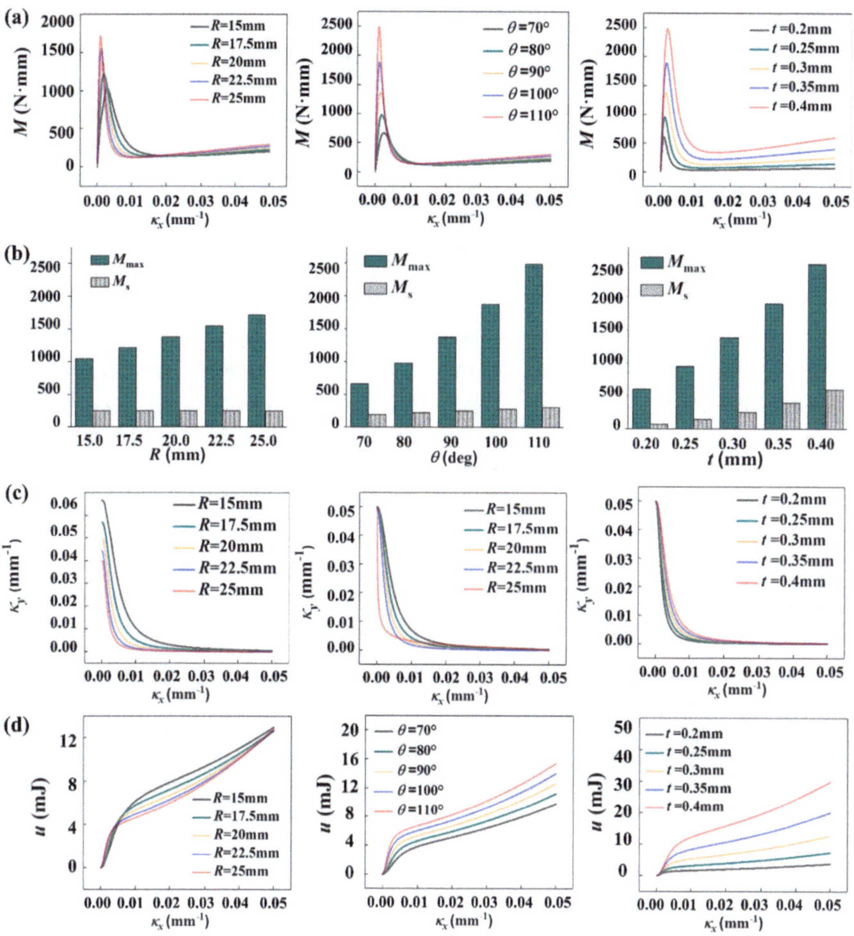

Fig. 5.8 Effect of geometric parameters on the folding behavior: (**a**) Folding moment versus longitudinal curvature curve. (**b**) Peak moment and propagation moment. (**c**) Transverse curvature and longitudinal curvature curve. (**d**) Strain energy versus longitudinal curvature curve

thickness increase, the strain energy stored per unit length in the composite tape spring also increases.

In summary, the folding performance of the composite tape spring is significantly influenced by geometric parameters such as cross-sectional radius, cross-sectional angle, and thickness. The cross-sectional radius primarily affects the peak moment, while cross-sectional angle and thickness influence both the peak and propagation moments. Additionally, these parameters directly affect the rate at which the transverse curvature stabilizes and the strain energy stored per unit length in the composite tape spring.

Table 5.5 The influence of geometric parameters on the peak and propagation moments

Geometric parameters	R (mm)	θ (°)	t (mm)	M_{max} (N mm)	M_s (N mm)
Groups of different R	15	90	0.3	1049.83	252.45
	17.5	90	0.3	1216.96	252.45
	20	90	0.3	1381.89	252.45
	22.5	90	0.3	1549.19	252.45
	25	90	0.3	1716.99	252.45
Groups of different θ	20	70	0.3	666.12	196.34
	20	80	0.3	979.51	224.39
	20	90	0.3	1381.89	252.45
	20	100	0.3	1880.57	280.49
	20	110	0.3	2488.79	308.54
Groups of different t	20	90	0.2	606.35	74.80
	20	90	0.25	951.91	146.09
	20	90	0.3	1381.89	252.45
	20	90	0.35	1894.69	400.88
	20	90	0.4	2489.27	598.40

References

1. Wuest W (1954) Einige anwendungen der theorie der zylinderschale. J Appl Math Mech/Z Angew Math Mech 34(12):444–454
2. Seffen KA, Pellegrino S (1983) Deployment dynamics of tape-springs. Proc Royal Soc Lond Ser A: Math Phys Eng Sci 1999(455):1003–1048
3. Liu TW, Bai JB, Fantuzzi N (2023) Analytical models for predicting folding behaviour of thin-walled tubular deployable composite boom for space applications. Acta Astronaut 208:167–178
4. Liu TW, Bai JB, Fantuzzi N et al (2023) Folding behaviour of thin-walled tubular deployable composite boom for space applications: experiments and numerical simulation. Acta Astronaut 209:159–171
5. Liu TW, Bai JB, Fantuzzi N et al (2024) Thin-walled deployable composite structures: A review. Prog Aerosp Sci 146:100985
6. Liu TW, Bai JB, Fantuzzi N (2024) New insights into folding tape-spring. Compos Struct 341:118225
7. Yao XF, Ma YJ, Yin YJ et al (2011) Design theory and dynamic mechanical characterization of the deployable composite tube hinge. Sci China Phys Mech Astron 54:633–639
8. Iqbal K, Pellegrino S, Daton-Lovett A (2000) Bistable composite slit tubes. In: IUTAM-IASS symposium on deployable structures. Cambridge, UK, September 6–9, 1998. Springer Netherlands, Dordrecht
9. Yee JCH, Soykasap O, Pellegrino S (2004) Carbon fibre reinforced plastic tape-springs. In: 45th AIAA/ASME/ASCE/AHS/ASC Structures, Structural Dynamics and Materials Conference, Palm Springs, California, April 19-22, 2004
10. Zuo Y, Jin G, Xie P (2018) Calculative and experimental study of the CFRP tape-spring. J Mech Sci Technol 32:3603–3609

Chapter 6
Compressive Folding of Composite Helical Structures

Abstract This chapter proposes an analytical model considering geometric nonlinearity to address inaccuracies in previous studies neglecting dynamic changes in helix diameter, angle, and active coil number during compression. Using Frenet unit vectors for force analysis, it derives compressive stiffness via energy principles and solves compressive strength with the Tsai-Hill criterion, using iteration for geometric parameter updates. Experiments on carbon fiber unidirectional and glass fiber woven composite helical structures validate the model. Parametric studies show compressive stiffness decreases from 1280 N/m to 504 N/m as helix angle increases to 45°, and from 886 N/m to 7 N/m as diameter increases to 500 mm, highlighting significant geometric nonlinearity effects in large-angle/helix diameter structures.

Composite helical structures can store and release strain energy through elastic deformation, making them highly promising for applications in aerospace, automotive, and other fields, such as deployable helical antennas and shock-absorbing springs. The compressive performance, including stiffness, strength, and load versus displacement behavior, is one of the most important basic mechanical properties of composite helical structures. A review of the literature indicates that most previous studies did not account for the effects of geometric nonlinearity during loading. Consequently, analytical methods or finite element methods have been unable to effectively predict the compressive performance of composite helical structures. To address this limitation, this chapter proposes a novel analytical model that incorporates geometric nonlinear effects to accurately predict the compressive stiffness and strength of composite helical structures. The predictions of this model show good agreement with existing experimental and numerical results. Based on the proposed model, further analysis was conducted to investigate the influence of the helix angle and helix diameter on the compressive performance of composite helical structures. The results reveal that composite helical structures with large helical angles exhibit significant geometric nonlinear effects.

J.-B. Bai et al., *Folding Mechanics of Deployable Composite Structures*, Solid Mechanics and Its Applications 283,
https://doi.org/10.1007/978-3-031-99254-4_6

6.1 Analytical Model

6.1.1 Geometric Model

The overall appearance of the composite helical structure is helical shape, and the cross section of helix is circular. The composite helical structure has the ability of elastic deformation along the direction of helix axis. When the compressive load is applied to the composite helical structure along the direction of the helix axis, the helix angle and height decrease with increase of compressive load. After the compressive load is removed, the composite helical structure can recover to original geometric configuration by releasing elastic strain energy.

In order to establish an analytical model for predicting the compressive behavior of the composite helical structure [1–3], basic assumptions are made as follows:

(1) The helix angle, helix height, helix diameter, and number of active coils of the composite helical structure change with the increase of compressive deformation, and the overall appearance of the composite helical structure is always helical shape.
(2) During compressive deformation process, the cross section of helix retains the circular shape with fixed inner diameter d_0 and outer diameter d_1, and wire length of helix remains unchanged.
(3) The compression of the composite helical structure is idealized as the quasi-static elastic compressive deformation mode.
(4) The focus of the analytical model is to investigate the stiffness, strength, and load versus displacement curve of the composite helical structure, without considering compressive stability.

According to assumption (1), the composite helical structure can be equivalent to the simplified model shown in Fig. 6.1a, and helix equations can be expressed as

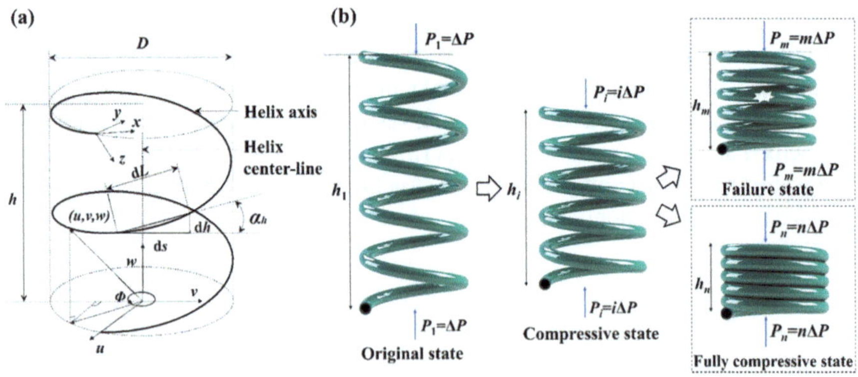

Fig. 6.1 The geometric model of the composite helical structure: (**a**) Simplified model. (**b**) Compressive deformation process

$$\begin{cases} u = \frac{1}{2} D \cos\phi \\ v = \frac{1}{2} D \sin\phi \\ w = \frac{1}{2} D\phi \tan\alpha_{\rm h} \end{cases} \tag{6.1}$$

$$L = \int_0^{\pi ND} \frac{1}{\cos\alpha_{\rm h}} {\rm d}s = \frac{\pi ND}{\cos\alpha_{\rm h}} \tag{6.2}$$

$$h = \int_0^{\pi ND} \tan\alpha_{\rm h} {\rm d}s = \pi ND \tan\alpha_{\rm h} \tag{6.3}$$

where D, $\alpha_{\rm h}$, N, and h are the helix diameter, helix angle, number of active coils, and height of the composite helical structure, respectively.

By combining Eqs. (6.2) and (6.3) with trigonometric function $\frac{1}{\cos^2\alpha_{\rm h}} = 1 + \tan^2\alpha_{\rm h}$, $\alpha_{\rm h}$ can be eliminated, and the relationship between D, L, N, and h can be deduced as

$$D = \sqrt{\frac{L^2 - h^2}{N^2\pi^2}} \tag{6.4}$$

By combining Eq. (6.3) with trigonometric functions $\sin\alpha_{\rm h} = \sqrt{\frac{\tan^2\alpha_{\rm h}}{1+\tan^2\alpha_{\rm h}}}$ and $\cos\alpha_{\rm h} = \sqrt{\frac{1}{1+\tan^2\alpha_{\rm h}}}$, $\sin\alpha_{\rm h}$ and $\cos\alpha_{\rm h}$ can be, respectively, described as

$$\sin\alpha_{\rm h} = \sqrt{\frac{h^2}{h^2 + N^2\pi^2D^2}} \tag{6.5}$$

$$\cos\alpha_{\rm h} = \sqrt{\frac{N^2\pi^2D^2}{h^2 + N^2\pi^2D^2}} \tag{6.6}$$

According to assumption (2), cross-section area, inertia moment, polar moment of inertia, and section modulus of helix cross section can be, respectively, written as

$$A = \frac{\pi\left(d_1^2 - d_0^2\right)}{4} \tag{6.7}$$

$$I_{\rm p} = \frac{\pi\left(d_1^4 - d_0^4\right)}{32} \tag{6.8}$$

$$I = \frac{\pi\left(d_1^4 - d_0^4\right)}{64} \tag{6.9}$$

$$W = \frac{\pi\left(d_1^3 - d_0^3\right)}{32} \tag{6.10}$$

Actually, in the compressive process of the composite helical structure, geometric parameters change with the increase of compressive load (shown in Fig. 6.1b). According to assumption (2), geometric parameters of the composite helical structure change continuously. In turn, this change can affect compressive behavior of the composite helical structure. Therefore, this chapter proposes an analytical model for predicting compressive behavior of the composite helical structure considering the change of geometric parameters (i.e., geometric nonlinearity). It is assumed that the composite helical structure can be compressed to the maximum compressive displacement (i.e., the two adjacent coils are closely contacted), and this process is divided into n steps; in each step, a small compressive load increment ΔP is applied to the composite helical structure, which leads to the geometrical change of helix diameter, helix angle, height, and number of active coils; in the next step, a small compressive load increment ΔP is applied to the composite helical structure with new geometric parameters, and the geometrical change is calculated again; the same iterative calculation is performed n times until reaching the maximum compressive displacement of the composite helical structure; during the whole compressive process, the Tsai-Hill criterion is used to determine whether failure of the composite helical structure occurs in each step; if the failure occurs, this step is recorded as the mth step.

Therefore, geometric parameters of the composite helical structure in the $(i + 1)$th step can be, respectively, described as

$$h_{i+1} = h_i - \Delta h_i \tag{6.11}$$

$$D_{i+1} = \sqrt{\frac{L^2 - h_{i+1}{}^2}{N_{i+1}{}^2 \pi^2}} \tag{6.12}$$

$$\sin \alpha_{\mathrm{h}}^{i+1} = \sqrt{\frac{h_{i+1}{}^2}{h_{i+1}{}^2 + N_{i+1}{}^2 \pi^2 D_{i+1}{}^2}} \tag{6.13}$$

$$\cos \alpha_{\mathrm{h}}^{i+1} = \sqrt{\frac{N_{i+1}{}^2 \pi^2 D_{i+1}{}^2}{h_{i+1}{}^2 + N_{i+1}{}^2 \pi^2 D_{i+1}{}^2}} \tag{6.14}$$

where i denotes the ith step ($i = 1, 2, 3 \ldots m$ or $i = 1, 2, 3 \ldots n$). h_i and Δh_i are height and height variation of the composite helical structure in the ith step, respectively.

The number of active coils of the composite helical structure in the $(i + 1)$th step can be expressed as [4]

$$N_{i+1} = N_i \frac{(1+\nu_{21})\cos \alpha_{\mathrm{h}}^{i+1} \cos \alpha_{\mathrm{h}}^{i} + \sin \alpha_{\mathrm{h}}^{i+1} \sin \alpha_{\mathrm{h}}^{i}}{(1+\nu_{21})\cos^2 \alpha_{\mathrm{h}}^{i+1} + \sin^2 \alpha_{\mathrm{h}}^{i+1}} \tag{6.15}$$

Equation (6.15) is based on the assumption that the composite helical structure retains helical shape and the wire length remains unchanged under compressive load. N_i is the number of active coils in the ith step.

The maximum compressive displacement can be obtained from the initial geometric configuration of the composite helical structure:

$$\delta_{\max} = h_1 - N_1 d_1 \tag{6.16}$$

6.1.2 Analytical Solution for Compressive Stiffness

Frenet unit vectors are used to describe the stress resultants of the composite helical structure under the axial compressive load increment ΔP. The stress resultants of the composite helical structure on the any cross section A–A are shown in Fig. 6.2, including the twisting moment increment around the direction of x-axis in the ith step, the bending moment increment around the direction of z-axis in the ith step, the axial force increment along the direction of x-axis in the ith step, and the shear force increment along the direction of z-axis in the ith step:

$$\Delta T_i = \frac{\Delta P D_i}{2} \cos \alpha_{\mathrm{h}}^i \tag{6.17}$$

$$\Delta M_i = \frac{\Delta P D_i}{2} \sin \alpha_{\mathrm{h}}^i \tag{6.18}$$

$$\Delta F_i = -\Delta P \sin \alpha_{\mathrm{h}}^i \tag{6.19}$$

$$\Delta Q_i = \Delta P \cos \alpha_{\mathrm{h}}^i \tag{6.20}$$

The strain energy expression of the helix composite structure in the ith step can be expressed as

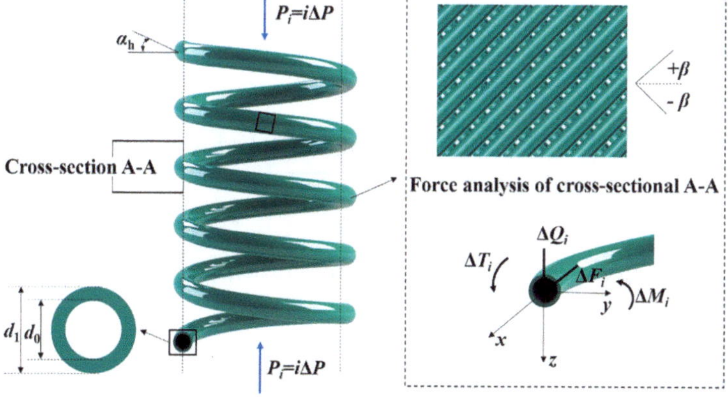

Fig. 6.2 Geometric configuration and force analysis of helix composite structure

$$\frac{\Delta P \Delta h_i}{2} = \int_0^L \frac{\Delta T_i^2}{2G_{xy}I_\mathrm{p}}\mathrm{d}l + \int_0^L \frac{\Delta M_i^2}{2E_xI}\mathrm{d}l + \int_0^L \frac{\Delta F_i^2}{2E_xA}\mathrm{d}l + \int_0^L \frac{k_\mathrm{S}\Delta Q_i^2}{2G_{xy}A}\mathrm{d}l \tag{6.21}$$

where E_x and G_{12} are the elastic modulus of composite laminate in the tangential direction and shear modulus of the composite helix structure in the x–y direction, respectively. The k_S is shape shear factor of circular cross section, and the expression is listed as follows:

$$k_S = \frac{A}{I^2}\,\overset{\text{,}}{_{Ayz}}\,\frac{S(z)^2}{b(z)^2}\mathrm{d}A \tag{6.22}$$

$$S(z) = \begin{cases} \int_z^{\frac{d_0}{2}} 2z\left(\sqrt{\left(\frac{d_1}{2}\right)^2 - z^2} - \sqrt{\left(\frac{d_0}{2}\right)^2 - z^2}\right)\mathrm{d}z \\ \quad + \int_{\frac{d_0}{2}}^{\frac{d_1}{2}} 2z\sqrt{\left(\frac{d_1}{2}\right)^2 - z^2}\,\mathrm{d}z \left(0 < z < \frac{d_0}{2}\right) \\ \int_z^{\frac{d_1}{2}} 2z\sqrt{\left(\frac{d_1}{2}\right)^2 - z^2}\,\mathrm{d}z \qquad \left(\frac{d_0}{2} < z < \frac{d_1}{2}\right) \end{cases} \tag{6.23}$$

$$b(z) = \begin{cases} \left(\sqrt{\left(\frac{d_1}{2}\right)^2 - z^2} - \sqrt{\left(\frac{d_0}{2}\right)^2 - z^2}\right) \left(0 < z < \frac{d_0}{2}\right) \\ \sqrt{\left(\frac{d_1}{2}\right)^2 - z^2} \qquad \left(\frac{d_0}{2} < z < \frac{d_1}{2}\right) \end{cases} \tag{6.24}$$

Substituting Eqs. (6.17)–(6.20) into Eq. (6.21), the height variation (i.e., compressive deformation increment) of the composite helix structure in the ith step can be obtained as

$$\Delta h_i = \frac{8N_i \Delta P D_i^3}{\cos\alpha_\mathrm{h}^i\left(d_1^4 - d_0^4\right)} \left[\begin{array}{l} \frac{\cos^2\alpha_\mathrm{h}^i}{G_{12}} + \frac{2\sin^2\alpha_\mathrm{h}^i}{E_x} \\ + \frac{\left(d_1^2 + d_0^2\right)\sin^2\alpha_\mathrm{h}^i}{2E_xD_i^2} + \frac{\left(d_1^2 + d_0^2\right)\cos^2\alpha_\mathrm{h}^i}{G_{12}D_i^2} \end{array}\right] \tag{6.25}$$

Therefore, compressive stiffness of the composite helix structure in the ith step can be deduced as

$$K_i = \frac{\dfrac{\cos\alpha_h^i\left(d_1^4 - d_0^4\right)}{8N_iD_i^3}}{\dfrac{\cos^2\alpha_h^i}{G_{12}} + \dfrac{2\sin^2\alpha_i}{E_x} + \dfrac{\left(d_1^2 + d_0^2\right)\sin^2\alpha_h^i}{2E_xD_i^2} + \dfrac{\left(d_1^2 + d_0^2\right)\cos^2\alpha_h^i}{G_{12}D_i^2}} \tag{6.26}$$

According to the Eq. (6.25), the height variation of the composite helix structure $\Delta h_i (i = 1,2,3\ldots m$ or $i = 1,2,3\ldots n)$ can be obtained. Thus, total height variation (compressive deformation) and total compressive load under k steps can be obtained by progressive accumulation of Δh_i and ΔP, as follows:

$$\delta_k = \sum_{i=1}^{k} \Delta h_i \tag{6.27}$$

$$P_k = k\Delta P \tag{6.28}$$

By fitting the data points $[\delta_k, P_k](k = 1,2,3\ldots m$ or $k = 1,2,3\ldots n)$, compressive load versus displacement curve of the composite helix structure under compressive load can be obtained until maximum compressive displacement of the composite helix structure $\delta_{\max}$ or compressive failure load of the composite helix structure P_m. Using the least squares method to linear fit these data points, the compressive stiffness of the composite helix structure can be obtained as

$$K = \begin{cases} \dfrac{\sum_{k=1}^{m}\left(\delta_k - \overline{\delta}\right)\left(P_k - \overline{P}\right)}{\sum_{k=1}^{m}\left(\delta_k - \overline{\delta}\right)^2} & (\text{Failure}) \\[2ex] \dfrac{\sum_{k=1}^{n}\left(\delta_k - \overline{\delta}\right)\left(P_k - \overline{P}\right)}{\sum_{k=1}^{n}\left(\delta_k - \overline{\delta}\right)^2} & (\text{Reaching}\, \delta_{\max}) \end{cases} \tag{6.29}$$

where $\overline{\delta}$ and $\overline{P}$ are the average compressive deformation and average compressive load of the composite helix structure in the direction of helix axis, respectively; it is possible to show as follows:

$$\overline{\delta} = \begin{cases} \left(\sum_{k=1}^{m}\delta_k\right)/m & (\text{Failure}) \\ \left(\sum_{k=1}^{n}\delta_k\right)/n & (\text{Reaching}\, \delta_{\max}) \end{cases} \tag{6.30}$$

$$\overline{P} = \begin{cases} \left(\sum_{k=1}^{m} P_k \right) / m & (\text{Failure}) \\ \left(\sum_{k=1}^{n} P_k \right) / n & (\text{Reaching}\, \delta_{\max}) \end{cases} \tag{6.31}$$

In the compressive deformation process, the composite helix structure may not fail, and accumulative height variation can reach the maximum compressive displacement (i.e., the two adjacent coils of composite helix structure are closely contacted):

$$\delta_{\max} = \sum_{i=1}^{n} \Delta h_i \tag{6.32}$$

Substituting Eqs. (6.16)–(6.25) into (6.32), it is possible to have

$$h_1 - N_1 d_1 = \sum_{i=1}^{n} \frac{8N_i \Delta P D_i^3}{\cos\alpha_{\mathrm{h}}^i \left(d_1^4 - d_0^4\right)} \left[\frac{\cos^2\alpha_i}{G_{12}} + \frac{2\sin^2\alpha_{\mathrm{h}}^i}{E_x} + \frac{\left(d_1^2 + d_0^2\right)\sin^2\alpha_{\mathrm{h}}^i}{2E_x D_i^2} + \frac{\left(d_1^2 + d_0^2\right)\cos^2\alpha_{\mathrm{h}}^i}{G_{12} D_i^2} \right] \tag{6.33}$$

Then, the compressive load corresponding to the maximum displacement of the composite helix structure can be determined as

$$\begin{aligned} P_n &= n\Delta P \\ &= h_1 - N_1 d_1 \left\{ \sum_{i=1}^{n} \frac{8N_i D_i^3}{\cos\alpha_{\mathrm{h}}^i \left(d_1^4 - d_0^4\right)} \left[\frac{\cos^2\alpha_{\mathrm{h}}^i}{G_{12}} + \frac{2\sin^2\alpha_{\mathrm{h}}^i}{E_x} + \frac{\left(d_1^2 + d_0^2\right)\sin^2\alpha_{\mathrm{h}}^i}{2E_x D_i^2} + \frac{\left(d_1^2 + d_0^2\right)\cos^2\alpha_{\mathrm{h}}^i}{G_{12} D_i^2} \right] \right\}^{-1} \end{aligned} \tag{6.34}$$

6.1.3 Analytical Solution for Compressive Strength

In the compression process of the composite helical structure, failure may occur. Therefore, this section calculates the compressive strength P_m of the composite helical structure. When the composite helical structure is subjected to a compressive load along the helical axis, the shear stress on the cross section is composed of two shear stress components: one generated by transverse shear force and the other by torque. Based on the theory of bar torsion, the shear stress at any cross section is proportional to the distance from the center, resulting in the maximum shear stress on the outermost layer. The outermost ply of the composite material is shown in Fig. 3.1, with the total shear stress in the outer layer denoted as τ_{xy}. When the composite helical structure bears compressive load, torque, bending moment, shear force, and axial force, it cannot produce normal stress in the y direction, so $\sigma_y = 0$. By using the coordinate transformation relation for stress components (Eq. 3.17), simplification and arrangement yield

$$\begin{bmatrix} \sigma_1 \\ \sigma_2 \\ \tau_{12} \end{bmatrix} = \begin{bmatrix} \sigma_x \cos^2 \beta + 2\tau_{xy} \cos \beta \sin \beta \\ \sigma_x \sin^2 \beta - 2\tau_{xy} \cos \beta \sin \beta \\ -\sigma_x \sin \beta \cos \beta + \tau_{xy} \left(\cos^2 \beta - \sin^2 \beta \right) \end{bmatrix} \tag{6.35}$$

According to the Tsai-Hill criterion [5], the composite helical structure fails if the following condition is met:

$$\frac{\sigma_1^{\ 2}}{X^2} - \frac{\sigma_1 \sigma_2}{X^2} + \frac{\sigma_2^{\ 2}}{Y^2} + \frac{\tau_{12}^{\ 2}}{S^2} = 1 \tag{6.36}$$

Substituting Eq. (6.35) into (6.36), it is possible to have

$$\begin{aligned} & X^{\ 2}\left[\sigma_x^{\ 2}\left(\cos^4 \beta - \sin^2 \beta \cos^2 \beta\right) + 8\tau_{xy}^{\ 2} \sin^2 \beta \cos^2 \beta \right. \\ & \left. \quad + \sigma_x \tau_{xy} \left(6 \sin \beta \cos^3 \beta - 2 \sin^3 \beta \cos \beta\right)\right] \\ & + Y^{\ 2}\left[\sigma_x^{\ 2} \sin^4 \beta + 4\tau_{xy}^{\ 2} \sin^2 \beta \cos^2 \beta - 4\sigma_x \tau_{xy} \sin^3 \beta \cos \beta\right] \\ & S^{\ 2}\left[\sigma_x^{\ 2} \sin^2 \beta \cos^2 \beta + \tau_{xy}^{\ 2}\left(\sin^4 \beta + \cos^4 \beta - 2 \sin^2 \beta \cos^2 \beta\right)\right] = 1 \end{aligned} \tag{6.37}$$

The maximum normal stress $\sigma_{x,m}$ in the outermost layer of the composite helical structure at m step can be expressed as

$$\sigma_{x,m} = \sum_{i=1}^{m} \left(\frac{\Delta F_i}{A} - \frac{\Delta M_i}{W} \right) \tag{6.38}$$

Substituting Eqs. (6.7), (6.10), (6.18), and (6.19) into Eq. (6.38) yields

$$\sigma_{x,m} = \sum_{i=1}^{m} \left[-\frac{4\Delta P \sin \alpha_{\mathrm{h}}^{i}}{\pi\left(d_1^{\ 2} - d_0^2\right)} - \frac{16\Delta P D_i \sin \alpha_{\mathrm{h}}^{i}}{\pi\left(d_1^{\ 3} - d_0^{\ 3}\right)} \right] \tag{6.39}$$

The maximum shear stress $\tau_{xy,m}$ in the outermost layer of the composite helical structure at m step can be expressed as

$$\tau_{xy,m} = \sum_{i=1}^{m} \left(\frac{\Delta T_i d_1}{2I_p} + \frac{2\Delta Q_i}{A} \right) \tag{6.40}$$

Substituting Eqs. (6.7), (6.8), and (6.17) into Eq. (6.40) yields

$$\tau_{xy,m} = \sum_{i=1}^{m} \left[\frac{8\Delta P d_1 D_i \cos\alpha_h^i}{\pi\left(d_1^4 - d_0^4\right)} + \frac{8\Delta P \cos\alpha_h^i}{\pi\left(d_1^2 - d_0^2\right)} \right] \tag{6.41}$$

By substituting Eqs. (6.39) and (6.41) into Eq. (6.37), the compressive failure load of the composite helical structure can be determined:

$$\begin{aligned} P_m &= m\Delta P \\ &= \left\{ \begin{aligned} & X^{-2}\left[B_m^2\left(\cos^4\beta - \sin^2\beta\cos^2\beta\right) + 8C_m^2\sin^2\beta\cos^2\beta \right. \\ & \left. + B_m C_m\left(6\sin\beta\cos^3\beta - 2\sin^3\beta\cos\beta\right)\right] \\ & + Y^{-2}\left[B_m^2\sin^4\beta + 4C_m^2\sin^2\beta\cos^2\beta - 4B_m C_m\sin^3\beta\cos\beta \right] \\ & + S^{-2}\left[B_m^2\sin^2\beta\cos^2\beta + C_m^2\left(\sin^4\beta + \cos^4\beta - 2\sin^2\beta\cos^2\beta\right)\right] \end{aligned} \right\}^{-0.5} \end{aligned} \tag{6.42}$$

where B_m and C_m are intermediate variables at m step, specifically expressed as

$$B_m = \sum_{i=1}^{m} \left[-\frac{4\sin\alpha_h^i}{\pi\left(d_1^2 - d_0^2\right)} - \frac{16D_i\sin\alpha_h^i}{\pi\left(d_1^3 - d_0^3\right)} \right] \tag{6.43}$$

$$C_m = \sum_{i=1}^{m} \left[\frac{8d_1 D_i\cos\alpha_h^i}{\pi\left(d_1^4 - d_0^4\right)} + \frac{8\cos\alpha_h^i}{\pi\left(d_1^2 - d_0^2\right)} \right] \tag{6.44}$$

6.2 Model Validation

Based on the analytical model considering geometric nonlinearity proposed in the previous sections, compressive behavior (i.e., stiffness, strength, and load versus displacement relationship) can be predicted using Eqs. (6.25)–(6.27), (6.34), and (6.42). According to the above approach, iterative solution procedure was developed to analyze the compressive performance of the composite helical structure. Figure 6.3 shows the flowchart of the analytical procedure for calculating compressive strength and load versus displacement curve by MATLAB code. Compressive stiffness can be obtained by using the least squares method to linearly fit the data points of the compressive load versus displacement curve.

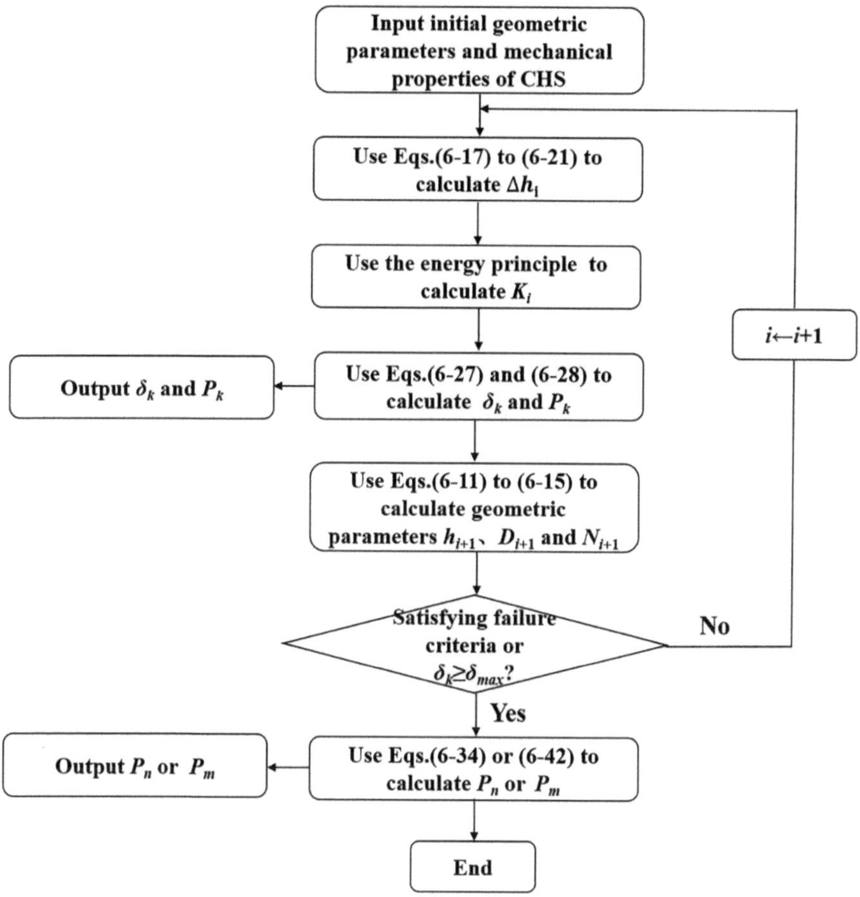

Fig. 6.3 Flowchart of the analytical procedures for calculating compressive strength and load versus displacement curve

In order to verify the effectiveness of the analytical model proposed in this section, its predictions are compared with experimental data from literatures [6, 7] as well as the classical Wahl model predictions [4]. In literature [6], the helix angle of the composite helical structure is 4°, and the composite helical structure is made from unidirectional reinforced composite, while in literature [7], the helix angle is 10° and the composite helical structure is made from glass plain weave fabric composite. These two groups of experimental data cover the cases of small and larger helix angles (i.e., 4° and 10°) as well as of different materials, which is convenient to validate the analytical model. The geometric parameters are listed in Table 6.1 [6, 7], and Table 6.2 provides the properties of unidirectional reinforced composite [6] and plain weave fabric composite [7].

As shown in Fig. 6.4 and Table 6.3, the experimentally measured compressive stiffness is 30,300 N/m in literature [6] and 1270 N/m in literature [7]. The

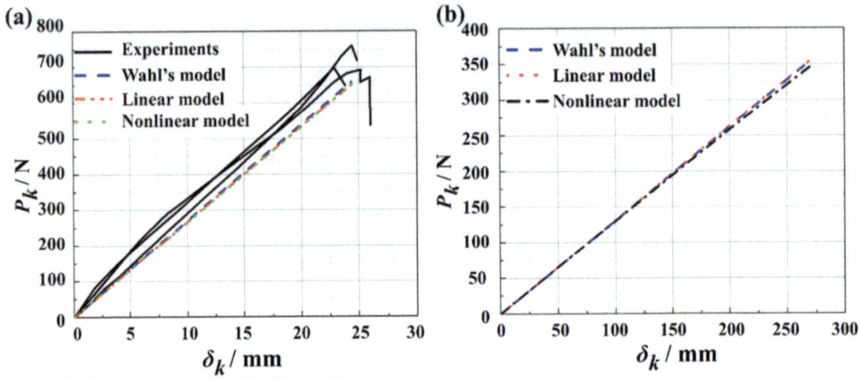

Fig. 6.4 Compressive load versus displacement curves: (**a**) Unidirectional reinforced composite helix structures. (**b**) Plain weave fabric composite helix structures

Table 6.1 Specifications and properties of the composite helix structure

Geometric parameters	Type of materials: M40/epoxy unidirectional reinforced composite [6]	Type of materials: E-glass/epoxy fabric plain-woven composite [7]
D(mm)	100	100
d_1(mm)	12	10
d_0 (mm)	8	8
α_h(°)	4	10
N	4	6
t_s (mm)	0.2	0.25
β (°)	±45	±45
V_f	0.61	N.A.

Table 6.2 Properties of composites and constituents

Property parameters	Type of materials: M40/epoxy648 unidirectional reinforced composite [6]	Type of materials: E-glass/epoxy fabric composite [7]
E_f (GPa)	397	N.A.
E_m (GPa)	2.5	N.A.
E_1 (GPa)	243.15*	28.00
E_2 (GPa)	6.35*	28.00
G_f (GPa)	6.00 [9]	N.A.
G_{12} (GPa)	2.10*	9.50
ν_f	0.2 [9]	N.A.
ν_{21}	0.2	0.28

Note: * represents the calculated value. Literature [6] does not directly give E_1, E_2, and G_{12} of M40/epoxy composite. E_1, E_2, and G_{12} of M40/epoxy composite can be calculated by the rule of mixtures and constitutive relationship. The values of shear modulus G_f and Poisson's ratio ν_f of M40 carbon fiber are from literature [9]

Table 6.3 Comparison between experiments and predictions for compressive stiffness

Compressive stiffness K (N/m)		Type of materials: M40/epoxy648 unidirectional reinforced composite [6]	E-glass/epoxy fabric composite [7]
Experiments		30,300	1270
Wahl model [4]	Predictions	27,000	1310
	Relative deviation	10.89%	3.15%
Linear model	Predictions	26,905	1314
	Relative deviation	11.20%	3.46%
Nonlinear model	Predictions	26,860	1280
	Relative deviation	11.35%	0.79%

analytical model proposed for predicting the compressive behavior of composite considering geometric nonlinearity can be degraded to a linear model by neglecting the geometric nonlinearity effect, namely the geometric parameters are constant during the whole compressive loading process. To further validate the model's effectiveness, the Wahl model, the linear model degraded from the nonlinear model, and the nonlinear model proposed were all used to predict the compressive stiffness and load versus displacement curve of the composite helical structure. There is a lack of necessary strength parameters (i.e., X, Y, and S) in literatures [6, 7]; failure during compression for the composite helical structure in literature [6] was measured by the average failure displacement from three groups of experimental data to predict compressive stiffness. In literature [7], when the compressive deformation of the composite helical structure reached the maximum compressive displacement, no failure occurred. So the above three analytical models are used to predict the compressive behavior; it is determined that the maximum compressive displacement can be reached, as are shown in Fig. 6.4 and Table 6.3.

From Fig. 6.4a and Table 6.3, for the composite helical structure with helix angle of 4° in literature [6], the compressive stiffness and load versus displacement curves predicted by all three models agree well with the experimental results, with relative deviations from the experimental stiffness of 10.89%, 11.20%, and 11.35%, respectively. The high consistency and similar accuracy of all three predictions indicate that the geometric nonlinearity has almost no influence on the prediction of compressive behavior for composite helical structures with small helix angles, and this effect can be ignored.

From Fig. 6.4b and Table 6.3, for the composite helical structure with helix angle of 10° in literature [7], the Wahl model, the linear model, and the nonlinear model all predict compressive stiffness and load versus displacement curves that align closely with the experimental results, with relative deviations of 3.15%, 3.46%, and 0.79%, respectively. Notably, the predicted compressive stiffness using the nonlinear model is more consistent with experimental results, and the prediction accuracy is significantly improved. It is shown that with the increase of the helix angle, geometric nonlinearity effect on the prediction of compressive stiffness increases, which cannot be ignored.

In summary, the analytical model proposed for predicting the compressive behavior considers the geometric nonlinearity effect and demonstrates good predictive accuracy for both small and large helix angles, which proves the validity and applicability of the nonlinear analytical model proposed.

6.3 Results and Discussion

When predicting compressive behavior of composite helical structures with small helix angles, the effects of internal axial force, shear force, bending moment, and geometric nonlinearity are minimal. Thus, both the Wahl model and the linear model proposed in this chapter can still achieve good predictive accuracy while neglecting these factors, making them relatively simple and convenient to use. However, neither the Wahl model nor the linear model can effectively predict the compressive performance of composite helical structures with large helix angles. In contrast, the nonlinear model proposed here achieves higher predictive accuracy for two main reasons: For composite helical structures with large helix angle under compressive load, (1) the strain energy is significantly generated by internal axial force, shear force, and bending moment, making these internal forces non-negligible, and (2) the mean diameter, height, helix angle, and number of active coils change significantly with increasing compressive load, necessitating the consideration of geometric nonlinearity.

The nonlinear model proposed considers both geometric nonlinearity (i.e., changes in helix diameter, height, helix angle, and number of active coils during compression) and comprehensive internal forces (i.e., torque, axial force, shear force, and bending moment), aligning better with the physical characteristics of the structure. Therefore, it demonstrates better predictive performance and a wider range of applicability. Omitting geometric nonlinearity in the analytical model evidently impacts prediction results, especially for large-helix-angle composite helical structures, where neglecting these changes during loading can lead to significant deviations from experimental results. In past studies, the effects of geometric nonlinearity during deformation were often ignored, limiting the effectiveness of analytical or finite element methods to predicting the compressive performance of only small-helix-angle composite helical structures. Hence, for accurate analytical predictions or numerical simulations of the compressive behavior of composite helical structures, accounting for geometric nonlinearity yields results that are both more accurate and more representative of the actual physical processes.

6.4 Parametric Study

6.4.1 The Effect of Helix Angle

As shown in Sect. 6.2, it is clear that the Wahl model, the linear model degraded from the nonlinear model, and the nonlinear model can effectively predict the compressive behavior of composite helical structures with small helix angle, while the predictions of the nonlinear model are more accurate. Currently, there are few investigations on the compressive behavior of composite helical structures with large helix angle, but composite helical structures with large helix angle are often needed in some fields. For example, due to the high storage ratio and electromagnetic requirements, the deployable antenna applied in the aerospace field needs to use the composite helical structures with large helix angle. Therefore, based on the geometric parameters and mechanical properties of the composite helical structure from literature [7], in which only the helix angle is varied while all other parameters remain constant, the nonlinear model proposed was used to analyze the compressive behavior of composite helical structures with helix angles of 10°, 15°, 20°, 25°, 30°, 35°, 40°, and 45°. Additionally, to demonstrate the effect of geometric nonlinearity on the prediction of compressive performance for composite helical structures with large helix angles, a comparison was made with the prediction results from the Wahl model and the linear model. Due to the lack of necessary strength parameters in literature [7], it was assumed that the compression deformation could reach its maximum, and the prediction results are shown in Table 6.4, Figs. 6.5 and 6.6.

The following can be observed: (1) The compressive stiffness of composite helical structures with different helix angles predicted by the Wahl model is 1310 N/m. The main reason is that the Wahl model only considers twisting moment while ignoring axial force, shear force, and bending moment. (2) The linear model derived from the nonlinear model proposed predicts that the compressive stiffness of composite helical structures decreases as the helix angle increases, from a maximum value of 1314 N/m to 1057 N/m, as it accounts for the strain energy contributions to axial force, shear force, bending moment, and twisting moment. (3) The compressive stiffness predicted by the nonlinear model significantly decreases with the

Table 6.4 Predicted compressive stiffness with the change of helix angle

	Compressive stiffness K (N/m)		
α_h (°)	Wahl model [4]	Linear model	Nonlinear model
10	1310	1314	1280
15	1310	1299	1205
20	1310	1278	1115
25	1310	1251	1006
30	1310	1216	884
35	1310	1172	764
40	1310	1122	627
45	1310	1057	504

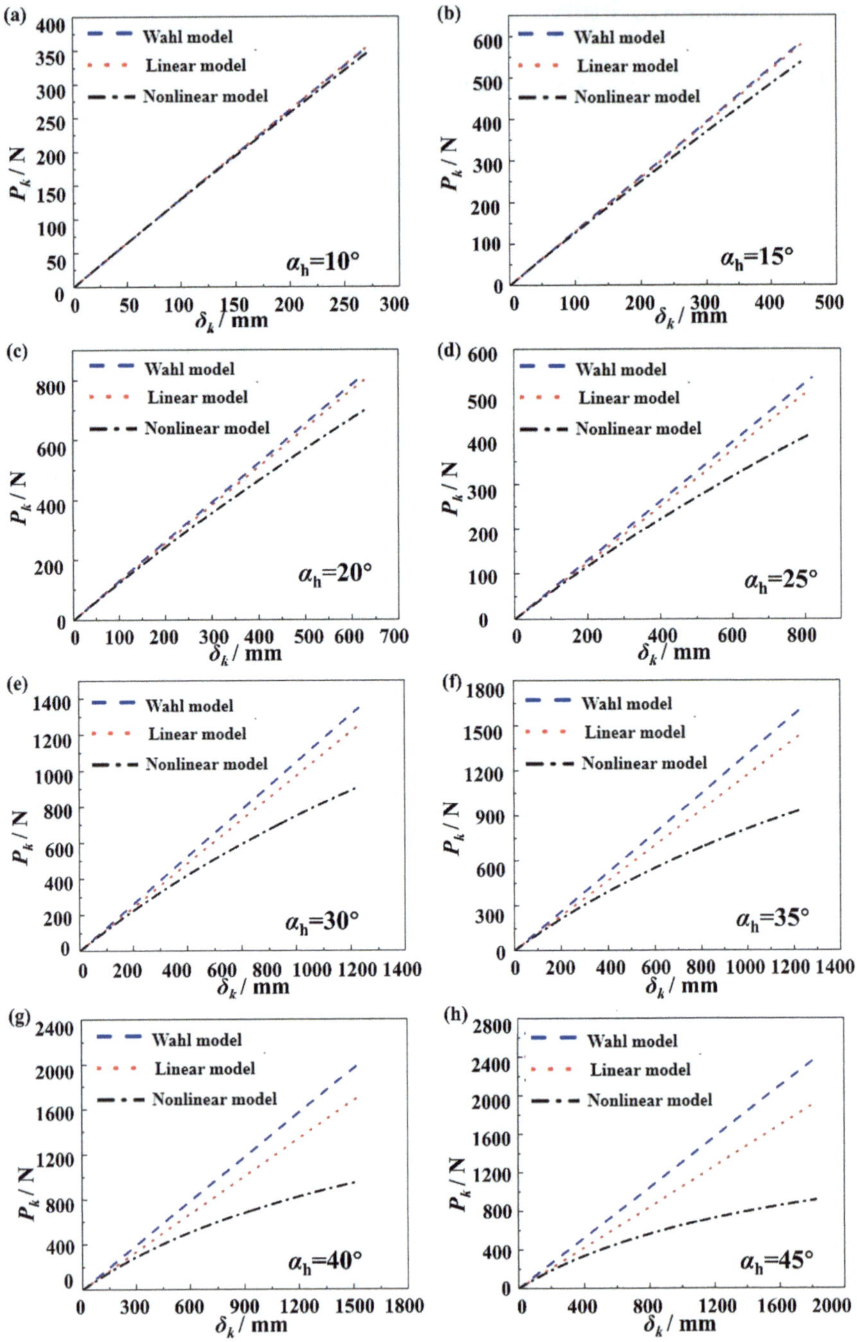

Fig. 6.5 Effect of helix angle on compressive load versus displacement curve of different analytical models: **(a)** $\alpha_h = 10°$. **(b)** $\alpha_h = 15°$. **(c)** $\alpha_h = 20°$. **(d)** $\alpha_h = 25°$. **(e)** $\alpha_h = 30°$. **(f)** $\alpha_h = 35°$. **(g)** $\alpha_h = 40°$. (h) $\alpha_h = 45°$

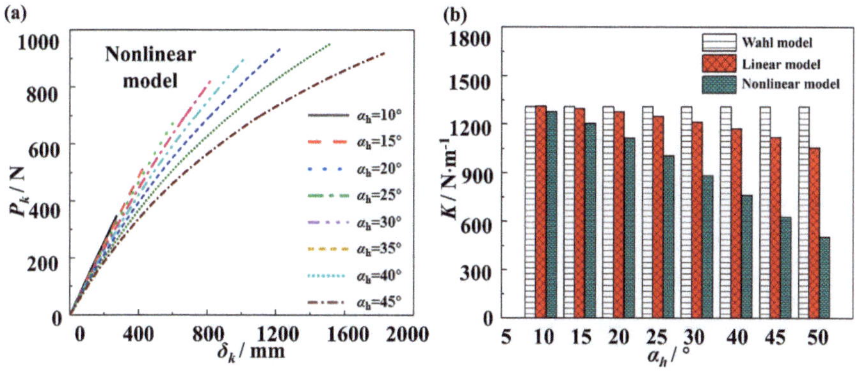

Fig. 6.6 Effect of helix angles on compressive behavior: (**a**) Compressive load versus displacement curves. (**b**) The relationship of compressive stiffness and helix angle

increase of helix angle, from 1280 N/m to 504 N/m. As the helix angle increases, the nonlinear characteristics of the load versus displacement curve become more pronounced, indicating that the helix angle is a key factor affecting the nonlinear characteristics of the load versus displacement curve of composite helical structures. Therefore, geometric nonlinearity must be considered when predicting the compressive behavior of composite helical structures with large helix angles.

Figure 6.7 and Table 6.5 show the maximum normal and shear stresses in the principal direction of composite helical structures with different helix angles predicted by the nonlinear model in the compressive deformation. The maximum normal stresses increase with compression, while the maximum shear stress initially increases and then decreases with increasing compression. Due to the lack of necessary strength parameters, typical strength values for glass plain weave fabric composites from literature [8] (X = 473 MPa, Y = 473 MPa, and S = 85 MPa) were selected as a reference. By substituting these strength values into Eq. (6.37), the compressive failure loads for composite helical structures with different helix angles were calculated (as shown in Table 6.5). From Table 6.5, it can be seen that when the helix angle reaches 20° or above, the load exceeds the failure load calculated using Eq. (6.37) when compression reaches the maximum displacement. This suggests that the composite helical structure is unlikely to fail at maximum compression displacement.

6.4.2 *The Effect of Helix Diameter*

Based on the analysis of the compressive properties of composite helical structures with different helix angles, a composite helical structure with a helix angle of 30° was selected. Only the value of the helix diameter was changed, while the other geometric parameters and performance parameters remained unchanged. The Wahl model, the linear model degraded from the nonlinear model proposed, and the

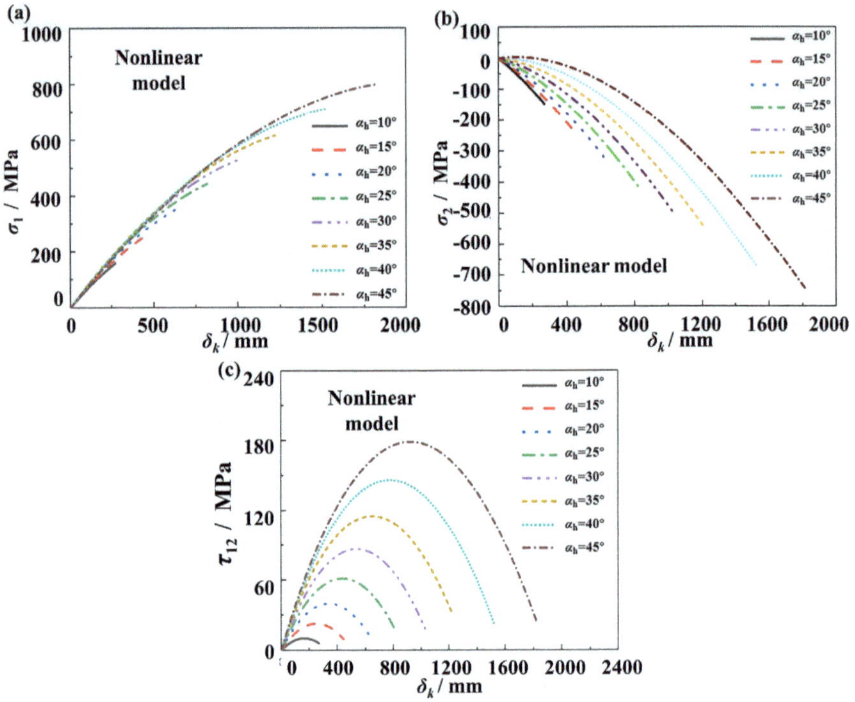

Fig. 6.7 Effect of helix angle on the maximum stress of the composite helical structure in the compressive deformation process: **(a)** σ_1. **(b)** σ_2. **(c)** τ_{12}

Table 6.5 Predicted compressive strength with change of helix angle using nonlinear model

			The maximum stress (MPa)		
α_h (°)	P_n (N)	P_m (N)	σ_1	σ_2	τ_{12}
10	346	N.F.	159	−147	10
15	537	N.F.	256	−237	23
20	699	529	351	−326	40
25	824	414	443	−412	61
30	907	330	534	−497	87
35	932	276	616	−552	115
40	955	240	711	−668	146
45	919	212	800	−753	179

Note: N.F. indicates that composite helix structure can be compressed to the maximum compressive displacement without failure, and the failure load cannot be calculated

nonlinear model were used to predict the compressive properties of composite helical structures with helix diameters of 100 mm, 200 mm, 300 mm, 400 mm, and 500 mm. The prediction results are shown in Fig. 6.8 and Table 6.6. As can be seen from Fig. 6.9 and Table 6.6, the compressive stiffness predicted by the Wahl model and the linear model is significantly higher than that predicted by the nonlinear

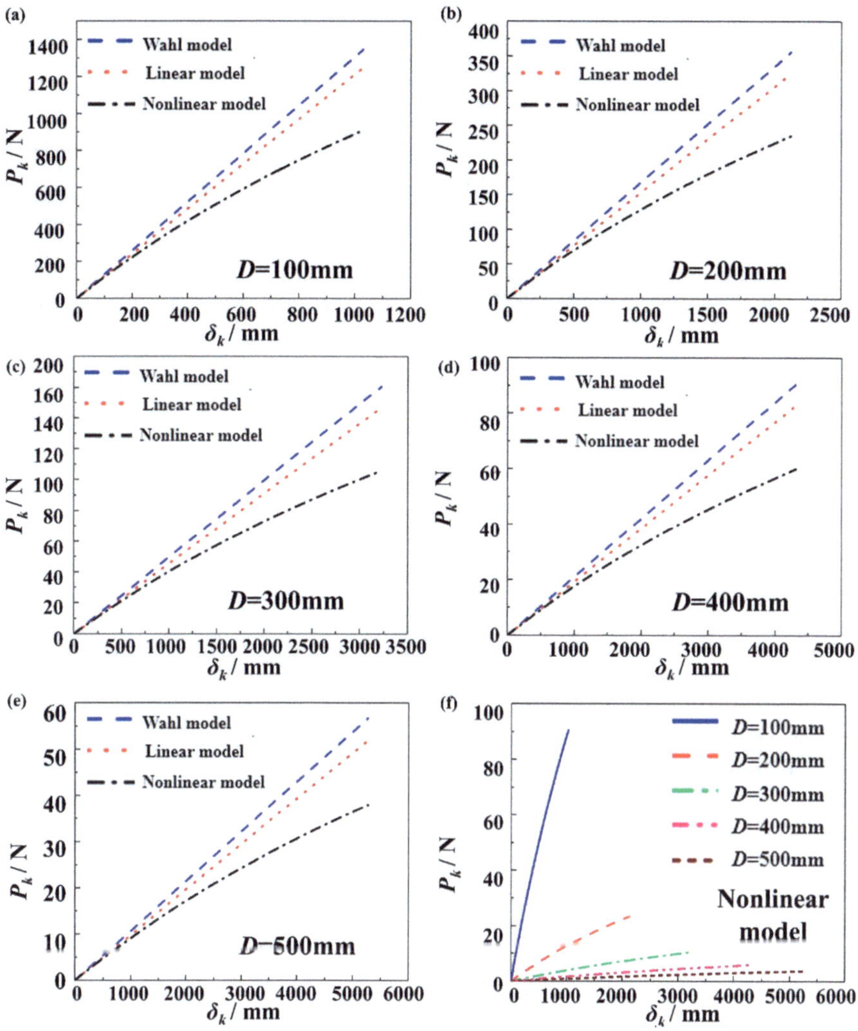

Fig. 6.8 Effect of helix diameter on compressive load versus displacement curve: (**a**) $D = 100$ mm. (**b**) $D = 200$ mm. (**c**) $D = 300$ mm. (**d**) $D = 400$ mm. (**e**) $D = 500$ mm. (f) The whole compressive load versus displacement curves

Table 6.6 Predicted compressive stiffness with the change of helix diameter

	Compressive stiffness K (N/m)		
D (mm)	Wahl model [4]	Linear model	Nonlinear model
100	1310	1216	886
200	168	154	110
300	50	46	33
400	21	20	14
500	11	10	7

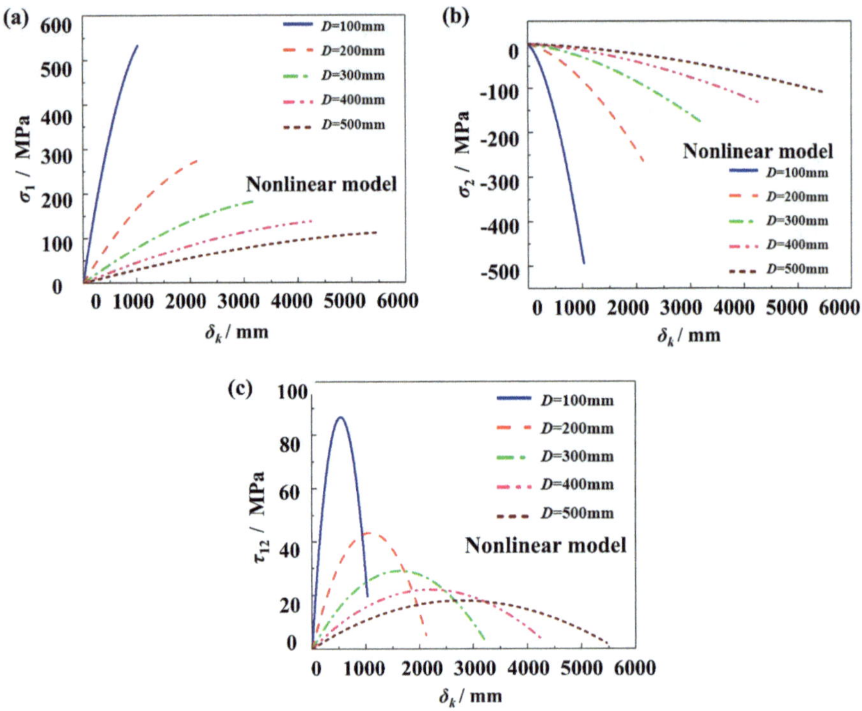

Fig. 6.9 Effect of helix diameter on the maximum stress in the compressive deformation process: (**a**) σ_1. (**b**) σ_2. (**c**) τ_{12}

Table 6.7 Predicted compressive strength with change of helix diameter using nonlinear model

			The maximum stress (MPa)		
D (mm)	P_n (N)	P_m (N)	σ_1	σ_2	τ_{12}
100	907	330	533	−494	87
200	236	N.F.	273	−263	43
300	107	N.F.	184	−179	29
400	61	N.F.	139	−134	22
500	40	N.F.	113	−108	18

model proposed, which is similar to the analysis results in Sect. 6.4.1; the prediction results of the nonlinear model proposed show that as the helix diameter increases, the compressive stiffness of the composite helical structure decreases significantly, indicating that the helix diameter is another key factor affecting the compressive stiffness of the composite helical structure.

Figure 6.9 and Table 6.7 show the maximum normal and shear stresses in the principal direction of composite helical structures with different helix diameters predicted by the nonlinear model in the compressive deformation. Similarly, the strength values of typical glass fabric composite materials from literature [8] were used as a reference ($X = 473$ MPa, $Y = 473$ MPa, and $S = 85$ MPa). By substituting the selected strength values into the Tsai-Hill failure criterion, the failure loads of composite helical structures with different helix diameters can be obtained (as shown in Table 6.7). As can be seen from Table 6.7, for composite helical structures with a helix diameter of 200 mm or above, when the maximum displacement is reached under compression, the Tsai-Hill failure criterion is still not satisfied, indicating that the composite helical structure is likely to reach the maximum displacement without failure.

References

1. Liu TW, Bai JB, Lin QH et al (2021) An analytical model for predicting compressive behaviour of composite helical Structures: considering geometric nonlinearity effect. Compos Struct 255:112908
2. Bai JB, Liu TW, Wang ZZ et al (2021) Determining the best practice-Optimal designs of composite helical structures using Genetic Algorithms. Compos Struct 268:113982
3. Bai JB, Li SL, Fantuzzi N et al (2024) Tensile performance of 3D-printed helical springs. Mech Adv Mater Struct:1–13
4. Wahl AW (1963) Mechanical springs. McGraw-Hill Book Company, NewYork
5. Shen GL, Hu GK (2006) Mechanics of composite materials. Tsinghua University Press, Beijing
6. Xiong ZY, Song RX, Kang ZX et al (2015) Analysis on rigidity of composite helical spring and its influence factors. Eng Mech 32(9):216–221. and 228
7. Gobbi M, Mastinu G (2001) On the optimal design of composite material tubular helical springs. Meccanica 36(5):525–553
8. Okutan B, Aslan Z, Karakuzu R (2001) A study of the effects of various geometric parameters on the failure strength of pin-loaded woven-glass-fiber reinforced epoxy laminate. Compos Sci Technol 61(10):1491–1497
9. Li YC, Song YP, Hu F (2020) Research on performance of TWF composites for space antenna. Chinese Space Sci Technol:1–10

Chapter 7
Coiling Folding of Lenticular Deployable Composite Booms

Abstract This chapter establishes a theoretical model for lenticular deployable composite booms based on equilibrium and energy principles, describing load-displacement relationships during flattening (compression/tension) and coiling, and deriving folding moments and ultimate coiling radii. Assuming ultra-thin-walled curved beams, it maintains neutral surface length during flattening, uses polar coordinate polynomials for coiling curvature, and solves stress levels via classical laminate theory and the maximum stress criterion for ultimate radius. Vacuum bag-fabricated samples tested on an INSTRON machine validate load-displacement curves, with stage I linear segment and full nonlinear tension flattening predicted within 22–23% error, and coiling strain predictions matching experimental trends.

Lenticular deployable composite booms serve as essential components for constructing deployable space structures. They offer advantages such as lightweight, high stiffness, and reusability in deployment, making them suitable as structural backbones for deployable parabolic mesh antennas, solar sails, and space probes. During the early work, the authors of this book [1] focused on the design, fabrication, and experiments of lenticular deployable composite booms. Subsequent studies have extensively assessed mechanical behavior of these booms, including (1) investigations into the thermal behavior in space environments [2]; (2) development of various analytical models to predict the nonlinear tensile response and neutral axis position during flattening under tension [3–6] as well as in-plane strain and interlaminar shear stress during folding [7–9]; (3) exploration of temperature effects on the buckling characteristics under axial compression [10, 11]; and (4) multi-objective optimization design studies of lenticular deployable composite booms [12].

This chapter mainly introduces analytical models for predicting the folding behavior of lenticular deployable composite booms. Geometric equations are first established to describe the folding deformation of these booms, and by employing equilibrium equations and functional relationships, analytical models are derived to predict the load versus displacement relationship for flattening under compression, tension, and coiling deformations. On this basis, analytical models for the folding moment and ultimate coiling radius of lenticular deployable composite booms are further developed. Samples were fabricated using the vacuum bagging method and

J.-B. Bai et al., *Folding Mechanics of Deployable Composite Structures*, Solid Mechanics and Its Applications 283,
https://doi.org/10.1007/978-3-031-99254-4_7

subjected to large deformation experiments, including compression, tension, coiling, and folding, to validate the predictive accuracy of analytical models.

7.1 Analytical Model

7.1.1 Geometric Model

The lenticular deployable composite boom is generally manufactured by bonding ultrathin-walled axisymmetrical curved composite shells through the vacuum bagging process, which can be folded into a roll for packaging by using a roller, which then automatically recovers its original shape in a free state due to the elastic deformation energy (shown in Fig. 7.1). In order to analyze its geometrical behavior, fundamental assumptions [6–9] are made as follows:

(1) The length of the lenticular deployable composite boom is much larger than any other dimension of the cross section, and the displacements are finite together with the material that remains linear-elastic within a small deflection range.
(2) The change in the thickness of lenticular deployable composite boom is negligibly small during folding deformation, and hence global deformation can be described with the changes in the shape and curvature of its mid-surface without local stretching.

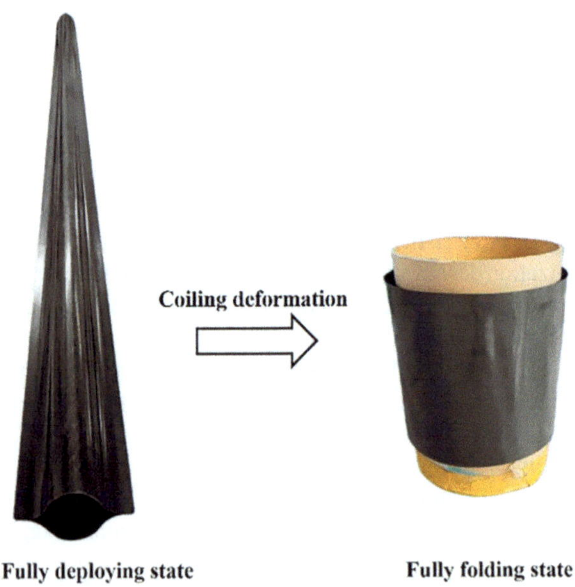

Fig. 7.1 Physical process of folding deformation of the lenticular deployable composite boom

(3) The folding deformation of the lenticular deployable composite boom is idealized as the linear superposition of both the quasi-static elastic flattening and coiling deformation modes, which remain identical along longitudinal and transverse directions, respectively.
(4) The lenticular deployable composite boom in folding deformation is approximated as the ultrathin-walled curved beam with a biaxisymmetrical lenticular cross section described by concave and convex tangential circular arcs containing the same central angle and curvature radius.
(5) The in-plane shear strain on the mid-surface of the lenticular deployable composite boom due to the change of normal stress (such as bending and torsional stresses) is negligibly small (Euler-Bernoulli hypothesis [13]), and the local buckling in the shell is neglected.

To describe the geometric behavior of the lenticular deployable composite boom during flattening under compression or tension, an appropriate polar coordinate system and minute element are selected, as shown in Fig. 7.2a, b. Taking flattening under compression as an example, the cross-sectional neutral axis in polar coordinates can be represented by the corresponding central angle φ and curvature radius r. Based on these fundamental assumptions, the force states of the lenticular deployable composite boom during compression or tension flattening can be derived, as depicted in Fig. 7.2c, d. Despite the existence of large deformations during the compression or tension flattening process, leading to significant changes in both geometric configuration and curvature radius, the corresponding geometric deformation can still be determined. Using the derived shape parameters φ and r, and applying force and moment equilibrium conditions, a load versus displacement relationship for the lenticular deployable composite boom under a vertical compressive load P_c is established as follows:

$$F = \frac{P_c}{2} \tag{7.1}$$

$$P_c = \frac{2M_x}{r\sin\varphi} \tag{7.2}$$

where P_c represents the compressive load, F denotes the reaction force at the bonded edge, and M_x is the bending moment in the upward x direction, with the detailed expressions available in [8].

Similar to the flattening deformation, the load versus displacement relationship of the lenticular deployable composite boom under a horizontal tensile load P_t can be derived as follows:

$$N = \frac{P_t}{2} \tag{7.3}$$

$$P_t = \frac{2M_x}{r(1-\cos\varphi)} \tag{7.4}$$

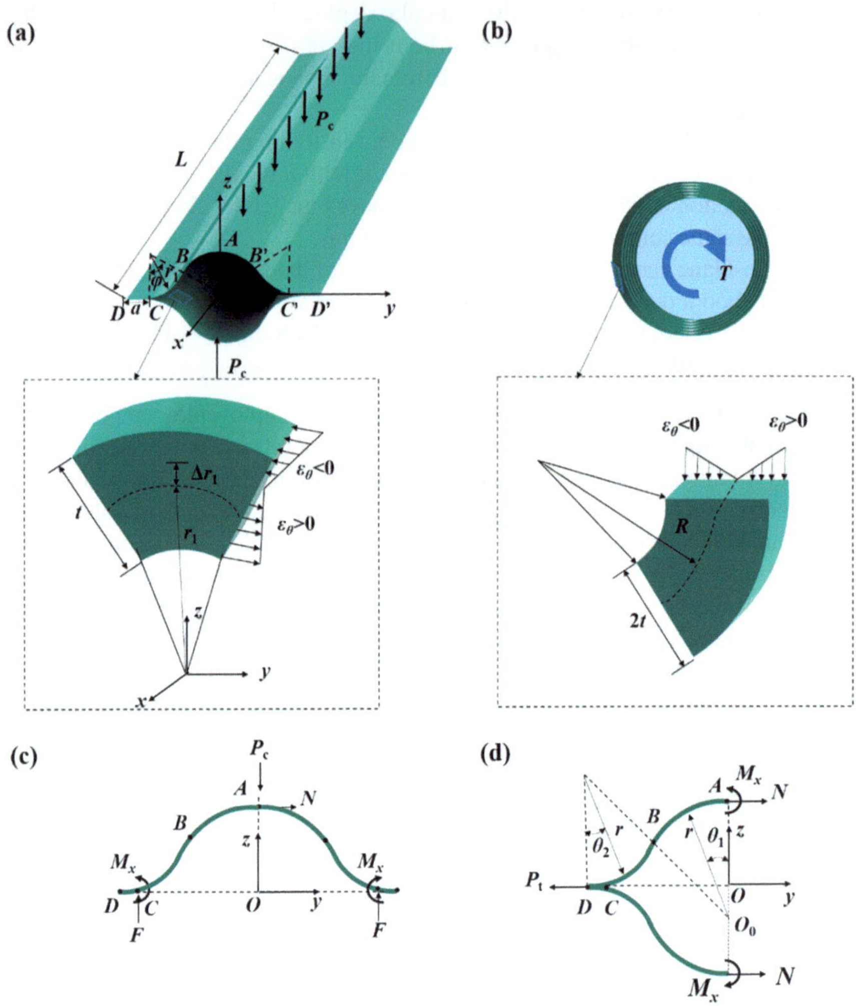

Fig. 7.2 Geometric model of the lenticular deployable composite boom: (**a**) Deployed state. (**b**) Folding state. (**c**) Forces for the flattened shell under compression. (**d**) Forces for the flattened shell under tension

where P_t represents the tensile load, and N is the reaction force at the apex edge.

7.1.2 Coiling Deformation Mode

The fundamental assumptions presented in Sect. 7.1 describe the geometric behavior of the lenticular deployable composite boom during the coiling deformation that occurs after it is completely flattened or elongated. The selected minute element is shown in Fig. 7.3a, b that illustrates that the coiling deformation of the lenticular deployable composite boom can be represented by the longitudinal curvature K (or longitudinal curvature radius ρ_2) of the neutral axis after complete flattening in the polar coordinate system. Based on this, the geometric model for the coiling deformation of the lenticular deployable composite boom is derived.

Based on Fig. 7.3 and assumption (3), the coiling deformation of the lenticular deployable composite boom after complete flattening can be described using a polynomial shape function in the polar coordinate system, as follows:

$$\rho_1 = a_1\alpha + a_0, \alpha_0 \le \alpha \le \alpha_1 \tag{7.5}$$

The coiling deformation of the lenticular deployable composite boom after complete flattening or elongation can be described using either the curvature radius ρ_2 (or curvature K) or the radian $\mathrm{d}\beta'$, alternatively:

$$K = \frac{1}{\rho_2} = \frac{2\left(\rho_1'\right)^2 + \rho_1^2}{\left[\left(\rho_1'\right)^2 + \rho_1^2\right]^{3/2}} \tag{7.6}$$

or

$$d\beta' = K\mathrm{d}s = K\sqrt{\left[\rho_1^2 + \left(\rho_1'\right)^2\right]}\mathrm{d}\alpha \tag{7.7}$$

For the coiling deformation of the lenticular deployable composite boom after complete flattening or elongation, interlaminar shear stress will be present at the interface of the bonded upper and lower semi-shells. By applying the force equilibrium condition along the β' direction on the minute element shown in Fig. 7.3b, we obtain

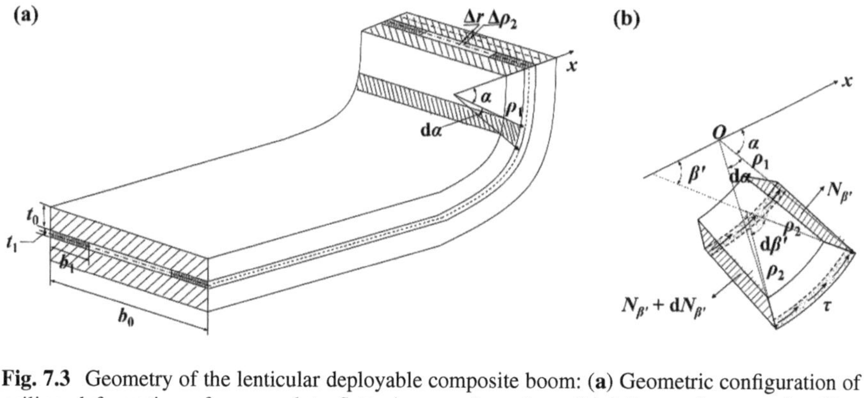

Fig. 7.3 Geometry of the lenticular deployable composite boom: (**a**) Geometric configuration of coiling deformation after complete flattening or elongation. (**b**) Minute element of coiling deformation

$$\left(N_{\beta'}+dN_{\beta'}\right)\cos\frac{d\beta'}{2}-N_{\beta'}\cos\frac{d\beta'}{2}-\tau\cdot 2b_1\cdot 2\rho_2\sin\frac{d\beta'}{2}=0 \tag{7.8}$$

where $N_{\beta'}$ represents the axial force tangential to the β' direction on the cross section of the minute element, and τ denotes the interlaminar shear stress at the bonded interface. Given that $d\beta'$ is infinitesimally small, an approximation can be applied, simplifying the equation to

$$\tau=\frac{1}{2b_1\sqrt{\left[\rho_1^2+\left(\rho_1^{'}\right)^2\right]}}\frac{dN_{\beta'}}{d\alpha} \tag{7.9}$$

Based on the basic mechanics of composite materials in Chap. 3, the interlaminar shear stress at the bonded interface during the coiling deformation of a fully flattened or compressed lenticular deployable composite boom can be derived as follows:

$$\tau=\frac{b_0\rho_1\rho_1^{'}\left[2\left(\rho_1^{'}\right)^2-6\rho_1^{'}-\rho_1^2\right]}{4n^2b_1\left[\left(\rho_1^{'}\right)^2+\rho_1^2\right]^3}\sum_{k=1}^{n}\bar{Q}_{11}^k\left[nt_0t_1+(2k-1)t_0^2\right] \tag{7.10}$$

The detailed derivation process for the above can be found in literature [8].

7.1.3 Folding Moment

This section further characterizes the mechanical properties of the lenticular deployable composite boom during the folding deformation process, based on the fundamental assumptions presented in Sect. 7.1.1. As shown in Fig. 7.2a, b, the flattening deformation of the lenticular deployable composite boom can be described by the central angle of the neutral axis in the cross section within a polar coordinate system. The coiling deformation, in turn, can be described by the longitudinal curvature of the neutral axis after the lenticular deployable composite boom has been completely flattened in polar coordinates. Based on these descriptions, the geometric model for the folding deformation of the lenticular deployable composite boom is established as follows [12].

According to assumptions (1) and (2) in Sect. 7.1.1, the geometrical constraint to be satisfied is

$$\cos\varphi=\frac{r_2}{r_1+r_2} \tag{7.11}$$

where r_1 and r_2 are the curvature radii of the concave and convex arcs, respectively, and φ is the central angle of neutral axis on cross section.

The half lenticular deployable composite boom is composed of two concave arcs, one convex arc, and two flat adhesive edges. Therefore, the width of the completely flattening lenticular deployable composite boom is

$$b = 2\varphi(r_1 + r_2) + 2a \tag{7.12}$$

where a is the width of bonding interface.

The cross-section area of the completely flattening lenticular deployable composite boom can be expressed as

$$A = 2bt = 4t\varphi(r_1 + r_2) + 4ta \tag{7.13}$$

The inertia moments of the lenticular deployable composite boom in initial state (i.e., completely deployed state) can be written as

$$\begin{aligned} I_y &= \int y^2 \mathrm{d}A \\ I_Z &= \int x^2 \mathrm{d}A \end{aligned} \tag{7.14}$$

Substituting geometric parameters into Eq. (7.14), it is possible to show as

$$\begin{aligned} I_y &= 4\int_0^{\varphi_0} (r_{2,0}\cos\beta)^2 \mathrm{d}\beta r_{2,0}t + 4\int_0^{\varphi_0} \left[r_{1,0}(1-\cos\alpha)\right]^2 \mathrm{d}\alpha r_{1,0}t + \frac{at^3}{3} \\ I_z &= 4\int_0^{\varphi_0} (r_{2,0}\sin\beta)^2 \mathrm{d}\beta r_{2,0}t + 4\int_0^{\varphi_0} (r_{1,0}\sin\varphi_0 + r_{2,0}\sin\varphi_0 - r_{1,0}\sin\alpha)^2 \mathrm{d}\alpha r_{1,0}t + \frac{ta^3}{3} \end{aligned} \tag{7.15}$$

where $r_{1,0}$ and $r_{2,0}$ are the initial curvature radii of the concave and convex arcs, respectively, and φ_0 is the initial central angle of neutral axis on cross section.

Integrating Eq. (7.15) will take the form below:

$$\begin{aligned} I_y &= r_{2,0}^3 t(2\varphi_0 + \sin 2\varphi_0) + r_{1,0}^3 t(6\varphi_0 - 8\sin\varphi_0 + \sin 2\varphi_0) + \frac{at^3}{3} \\ I_z &= (r_{1,0}^3 t + r_{2,0}^3 t)(2\varphi_0 - \sin 2\varphi_0) + 4r_{1,0}t\varphi_0\sin^2\varphi_0(r_{1,0} + r_{2,0})^2 \\ &\quad + 8r_{1,0}^2 t\sin\varphi_0(\cos\varphi_0 - 1)(r_{1,0} + r_{2,0}) + \frac{ta^3}{3} \end{aligned} \tag{7.16}$$

The folding deformation process of the lenticular deployable composite boom is shown in Fig. 7.4. The energy method is used to determine the folding moment required during the folding process. To establish the strain energy generated during the folding process, the process is divided into two steps, as illustrated in Fig. 7.4a–c. In the first step, a compressive load is applied at the apex to flatten the lenticular deployable composite boom. In this step, the external work is converted into

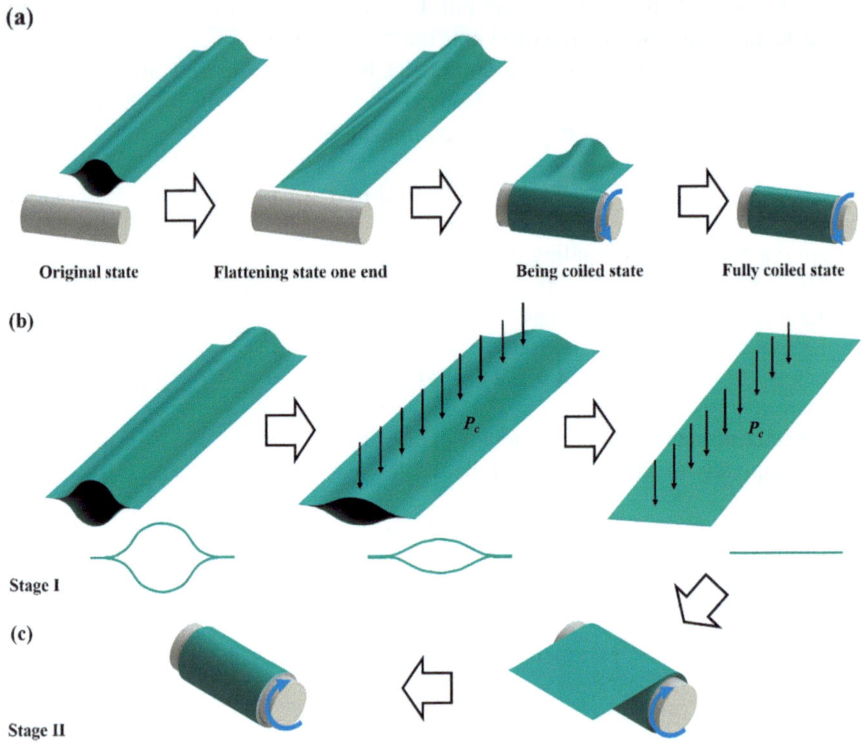

Fig. 7.4 Schematic diagram of folding process of the lenticular deployable composite boom: (**a**) Folding process. (**b**) Flattening process. (**c**) Coiling process

bending strain energy, as shown in Fig. 7.4b. In the second step, after the lenticular deployable composite boom is completely flattened, a bending moment is applied at both ends, changing the curvature in the x direction from 0 to $1/R$, as shown in Fig. 7.4c.

In the first step, the compressive load versus displacement curve of the semi-lenticular deployable composite boom can be expressed by a parametric equation involving φ, r_1, and r_2 [6]:

$$\begin{cases} \delta(\varphi, r_1, r_2) = (r_{1,0} + r_{2,0})(1-\cos\varphi_0) - (r_1 + r_2)(1-\cos\varphi) \\ P_c(\varphi, r_1, r_2) = \dfrac{4r_{1,0}BL(\varphi-\varphi_0)}{(r_1+r_2)\varphi_0 \sin\varphi} \\ (\varphi \in [\varphi_0, 0], r_1 \in [r_{1,0}, +\infty], r_2 \in [r_{2,0}, +\infty]) \end{cases} \tag{7.17}$$

where

$$B=\sum_{k=1}^{n}\left\{E_{\beta}^{k}\left[\left(\Delta r_{1}^{k}-\Delta r_{1}^{k-1}\right)-r_{1,0}\ln\frac{r_{1,0}+\Delta r_{1}^{k}}{r_{1,0}+\Delta r_{1}^{k-1}}\right]\right\} \tag{7.18}$$

where n is the number of composite plies and β is the ply angle. E_{β}^{k} is the off-axis elastic modulus of the kth ply in tangential direction of β.

According to assumption (2), although there is a large deformation in the flattening process and the geometry and curvature of the lenticular deployable composite boom have changed significantly, the total length of the neutral surface has not changed. Thus, it is assumed to have

$$\varphi\left(r_{1}+r_{2}\right)=\varphi_{0}\left(r_{1,0}+r_{2,0}\right) \tag{7.19}$$

Substituting Eq. (7.19) into (7.17) and eliminating $r_1 + r_2$, the parametric equation of φ can be obtained as

$$\begin{aligned}
&\delta(\varphi)=\left(r_{1,0}+r_{2,0}\right)\left(1-\cos\varphi_{0}\right)-\frac{\varphi_{0}}{\varphi}\left(r_{1,0}+r_{2,0}\right)\left(1-\cos\varphi\right),\left(\varphi\in\left[\varphi_{0},0\right]\right)\\
&P_{c}(\varphi)=\frac{4r_{1,0}\varphi BL\left(\varphi-\varphi_{0}\right)}{\left(r_{1,0}+r_{2,0}\right)\varphi_{0}^{2}\sin\varphi}
\end{aligned} \tag{7.20}$$

The external work of the lenticular deployable composite boom when compressed from the initial state to the completely flattening state is twice the definite integral of parametric Eq. (7.20), that is,

$$U_{1}=2\int_{\varphi_{0}}^{0}P_{c}(\varphi)\delta'(\varphi)\,\mathrm{d}\varphi \tag{7.21}$$

According to the energy principle, the strain energy of the lenticular deployable composite boom in the completely flattening state is equal to the external work. Substituting Eq. (7.20) into (7.21), the strain energy of the lenticular deployable composite boom in the completely flattening state can be obtained as

$$U_{1}=\int_{\varphi_{0}}^{0}\frac{8r_{1,0}BL\left(\varphi_{0}-\varphi\right)\left(\varphi\sin\varphi+\cos\varphi-1\right)}{\varphi\varphi_{0}\sin\varphi}\,\mathrm{d}\varphi \tag{7.22}$$

Utilizing the classical laminate theory, the elastic deformation of the lenticular deployable composite boom in the coiling process can be expressed. The completely flattening lenticular deployable composite boom is only subjected to the bending moment in the x direction in the coiling process, so it is possible to determine as

$$\begin{cases} \mathbf{N} = \mathbf{0} \\ M_y = 0 \\ M_{xy} = 0 \\ \kappa_y = 0 \\ \kappa_{xy} = 0 \end{cases} \tag{7.23}$$

The symmetric laminates are selected, $\boldsymbol{B} = \mathbf{0}$. Substituting Eq. (7.23) into (3.59) yields

$$\begin{bmatrix} M_x \\ 0 \\ 0 \end{bmatrix} = \begin{bmatrix} D_{11} & D_{12} & D_{16} \\ D_{12} & D_{22} & D_{26} \\ D_{16} & D_{26} & D_{66} \end{bmatrix} \begin{bmatrix} \kappa_x \\ 0 \\ 0 \end{bmatrix} \tag{7.24}$$

Since the lenticular deployable composite boom is in the pure bending state during the coiling process, the strain energy in the coiling process is

$$U_2 = \frac{bL}{2} \begin{bmatrix} \kappa_x & 0 & 0 \end{bmatrix} \begin{bmatrix} D_{11} & D_{12} & D_{16} \\ D_{12} & D_{22} & D_{26} \\ D_{16} & D_{26} & D_{66} \end{bmatrix} \begin{bmatrix} \kappa_x \\ 0 \\ 0 \end{bmatrix} \tag{7.25}$$

In the coiling process, the curvature in the x direction of the lenticular deployable composite boom can be written as

$$\kappa_x = \frac{1}{R} \tag{7.26}$$

where R is the coiling radius:

$$U_2 = \frac{D_{11} L \left(\varphi_0 r_{1,0} + \varphi_0 r_{2,0} + a\right)}{R^2} \tag{7.27}$$

In the practical folding process of the lenticular deployable composite boom, one end is flattened in the first step, and then the folding process is completed through a roller in the second step. The schematic diagram of the whole folding deformation process is shown in Fig. 7.4a. In the first step of the compressive deformation process, except the cross-section shape of the neutral cross section, the cross-section shapes of the other positions along the length direction change. The cross section from the neutral cross section to the compressive end will be flattened in the width direction, and the cross section from the neutral cross section to the free end will be shortened. The position of the neutral cross section of the lenticular deployable composite boom in the compressive deformation process is at 2/3 of the length direction, and the projection length of the circular arc section of the cross section is linearly distributed along the length direction.

At the compressive end ($x = 0$), the projection length of the circular arc section of the cross section is

$$c|_{x=0} = 2\left(r_{1,0} + r_{2,0}\right)\varphi_0 \tag{7.28}$$

At the neutral cross section ($x = \frac{2}{3}L$), the projection length of the circular arc section of the cross section is

$$c|_{x=\frac{2}{3}L} = 2\left(r_{1,0} + r_{2,0}\right)\sin\varphi_0 \tag{7.29}$$

Therefore, the relationship between $c(x)$ and x can be deduced as

$$c(x) = \frac{3}{L}\left(r_{1,0} + r_{2,0}\right)\left(\sin\varphi_0 - \varphi_0\right)x + 2\left(r_{1,0} + r_{2,0}\right)\varphi_0, \left(x \in [0, L]\right) \tag{7.30}$$

According to the geometric configuration, it is possible to have

$$\begin{aligned} & c(x) = 2\left[r_{1,1}(x) + r_{2,1}(x)\right]\sin\varphi_1(x), \ \left(x \in [0,L]\right) \\ & \left[r_{1,1}(x) + r_{2,1}(x)\right]\sin\varphi_1(x) = \varphi_0\left(r_{1,0} + r_{2,0}\right), \ \left(x \in [0, L]\right) \end{aligned} \tag{7.31}$$

Substituting Eq. (7.31) into (7.30) and eliminating $r_{1,1}(x) + r_{2,1}(x)$, the relationship between x and $\varphi_1(x)$ can be expressed as

$$\frac{\sin\varphi_1(x)}{\varphi_1(x)} = \frac{3\left(\sin\varphi_0 - \varphi_0\right)}{2\varphi_0 L}x + 1, \ \left(x \in [0, L]\right) \tag{7.32}$$

Obviously, Eq. (7.32) is an implicit function, which can be solved numerically by Newton iterative method.

According to Eq. (7.22) and the definition of double integral, the strain energy of the lenticular deployable composite boom in the flattening state at one end can be written as

$$U_0 = \int_0^L \mathrm{d}x \left| \int_{\varphi_0}^{\varphi_1(x)} \frac{8r_{1,0}B\left(\varphi_0 - \varphi\right)\left(\varphi\sin\varphi + \cos\varphi - 1\right)}{\varphi\varphi_0\sin\varphi} \mathrm{d}\varphi \right| \tag{7.33}$$

The total strain energy of the lenticular deployable composite boom in the folding process can be expressed as

$$U = U_1 + U_2 - U_0 \tag{7.34}$$

Substituting Eqs. (7.22), (7.27), and (7.33) into (7.34), it is possible to determine as

$$
\begin{aligned}
U = & \int_{\varphi_0}^{0} \frac{8r_{1,0}BL(\varphi_0-\varphi)(\varphi\sin\varphi+\cos\varphi-1)}{\varphi\varphi_0\sin\varphi}\mathrm{d}\varphi + \frac{D_{11}L(\varphi_0 r_{1,0}+\varphi_0 r_{2,0}+a)}{R^2} \\
& - \int_0^L \mathrm{d}x \left| \int_{\varphi_0}^{\varphi_1(x)} \frac{8r_{1,0}B(\varphi_0-\varphi)(\varphi\sin\varphi+\cos\varphi-1)}{\varphi\varphi_0\sin\varphi}\mathrm{d}\varphi \right|
\end{aligned} \tag{7.35}
$$

The external work of the lenticular deployable composite boom in the folding process can be written as

$$
W = \frac{TL}{R} \tag{7.36}
$$

where T is the folding moment of the lenticular deployable composite boom in the folding process.

According to the energy principle, the total strain energy of the lenticular deployable composite boom in the folding process is equal to the external work. Thus, the folding moment can be obtained by substituting Eq. (7.35) into (7.36), as follows:

$$
\begin{aligned}
T = & \int_{\varphi_0}^{0} \frac{8r_{1,0}BR(\varphi_0-\varphi)(\varphi\sin\varphi+\cos\varphi-1)}{\varphi\varphi_0\sin\varphi}\mathrm{d}\varphi + \frac{D_{11}(\varphi_0 r_{1,0}+\varphi_0 r_{2,0}+a)}{R} \\
& - \frac{R}{L}\int_0^L \mathrm{d}x \left| \int_{\varphi_0}^{\varphi_1(x)} \frac{8r_{1,0}B(\varphi_0-\varphi)(\varphi\sin\varphi+\cos\varphi-1)}{\varphi\varphi_0\sin\varphi}\mathrm{d}\varphi \right|
\end{aligned} \tag{7.37}
$$

7.1.4 Ultimate Coiling Radius

As the lenticular deployable composite boom is gradually being flattened, the curvatures of the concave and convex arcs change from $1/r_{1,0}$ and $1/r_{2,0}$ to 0. In the flattening process, the arc with greater curvature variation produces greater strain and then produces greater stress. Therefore, the curvature variation of the arc with smaller curvature radius is

$$
\kappa_y = \frac{1}{r} \tag{7.38}
$$

where

$$
r = \min\{r_{1,0}, r_{2,0}\} \tag{7.39}
$$

When the lenticular deployable composite boom is coiled with the ultimate coiling radius, the curvature variation in the x direction can be expressed as

$$\kappa_x = \frac{1}{R_{\text{ultimate}}} \tag{7.40}$$

where R_{ultimate} is the ultimate coiling radius of the lenticular deployable composite boom.

The stress-strain relationship of the kth ply ($k = 1, 2, 3, \ldots, n$) in the laminate is

$$\begin{bmatrix} \sigma_x \\ \sigma_y \\ \tau_{xy} \end{bmatrix}_k = \begin{bmatrix} \bar{Q}_{11} & \bar{Q}_{12} & \bar{Q}_{16} \\ \bar{Q}_{12} & \bar{Q}_{22} & \bar{Q}_{26} \\ \bar{Q}_{16} & \bar{Q}_{26} & \bar{Q}_{66} \end{bmatrix}_k \left\{ \begin{bmatrix} \kappa_x \Delta R_{\text{ultimate}} \\ \kappa_y \Delta r \\ 0 \end{bmatrix} \right\} \tag{7.41}$$

Substituting Eqs. (7.38) and (7.40) into (7.41), it is deduced as

$$\begin{aligned} \sigma_x^k &= \frac{\Delta R_{\text{ultimate}} \bar{Q}_{11}^k}{R_{\text{ultimate}}} + \frac{\Delta r \bar{Q}_{12}^k}{r} \\ \sigma_y^k &= \frac{\Delta R_{\text{ultimate}} \bar{Q}_{12}^k}{R_{\text{ultimate}}} + \frac{\Delta r \bar{Q}_{22}^k}{r} \\ \tau_{xy}^k &= \frac{\Delta R_{\text{ultimate}} \bar{Q}_{16}^k}{R_{\text{ultimate}}} + \frac{\Delta r \bar{Q}_{26}^k}{r} \end{aligned} \tag{7.42}$$

According to the coordinate transformation of stress components of the kth ply in the laminate, and substituting Eqs. (7.38) to (7.42) into (3.21), it is possible to determine as

$$\begin{aligned} \sigma_1^k = {} & \frac{\Delta R_{\text{ultimate}}}{R_{\text{ultimate}}} \left(\cos^2 \beta^k \bar{Q}_{11}^k + \sin^2 \beta^k \bar{Q}_{12}^k + 2 \sin \beta^k \cos \beta^k \bar{Q}_{16}^k \right) \\ & + \frac{\Delta r}{r} \left(\cos^2 \beta^k \bar{Q}_{12}^k + \sin^2 \beta^k \bar{Q}_{22}^k + 2 \sin \beta^k \cos \beta^k \bar{Q}_{26}^k \right) \end{aligned} \tag{7.43}$$

$$\begin{aligned} \sigma_2^k = {} & \frac{\Delta R_{\text{ultimate}}}{R_{\text{ultimate}}} \left(\sin^2 \beta^k \bar{Q}_{11}^k + \cos^2 \beta^k \bar{Q}_{12}^k - 2 \sin \beta^k \cos \beta^k \bar{Q}_{16}^k \right) \\ & + \frac{\Delta r}{r} \left(\sin^2 \beta^k \bar{Q}_{12}^k + \cos^2 \beta^k \bar{Q}_{22}^k - 2 \sin \beta^k \cos \beta^k \bar{Q}_{26}^k \right) \end{aligned} \tag{7.44}$$

$$\begin{aligned} \tau_{12}^k = {} & \frac{\Delta R_{\text{ultimate}}}{R_{\text{ultimate}}} \left[-\sin \beta^k \cos \beta^k \bar{Q}_{11}^k + \sin \beta^k \cos \beta^k \bar{Q}_{12}^k + \left(\cos^2 \beta^k - \sin^2 \beta^k \right) \bar{Q}_{16}^k \right] \\ & + \frac{\Delta r}{r} \left[-\sin \beta^k \cos \beta^k \bar{Q}_{12}^k + \sin \beta^k \cos \beta^k \bar{Q}_{22}^k + \left(\cos^2 \beta^k - \sin^2 \beta^k \right) \bar{Q}_{26}^k \right] \end{aligned} \tag{7.45}$$

The maximum stress criterion is utilized to predict the ultimate coiling radius of the lenticular deployable composite boom in the folding process. The critical stress state of the kth ply with tensile failure in the longitudinal (fiber) direction is

$$\sigma_1^k = X_{\text{t}} \tag{7.46}$$

Substituting Eq. (7.43) into (7.46), the ultimate coiling radius corresponding to the tensile failure of the kth ply in the longitudinal direction can be obtained, as follows:

$$R_{\text{ultimate}}^{1,k} = \frac{r\Delta R_{\text{ultimate}}\left(\cos^2\beta^k\overline{Q}_{11}^k + \sin^2\beta^k\overline{Q}_{12}^k + 2\sin\beta^k\cos\beta^k\overline{Q}_{16}^k\right)}{X_{\text{t}}r - \Delta r\left(\cos^2\beta^k\overline{Q}_{12}^k + \sin^2\beta^k\overline{Q}_{22}^k + 2\sin\beta^k\cos\beta^k\overline{Q}_{26}^k\right)} \tag{7.47}$$

The critical stress state of the kth ply with compressive failure in the longitudinal (fiber) direction is

$$\sigma_1^k = X_{\text{c}} \tag{7.48}$$

Substituting Eq. (7.43) into (7.48), the ultimate coiling radius corresponding to the compressive failure of the kth ply in the longitudinal direction can be obtained, as follows:

$$R_{\text{ultimate}}^{2,k} = \frac{r\Delta R_{\text{ultimate}}\left(\cos^2\beta^k\overline{Q}_{11}^k + \sin^2\beta^k\overline{Q}_{12}^k + 2\sin\beta^k\cos\beta^k\overline{Q}_{16}^k\right)}{X_{\text{c}}r - \Delta r\left(\cos^2\beta^k\overline{Q}_{12}^k + \sin^2\beta^k\overline{Q}_{22}^k + 2\sin\beta^k\cos\beta^k\overline{Q}_{26}^k\right)} \tag{7.49}$$

The critical stress state of the kth ply with tensile failure in the transverse (to the fiber) direction is

$$\sigma_2^k = Y_{\text{t}} \tag{7.50}$$

Substituting Eq. (7.44) into (7.50), the ultimate coiling radius corresponding to the tensile failure of the kth ply in the transverse (to the fiber) direction can be obtained, as follows:

$$R_{\text{ultimate}}^{3,k} = \frac{r\Delta R_{\text{ultimate}}\left(\sin^2\beta^k\overline{Q}_{11}^k + \cos^2\beta^k\overline{Q}_{12}^k - 2\sin\beta^k\cos\beta^k\overline{Q}_{16}^k\right)}{Y_{\text{t}}r - \Delta r\left(\sin^2\beta^k\overline{Q}_{12}^k + \cos^2\beta^k\overline{Q}_{22}^k - 2\sin\beta^k\cos\beta^k\overline{Q}_{26}^k\right)} \tag{7.51}$$

The critical stress state of the kth ply with compressive failure in the transverse (to the fiber) direction is

$$\sigma_2^k = Y_{\text{c}} \tag{7.52}$$

Substituting Eq. (7.44) into (7.52), the ultimate coiling radius corresponding to the compressive failure of the kth ply in the transverse (to the fiber) direction can be obtained, as follows:

$$R_{\text{ultimate}}^{4,k} = \frac{r\Delta R_{\text{ultimate}}\left(\sin^2\beta^k\bar{Q}_{11}^k + \cos^2\beta^k\bar{Q}_{12}^k - 2\sin\beta^k\cos\beta^k\bar{Q}_{16}^k\right)}{Y_c r - \Delta r\left(\sin^2\beta^k\bar{Q}_{12}^k + \cos^2\beta^k\bar{Q}_{22}^k - 2\sin\beta^k\cos\beta^k\bar{Q}_{26}^k\right)} \tag{7.53}$$

The critical stress state of the kth ply with shear failure is

$$t_{12}^k = S_{12} \tag{7.54}$$

Substituting Eq. (7.45) into (7.54), the ultimate coiling radius corresponding to the shear failure of the kth ply can be obtained, as follows:

$$R_{\text{ultimate}}^{5,k} = \frac{r\Delta R_{\text{ultimate}}\left[-\sin\beta^k\cos\beta^k\bar{Q}_{11}^k + \sin\beta^k\cos\beta^k\bar{Q}_{12}^k + \left(\cos^2\beta^k - \sin^2\beta^k\right)\bar{Q}_{16}^k\right]}{S_{12}r - \Delta r\left[-\sin\beta^k\cos\beta^k\bar{Q}_{12}^k + \sin\beta^k\cos\beta^k\bar{Q}_{22}^k + \left(\cos^2\beta^k - \sin^2\beta^k\right)\bar{Q}_{26}^k\right]} \tag{7.55}$$

The maximum value of the ultimate coiling radius corresponding to five failure modes and n plies is determined as the ultimate coiling radius of the lenticular deployable composite boom:

$$R_{\text{ultimate}} = \max R_{\text{ultimate}}^{j,k}, \begin{cases} j = 1,2,3,4,5 \\ k = 1,2,3\ldots n \end{cases} \tag{7.56}$$

7.2 Validation Analysis

7.2.1 Compression and Tensile Experiments

The experimental validation of the analytical solution for the compressive and tensile flattening deformation of the lenticular deployable composite boom proposed in Sect. 7.1.1 is carried out. The geometric configuration and dimensional parameters of the samples' cross section are as follows: the length is 65 mm, the thickness is 0.4 mm, the ply scheme is [45°/–45°/0°/–45°/45°], and the adhesive layer thickness is 0.1 mm. The material properties and specific parameters of the lamina are shown in Table 7.1. The lenticular deployable composite boom samples were fabricated using vacuum bagging and secondary adhesive bonding methods [1], with the specific preparation process shown in Fig. 7.5. Under dry conditions at room temperature, compressive flattening and tensile deformation tests of three samples were performed on an INSTRON-500N hydraulic servo experimental machine, with a loading rate of 2 mm/min. Figure 7.6a, b shows the loading boundary conditions and load directions for the tensile and compressive flattening deformation tests, respectively. During the testing process, no sliding of the loading boundary was observed visually on the samples.

Table 7.1 Specifications and properties of carbon/epoxy composite material [6]

Specifications and properties	Values
E_1 (GPa)	80.08
E_2 (GPa)	6.67
G_{12} (GPa)	2.93
G_{13} (GPa)	2.93
G_{23} (GPa)	2.50
ν_{12}	0.344
ν_{21}	0.036
X_t (MPa)	1063.48
Y_t (MPa)	65.99
X_c (MPa)	662.69
Y_c (MPa)	144.54
S_{12} (MPa)	126.76
ρ_d (g/mm^3)	1.60×10^{-3}
t_0 (mm)	0.08

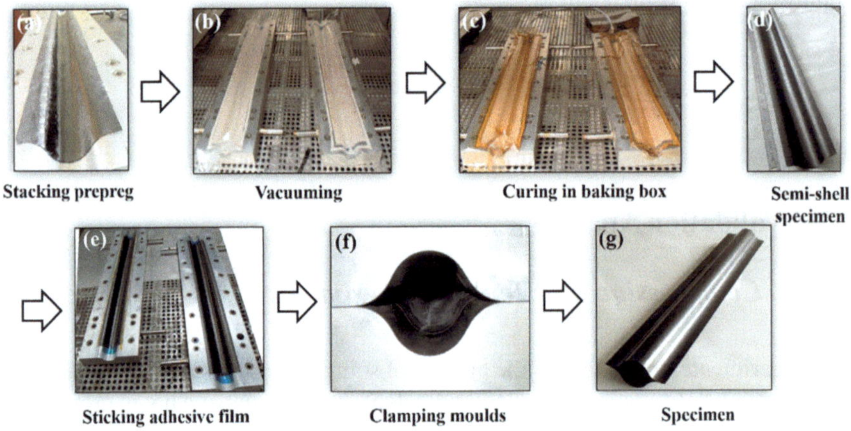

Fig. 7.5 The preparation of the lenticular deployable composite boom samples

Meanwhile, the load versus displacement curves for the tensile flattening deformation and compressive flattening deformation were recorded, as shown in Fig. 7.6c, d. After the completion of the tensile and compressive flattening deformation tests, all tested samples were able to recover their initial shape, and no failure was observed. From Fig. 7.6c, d, it is clear that the measured load versus displacement curves are almost identical, with very little dispersion. Additionally, both the tensile and compressive flattening deformation load versus displacement curves exhibit an approximately bilinear nature, with two linear segments and a small nonlinear transition segment between them.

Using the analytical method, the predicted load versus displacement curves for the compressive and tensile flattening deformation processes are obtained and

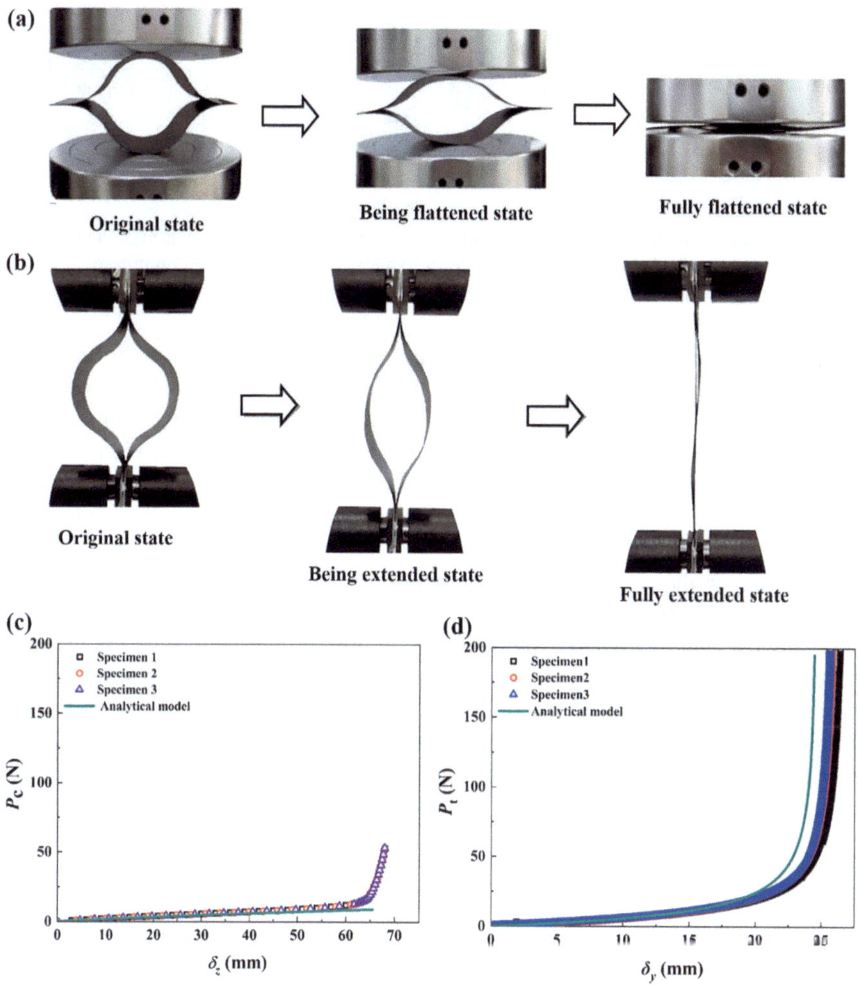

Fig. 7.6 Validation of the analytical model for compressive and tensile flattening of the lenticular deployable composite boom samples: (**a**) Compressive deformation. (**b**) Tensile deformation. (**c**) Load versus displacement curve for compressive flattening. (**d**) Load versus displacement curve for tensile flattening

compared with the experimental results. From Fig. 7.6c, d, it can be seen that the analytical model presented in this section provides a good match with the experimental results. The analytical model successfully predicts the first linear segment of the compressive flattening deformation load versus displacement curve and the entire nonlinear load versus displacement curve of the tensile flattening deformation.

In analytical models for compressive and tensile flattening, membrane strain is neglected, assuming that strain during the flattening process results solely from bending. Consequently, the strain corresponding to the fully flattened state in both

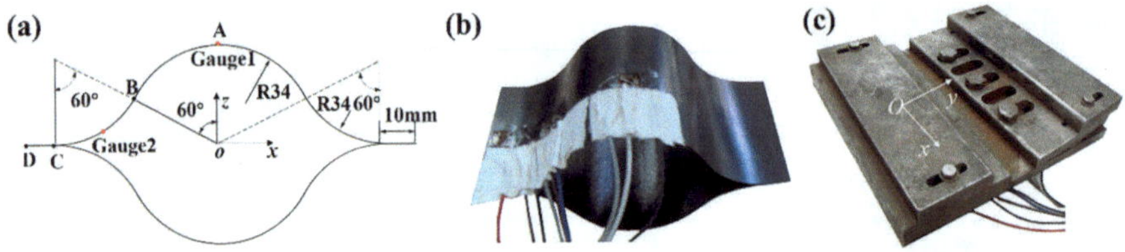

Fig. 7.7 Strain measurement experiments for compressive flattening deformation of the lenticular deployable composite boom samples: (**a**) Strain gauges' arrangement for compressive test. (**b**) Strain gauges attached to the sample. (**c**) Fully compressed test

compressive and tensile flattening processes is consistent. To validate the analytical solution, the strain at specific points on the fully compressed samples was measured. A sample (identical to those used in previous tests) was fully compressed by pressing blocks, with strain gauges placed at two locations on the surface: at point A on the apex of arc $\widehat{BAB'}$ and at the midpoint of arc $\widehat{BC}$ (as shown in Fig. 7.7a, b). The in-plane strains at these locations were recorded for three fully compressed samples, with results presented in Table 7.2. According to Table 7.2, the mean in-plane strains in the tangential direction at point A on arc $\widehat{BAB'}$ and at the midpoint on arc $\widehat{BC}$ were −0.7571% and 0.4826%, respectively.

Using the analytical model, the strain along the θ direction for the fully flattened state (compressed or stretched) at point A on arc $\widehat{BAB'}$ and the midpoint of arc $\widehat{BC}$ is predicted to be −0.5917% and 0.5917%, respectively. Comparing these analytical strain values with the experimental measurements (−0.7571% and 0.4826%), the relative errors between the analytical and experimental results are 22% and 23%, which fall within an acceptable range of variation. Thus, the analytical model proposed in this section provides reasonable accuracy in predicting the strain behavior of the lenticular deployable composite boom during tensile or compressive flattening deformation.

7.2.2 *Folding Experiments*

The analytical solution for the coiling deformation of the fully compressed and the lenticular deployable composite boom was further validated through experiments. A custom coiling deformation test setup, as shown in Fig. 7.8a (with a roller diameter of 160 mm), was used. The external constraint axis of the apparatus is spring-mounted to ensure that the lenticular deployable composite boom is tightly wound during coiling. Three samples were tested for coiling deformation at room temperature in a dry environment (Fig. 7.8b). The cross-sectional geometry and dimensions of the samples are as follows: the length of 1020 mm, the thickness of 0.4 mm, ply scheme of [45°/−45°/0°/−45°/45°], and adhesive layer thickness of 0.1 mm; additional parameters are provided in Table 7.1. The samples were fabricated using vacuum bagging and secondary bonding techniques [1]. To capture strain changes

Table 7.2 Experimentally measured in-plane strain of the lenticular deployable composite boom in fully compressed state

	Strain gauge 1 $\varepsilon_{\theta}^{*}/\%$	Strain gauge 2 $\varepsilon_{\theta}^{*}/\%$
1	−0.7605	0.4800
2	−0.7362	0.4817
3	−0.7746	0.4861
Average	−0.7571	0.4826
Prediction	−0.5917	0.5917
Error	22%	23%

Note: Strain gauge 1 corresponds to the strain at point A on arc $\widehat{BAB'}$, and strain gauge 2 corresponds to the strain at the midpoint on arc $\widehat{BC}$

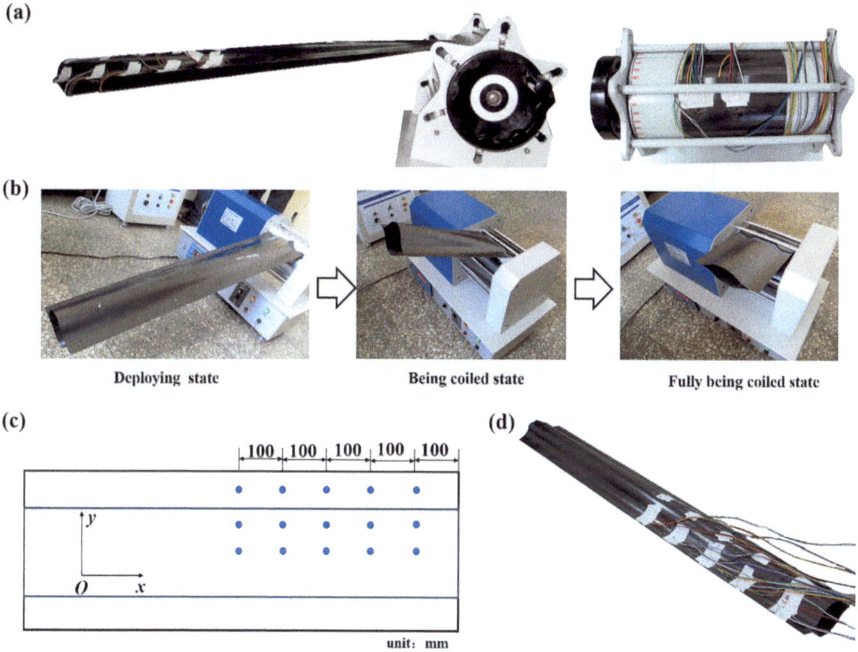

Fig. 7.8 Folding deformation tests of the lenticular deployable composite boom samples: (**a**) Custom coiling deformation test apparatus. (**b**) Coiling test. (**c**) Strain gauge layout for coiling test. (**d**) Strain gauges attached for coiling test

during coiling deformation, strain gauges were placed on the free surface of the lenticular deployable composite boom, near the endpoint of the nearly fully coiled state, as shown in Fig. 7.8c, d. During testing, the strain experienced by the three samples while being coiled around the reel was recorded. The strain variation with polar angle α at the apex along the β' and θ directions on arc $\widehat{AB}$ is shown in Fig. 7.9a, b, respectively. Similarly, the strain variation at the midpoint along the $\beta'$and θ directions on arc $\widehat{BC}$ is depicted in Fig. 7.9c, d.

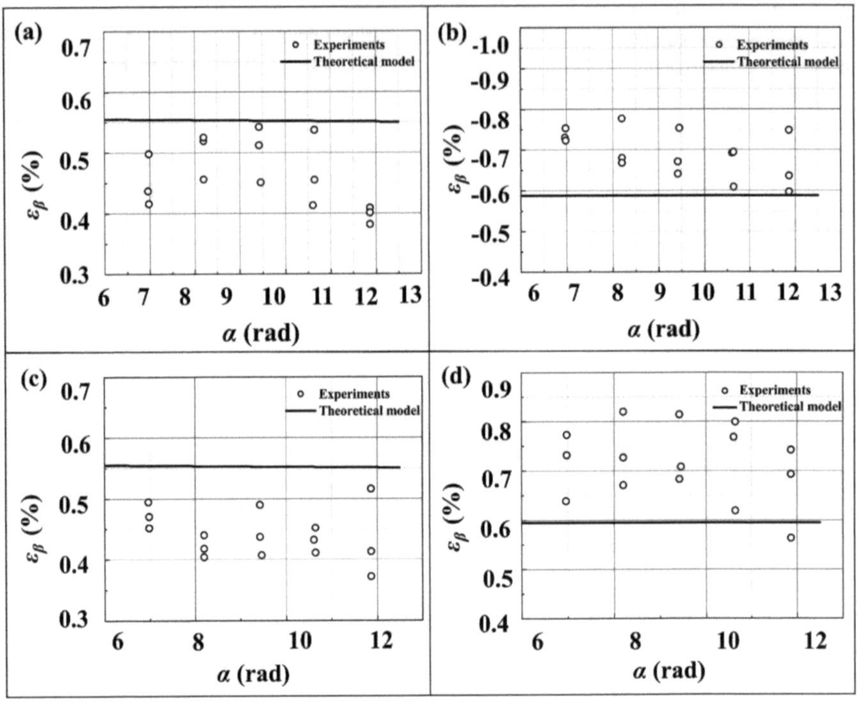

Fig. 7.9 Strain patterns of coiled lenticular deployable composite booms: (**a**) Strain pattern at the apex in the direction of β' with the increasing of α. (**b**) Strain pattern at the apex in the direction of θ with the increasing of α. (**c**) Strain pattern at the middle point of cross-sectional circular arc $\overset{\frown}{BC}$ in the direction of β' with the increasing of α. (**d**) Strain pattern at the middle point of cross-sectional circular arc $\overset{\frown}{BC}$ in the direction of θ with the increasing of α

All tested samples were able to fully recover to their initial shape after the coiling test, and no damage was observed during testing. Figure 7.9a–d shows some variability in strain measurements at the same location on the samples, with relative error ranging from 4% to 28%. This variability is mainly attributed to the inherent material properties' dispersion and uncertainties during sample preparation and testing processes, such as undetectable manufacturing defects, strain gauge positioning deviations, and control discrepancies in the coiling deformation test.

Using the analytical model and the parameters and properties listed in Table 7.1, the strain of the samples under fully flattened or compressed coiling deformation was calculated. As shown in Fig. 7.9a–d, the analytical predictions align well with the experimental data, displaying consistent trends and a relative error range of 10–38%, which falls within acceptable limits. Therefore, the analytical model presented in Sect. 7.1.1 is validated as reliable and effective in predicting the in-plane

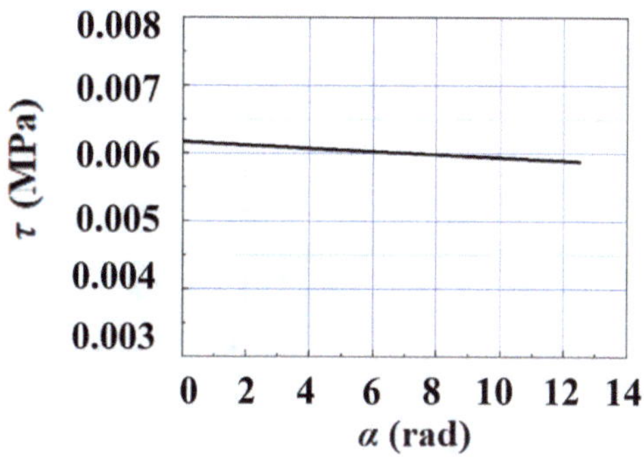

Fig. 7.10 Analytical solution of interlaminar shear stress on bonding interface of the coiled lenticular deployable composite boom

strain of lenticular deployable composite booms under fully flattened or compressed coiling deformation.

Based on the analytical model in Sect. 7.1.2 and the properties and parameters listed in Table 7.1, the interlaminar shear stress at the adhesive interface during coiling deformation of the lenticular deployable composite boom can be determined (as shown in Fig. 7.10). From Fig. 7.10, it can be seen that the interlaminar shear stress at the adhesive interface decreases with the increase of the polar angle. Although current experimental methods cannot validate this, the trend predicted by Eq. (7.10) is consistent with the real physical phenomenon, indicating that the analytical approach is reasonable.

7.2.3 *Folding Behavior Analysis*

To validate the effectiveness of the analytical models proposed in Sects. 7.1.3 and 7.1.4, the predicted results are compared with experimental results from literature [6]. The geometric parameters of the lenticular deployable composite boom are as follows: the initial curvature radii of the concave arc and convex arc are 34 mm, the length is 1020 mm, the thickness is 0.4 mm, the ply scheme is [45°/−45°/0°/−45°/45°], the adhesive layer thickness is 0.1 mm, the coiling radius is 80 mm, and other parameters are as listed in literature [6]. Figure 7.11 shows the folding moment versus torsional displacement curve for both the analytical model and experimental results. The experimentally measured folding moment increases with the torsional displacement within the initial range, eventually reaching a stable level, indicating that the folding deformation stabilizes. It is evident that the folding moment versus torsional displacement curve predicted by the analytical model closely matches the experimental results, where the measured folding moment is approximately 1 N m, and the analytical model predicts a folding moment of 1.289 N m. In comparison,

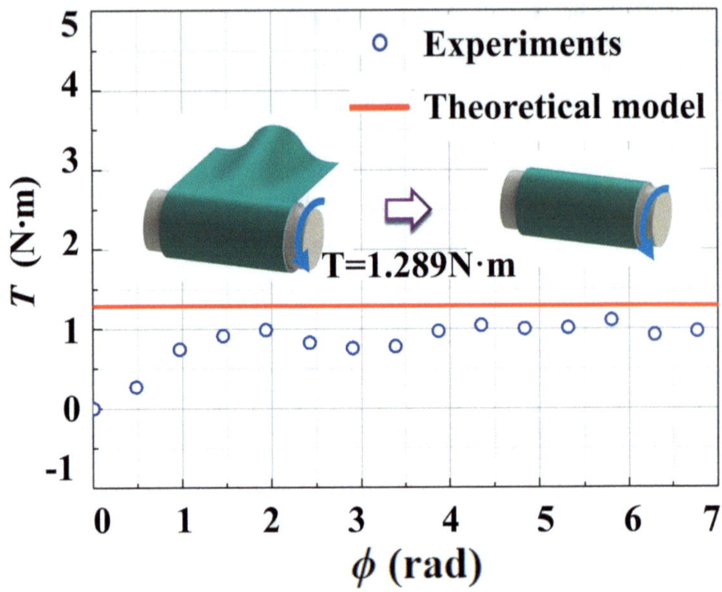

Fig. 7.11 Folding moment versus rotational displacement curves of the lenticular deployable composite boom in the folding process

the analytical model presented in Sect. 7.1.3 provides a good prediction of the folding moment for the lenticular deployable composite boom.

Similarly, using the same material properties and geometric parameters, the analytical model predicts the limit coiling radius of the lenticular deployable composite boom. In literature [6], the coiling radius of the lenticular deployable composite boom is 80 mm, and it can achieve multiple folding and deploying cycles without failure. The predicted coiling radius from the analytical model is 37.44 mm, which is <80 mm. Therefore, the lenticular deployable composite boom can be folded with a coiling radius of 80 mm, which is consistent with the experimental results. This confirms the validity of the analytical model for predicting the ultimate coiling radius of the lenticular deployable composite boom presented in Sect. 7.1.4.

7.3 Results and Discussion

Based on the results of model validation and experimental comparison in Sect. 7.2, the load versus displacement curves during the compression and tension experiments of the lenticular deployable composite boom show good agreement with the analytical model predictions. This confirms that the analytical method can effectively predict the load versus displacement curve for the first linear segment of compression deformation and the complete nonlinear load versus displacement curve for stretching deformation. Although the relative error between the analytical and experimental results ranges from 10% to 30%, the dispersion is within an acceptable range, indicating that the established analytical models provide reasonable accuracy for predicting the compression and tensile deformation of lenticular deployable composite booms.

From the strain results of lenticular deployable composite booms during the coiling deformation, the following conclusions can be drawn: the data dispersion is primarily due to the variability in the initial material properties as well as uncertainties in the preparation and experiment processes (e.g., undetectable defects during manufacturing, positioning errors of the strain gauges, and control deviations in the coiling deformation tests). Additionally, other potential factors, such as the initial curvature radius error, torsion deformation, warping deformation, transverse shear deformation, and elastic coupling effect, may also influence the experimental results. From an engineering perspective, these factors can be neglected in large deformation experiments, with a focus on the key factors.

The folding behavior from analytical solutions of the lenticular deployable composite boom during the folding process shows that the predicted folding moment and ultimate coiling radius are well consistent with the experimental results, demonstrating that the proposed analytical models exhibit good predictive accuracy.

References

1. Bai JB, Xiong JJ, Gao JP et al (2011) Fabrication and validation of collapsible composite lenticular tubes. Acta Aeronaut Astronaut Sin 32(7):1217–1223
2. Bai JB, Shenoi RA, Xiong JJ (2017) Thermal analysis of thin-walled deployable composite boom in simulated space environment. Comp Struct 173:210–218
3. Bai JB, Chen D, Xiong JJ et al (2017) A corrugated flexible composite skin for morphing applications. Comp Part B: Eng 131:134–143
4. Bai JB, Chen D, Xiong JJ et al (2019) A semi analytical model for predicting nonlinear tensile behaviour of corrugated flexible composite skin. Comp Part B: Eng 168:312–319
5. Liu TW, Bai JB, Li SL et al (2023) Large deformation and failure analysis of the corrugated flexible composite skin for morphing wing. Eng Struct 278:115463
6. Bai JB, Chen D, Xiong JJ, Shenoi RA (2019) Folding analysis for thin-walled deployable composite boom. Acta Astronaut 159:622–636
7. Wang LW, Bai JB, Shi Y (2021) Simplified analytical model for predicting neutral cross-section position of lenticular deployable composite boom in tensile deformation. Materials 14(24):7809
8. Bai JB, Xiong JJ, Gao JP et al (2013) Analytical solutions for predicting in-plane strain and interlaminar shear stress of ultra-thin-walled lenticular collapsible composite tube in fold deformation. Comp Struct 97:64–75
9. Bai JB, Chen D, Xiong JJ et al (2019) Folding analysis for thin-walled deployable composite boom. Acta Astronaut 159:622–636
10. Bai JB, Xiong JJ (2014) Temperature effect on buckling properties of ultra-thin-walled lenticular collapsible composite tube subjected to axial compression. Chinese J Aeronaut 27(5):1312–1317

11. Lin QH, Bai JB, Cong Q (2019) Novel experimental method for determining axial compression buckling behaviour of ultra-length foldable thin-walled lenticular composite tube. Comp Struct Manuf 62(4):51–55
12. Liu TW, Bai JB, Fantuzzi N (2022) Folding behaviour of the thin-walled lenticular deployable composite boom: analytical analysis and many-objective optimization. Mech Adv Mater Struct:1–19
13. Norman AD, Seffen KA, Guest SD (2009) Morphing of curved corrugated shells. Int J Solids Struct 46:1624–1633

Chapter 8
Coiling Folding of Tubular Deployable Composite Booms

Abstract This chapter develops an analytical model for tubular deployable composite booms using Archimedean spiral geometry and energy principles to predict coiling folding moment-rotation angle curves. The folding process is divided into two stages (with/without free regions), solving implicit equations via Newton iteration and considering ploy region strain energy. A custom folding mechanism tests six T700/epoxy samples, showing good agreement between theoretical and experimental external work, with failure indices <1. Parametric studies show cross-sectional radius minimally affects moments, while angle and thickness increase moments, with thickness enhancing failure risk.
Tubular deployable composite booms, as a typical large-deformation flexible composite structure, can be not only folded and compressed for storage but also deployed by utilizing the stored strain energy [1–6]. To further illustrate the large-deformation mechanisms of tubular deployable composite booms, this chapter establishes geometric equations to describe the coiling and folding behavior. By combining equilibrium equations and the energy principle, an analytical model is proposed to predict the folding moment versus displacement curve during the coiling and folding process. Failure mechanisms during the coiling and folding process are examined using the Tsai-Hill criterion and the maximum stress criterion. Samples were prepared, and large deformation experiments were conducted. The study shows that the predicted results from the analytical model are in good agreement with the experiments. Furthermore, the impact of geometric parameters on the coiling and folding behavior is also analyzed.

J.-B. Bai et al., *Folding Mechanics of Deployable Composite Structures*, Solid Mechanics and Its Applications 283,
https://doi.org/10.1007/978-3-031-99254-4_8

8.1 Analytical Model

8.1.1 Geometric Model

The deployable composite cylindrical shell can achieve a coiling and folding function and can be used as a monostable deployable composite boom in space deployable structures, with its cross-sectional angle typically >180°. This type of deployable composite cylindrical shell can be categorized into two forms: open and closed, differing only in the size of the cross-sectional angle (when the cross-sectional angle is >360°, it is in a closed form; otherwise, it is in an open form). However, there is no distinction in the analytical modeling methods for coiling and folding deformation. For clarity, the illustrations in this chapter adopt the open form. Similar to bending folding, the tubular deployable composite boom also has two types of coiling and folding modes, as shown in Fig. 8.1. When the coiling direction of the tubular deployable composite boom faces the opening direction, it is referred to as "equal-sense folding," as shown in Fig. 8.1a; when the coiling

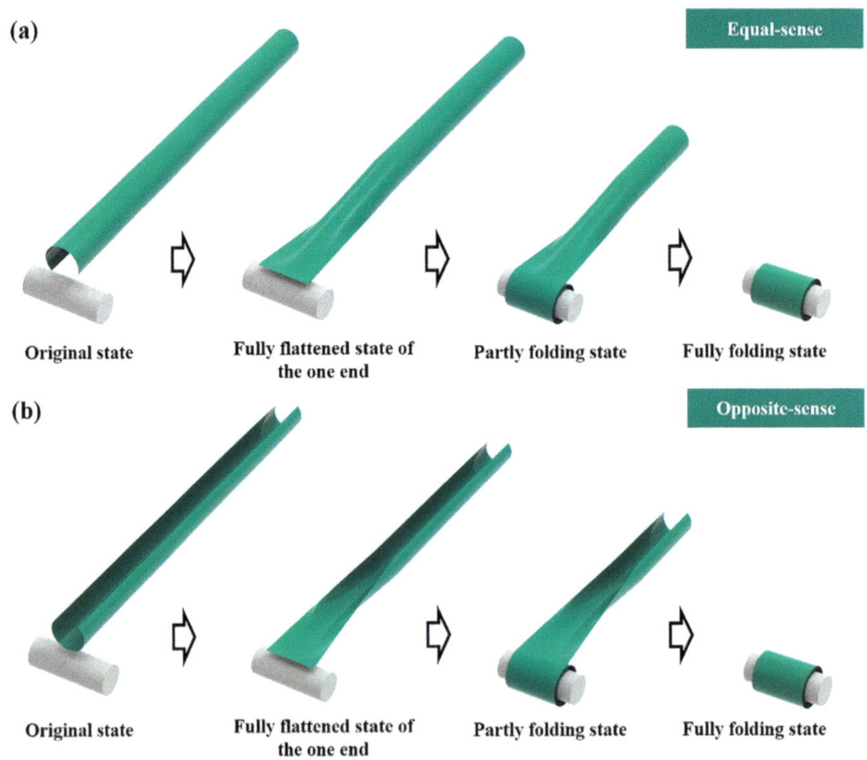

Fig. 8.1 Schematic description of the folding process of the tubular deployable composite boom: (**a**) Equal-sense folding. (**b**) Opposite-sense folding

direction faces away from the opening direction, it is referred to as "opposite-sense folding," as shown in Fig. 8.1b. To represent the geometric model of the tubular deployable composite boom during the equal-sense and opposite-sense folding processes, the following basic assumptions are made [1, 2]:

(1) The entire folding process of the tubular deployable composite boom is divided into two stages. The first stage includes the coiled region, ploy region, and free region. When the deployable composite boom has folded to a certain extent, the free region disappears, and the folding process enters the second stage, which only includes the coiling region and ploy region, as shown in Fig. 8.2 a–c.
(2) The longitudinal centerline of the tubular deployable composite boom in the folded state is an Archimedes' helix, as shown in Fig. 8.2d.
(3) The overall deformation can be described with the changes of the shape and curvature of the neutral surface without local extension.
(4) The folding deformation of the tubular deployable composite boom is idealized as a quasi-static elastic folding process.

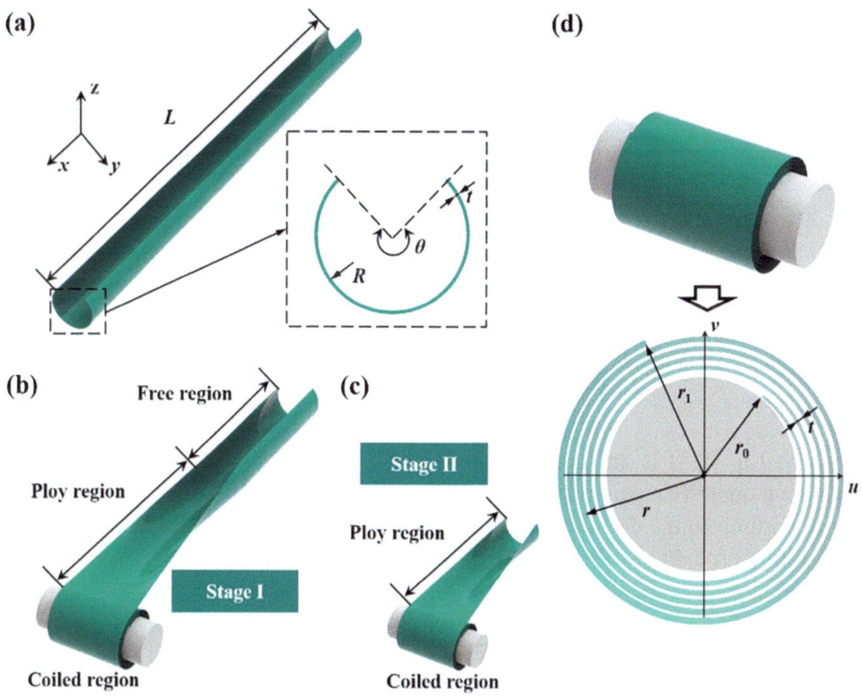

Fig. 8.2 Geometric configurations of the tubular deployable composite boom during folding process: (**a**) Deployed state. (**b**) Stage I of the partly folded state. (**c**) Stage II of partly folded state. (**d**) Longitudinal cross section in the fully folded state

Based on the assumption (2) and Fig. 8.2d, the coiled deformation of the tubular deployable composite boom can be described by a polynomial shape function in polar coordinates (ρ,ϕ), as given by

$$\rho = a\phi + b, \phi \in [\phi_0, \phi_1] \tag{8.1}$$

According to the geometric relationships, Eq. (8.1) should satisfy boundary conditions, as follows:

$$\begin{cases} a\phi_0 + b = r_0 \\ a\phi_1 + b = r_1 \end{cases} \tag{8.2}$$

where r_0 and r_1 are the polar radii at the start point and end point of the deployable composite boom in the fully folded state.

Generally, when $\phi_0 = 0$, then Eq. (8.2) can be written as

$$\begin{cases} b = r_0 \\ \phi_1 = \dfrac{r_1 - r_0}{a} \end{cases} \tag{8.3}$$

According to the assumption (3), it is possible to obtain

$$\int_{\phi_0}^{\phi_1} \sqrt{\rho^2 + (\rho')^2}\, \mathrm{d}\phi = L \tag{8.4}$$

Substituting Eqs. (8.1)–(8.3) into Eq. (8.4) and integrating give the following [3]:

$$\begin{aligned} & a\left[\frac{r_1}{2a}\sqrt{\left(\frac{r_1}{a}\right)^2 + 1} + \frac{1}{2}\ln\left(\frac{r_1}{2a} + \sqrt{\left(\frac{r_1}{a}\right)^2 + 1}\right)\right] \\ & -a\left[\frac{r_0}{2a}\sqrt{\left(\frac{r_0}{a}\right)^2 + 1} + \frac{1}{2}\ln\left(\frac{r_0}{2a} + \sqrt{\left(\frac{r_0}{a}\right)^2 + 1}\right)\right] = L \end{aligned} \tag{8.5}$$

Clearly, Eq. (8.5) is an implicit function that can be numerically solved by the unknown quantity a using the Newton-Raphson method.

According to the definition of curvature, the curvature in the x direction during the coiling process of the tubular deployable composite boom can be expressed as

$$\kappa_x = \frac{2(\rho')^2 + \rho^2}{\left[(\rho')^2 + \rho^2\right]^{\frac{3}{2}}} = \frac{2a^2 + \rho^2}{\left(a^2 + \rho^2\right)^{\frac{3}{2}}}, \phi \in \left[0, \frac{r_1 - r_0}{a}\right] \tag{8.6}$$

8.1.2 Analytical Solution for Folding Moment

The energy method is utilized to determine the folding moment for the tubular deployable composite boom during equal-sense and opposite-sense folding processes. To obtain the strain energy in the tubular deployable composite boom, the equal-sense and opposite-sense folding processes are divided into two steps, as shown in Fig. 8.3a, b. First, a bending moment M_y is applied at the edge to change the transverse curvature of the tubular deployable composite boom in the y direction

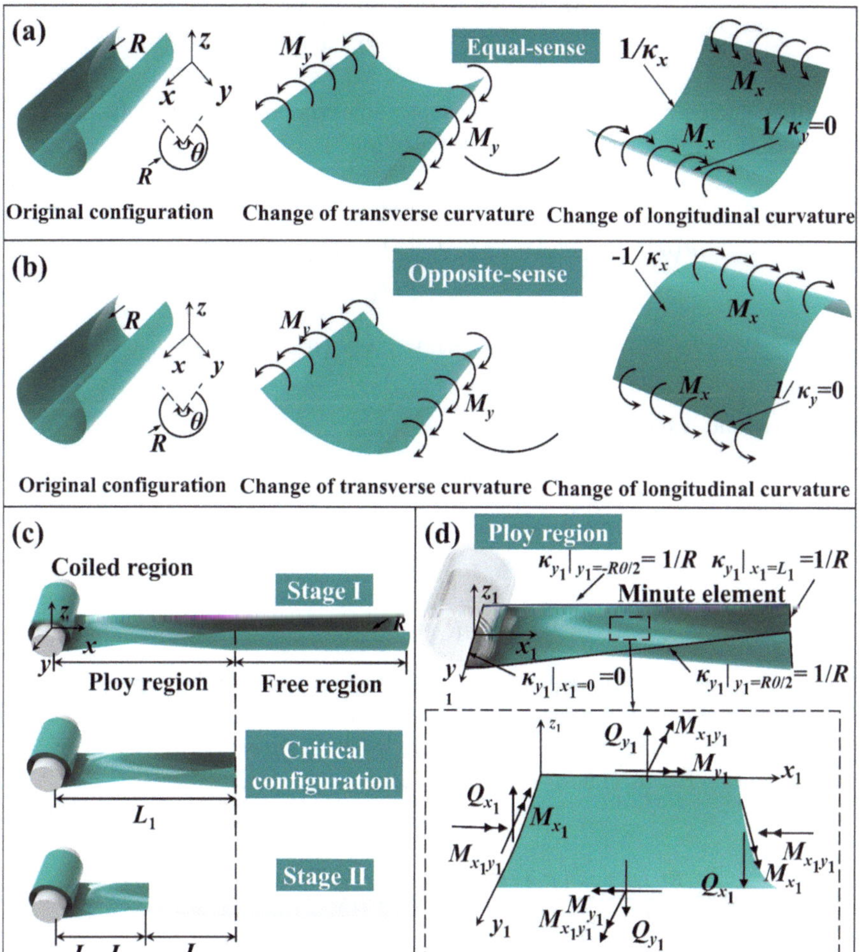

Fig. 8.3 Stress analysis and configuration of the tubular deployable composite boom during the folding process: (**a**) Equal-sense folding process in the first stage. (**b**) Opposite-sense folding process in the first stage. (**c**) Overall equal-sense folding process. (**d**) Loading conditions of any minute element in the ploy region

from $1/R$ to 0. Next, a bending moment M_x is applied at both edges of the tubular deployable composite boom to change its curvature in the x direction from 0 to κ_x. The external work is converted into bending strain energy in two stages.

Utilize an ABD matrix of classical laminate theory to describe the elastic deformation of the tubular deployable composite boom during the folding process. To avoid bending-twisting coupling effects, a symmetric laminate is selected. Furthermore, due to the restriction of the constraint mechanism, Eq. (3.59) can be simplified as

$$\begin{bmatrix} M_x \\ M_y \\ 0 \end{bmatrix} = \begin{bmatrix} D_{11} & D_{12} & D_{16} \\ D_{12} & D_{22} & D_{26} \\ D_{16} & D_{26} & D_{66} \end{bmatrix} \begin{bmatrix} \Delta\kappa_x \\ \Delta\kappa_y \\ 0 \end{bmatrix} \tag{8.7}$$

The strain energy per unit length of the deployable composite boom in the first stage from free state to coiled state can be deduced as [7].

$$\begin{aligned} u(\phi) &= \frac{R\theta}{2}\begin{bmatrix} \Delta\kappa_x & \Delta\kappa_y & 0 \end{bmatrix} \begin{bmatrix} D_{11} & D_{12} & D_{16} \\ D_{12} & D_{22} & D_{26} \\ D_{16} & D_{26} & D_{66} \end{bmatrix} \begin{bmatrix} \Delta\kappa_x \\ \Delta\kappa_y \\ 0 \end{bmatrix} \\ &= \frac{R\theta}{2}\left(D_{11}\Delta\kappa_x^2 + 2D_{12}\Delta\kappa_x\Delta\kappa_y + D_{22}\Delta\kappa_y^2\right) \end{aligned} \tag{8.8}$$

where $\Delta\kappa_x$ and $\Delta\kappa_y$ can be expressed as

$$\begin{cases} \Delta\kappa_x = \pm\dfrac{2a^2+\rho^2}{\left(a^2+\rho^2\right)^{\frac{3}{2}}}, \phi \in \left[0, \dfrac{r_1 - r_0}{a}\right] \\ \Delta\kappa_y = -\dfrac{1}{R} \end{cases} \tag{8.9}$$

Notably, the value of $\Delta\kappa_x$ depends on the folding direction of the deployable composite boom. For equal-sense folding, $\Delta\kappa_x$ takes a positive value, whereas for opposite-sense folding, $\Delta\kappa_x$ takes a negative value, as shown in Fig. 8.3a, b.

Substituting Eq. (8.9) into (8.8) yields

$$u(\phi) = \frac{R\theta D_{11}}{2}\left\{\frac{2(\rho')^2+\rho^2}{\left[(\rho')^2+\rho^2\right]^{\frac{3}{2}}}\right\}^2 \pm \theta D_{12}\left\{\frac{2(\rho')^2+\rho^2}{\left[(\rho')^2+\rho^2\right]^{\frac{3}{2}}}\right\} + \frac{\theta D_{22}}{2R} \tag{8.10}$$

In the practical folding process of the tubular deployable composite boom, one end is flattened in the first step, and then the folding process is completed through a roller in the second step, as shown in Fig. 8.3c. Therefore, there is a ploy region between the coiled region and the free region. Yang et al. [8] refined the ploy region into two specific regions and developed an analytical model to predict its length in their study. To provide a more convenient and efficient prediction of the ploy region length, this chapter simplifies the original model. For any minute element within the ploy region, there are internal moments and out-of-plane shear forces at the edges, as illustrated in Fig. 8.3d. The x_1, y_1, and z_1 are the local coordinate systems along the centerline of the ploy region. The equilibrium equations for the resultant forces and moments of the ploy region are given by

$$\begin{cases} \dfrac{\partial Q_{x_1}}{\partial x_1} + \dfrac{\partial Q_{y_1}}{\partial y_1} + q = 0 \\ \dfrac{\partial M_{x_1}}{\partial x_1} + \dfrac{\partial M_{x_1 y_1}}{\partial y_1} = Q_x \\ \dfrac{\partial M_{x_1 y_1}}{\partial x_1} + \dfrac{\partial M_{y_1}}{\partial y_1} = Q_y \end{cases} \tag{8.11}$$

The main deformation of the tubular deployable composite boom is the change of curvature caused by bending. Since the influence of internal forces is small, Eq. (8.11) can be simplified as

$$\frac{\partial^2 M_{x_1}}{\partial x_1^2} + 2\frac{\partial^2 M_{x_1 y_1}}{\partial x_1 \partial y_1} + \frac{\partial^2 M_{y_1}}{\partial y_1^2} = 0 \tag{8.12}$$

According to classical laminate theory, the relationship between the internal moment and curvature change is given by

$$\begin{bmatrix} M_{x_1} \\ M_{y_1} \\ M_{x_1 y_1} \end{bmatrix} = \begin{bmatrix} D_{11} & D_{12} & D_{16} \\ D_{12} & D_{22} & D_{26} \\ D_{16} & D_{26} & D_{66} \end{bmatrix} \begin{bmatrix} \Delta\kappa_{x_1} \\ \Delta\kappa_{y_1} \\ \Delta\kappa_{x_1 y_1} \end{bmatrix} \tag{8.13}$$

where

$$\begin{bmatrix} \Delta\kappa_{x_1} \\ \Delta\kappa_{y_1} \\ \Delta\kappa_{x_1 y_1} \end{bmatrix} = \begin{bmatrix} \kappa_{x_1} - \kappa_{x_1,0} \\ \kappa_{y_1} - \kappa_{y_1,0} \\ \kappa_{x_1 y_1} - \kappa_{x_1 y_1,0} \end{bmatrix} \tag{8.14}$$

where $\kappa_{x_1,0}$, $\kappa_{y_1,0}$, and $\kappa_{x_1 y_1,0}$ are the initial curvatures of the deployable composite boom in the deployed state:

$$\begin{cases} \kappa_{x_1,0} = 0 \\ \kappa_{y_1,0} = \dfrac{1}{R} \\ \kappa_{x_1y_1,0} = 0 \end{cases} \tag{8.15}$$

The κ_{x_1} is zero in the ploy region, and the deformation compatibility equation between the bending curvature and twisting curvature is given as

$$\frac{\partial \kappa_{y_1}}{\partial x_1} = \frac{\partial \kappa_{x_1y_1}}{\partial y_1} \tag{8.16}$$

Substituting Eqs. (8.13)–(8.16) into (8.12) as well as simplifying, it is possible to have

$$\frac{\partial^2 \kappa_{y_1}}{\partial x_1^2} + \frac{D_{22}}{D_{12} + 2D_{66}} \frac{\partial^2 \kappa_{y_1}}{\partial y_1^2} = 0 \tag{8.17}$$

Clearly, Eq. (8.17) is a second-order partial differential equation, and the boundary conditions are

$$\begin{cases} \kappa_{y_1}\big|_{x_1=0} = 0 \\ \kappa_{y_1}\big|_{x_1=L_1} = 1/R \\ \kappa_{y_1}\big|_{y_1=\pm R\theta/2} = 1/R \end{cases} \tag{8.18}$$

where L_1 is the length of the ploy region (as shown in Fig. 8.3c).

The approximate solution of the partial differential equation satisfying the boundary conditions is obtained by using the method of separation of variables, as follows:

$$\kappa_{y_1} = \left\{ \sum_{n=1}^{\infty} \left(W_1 e^{-W_3W_4x_1} + W_2 e^{W_3W_4x_1} \right) \cos W_4 y_1 \right\} + \frac{1}{R} \tag{8.19}$$

where W_1, W_2, W_3, and W_4 are intermediate variables, which are defined as

$$W_1 = -\frac{4(-1)^{n-1} e^{W_3W_4L_1}}{\pi R(2n-1)\left(e^{W_3W_4L_1} - e^{-W_3W_4L_1}\right)} \tag{8.20}$$

$$W_1 = \frac{4(-1)^{n-1} e^{-W_3W_4L_1}}{\pi R(2n-1)\left(e^{W_3W_4L_1} - e^{-W_3W_4L_1}\right)} \tag{8.21}$$

$$W_3 = \sqrt{\frac{D_{22}}{D_{12} + 2D_{66}}} \tag{8.22}$$

$$W_4 = \frac{(2n-1)\pi}{R\theta} \tag{8.23}$$

Substituting Eqs. (8.17) to (8.23) into (8.16) yields

$$\kappa_{x_1y_1} = \sum_{n=1}^{\infty}\left[-W_1W_3\left(e^{-W_3W_4x_1} - e^{-W_3W_4L_1}\right) + W_2W_3\left(e^{W_3W_4x_1} - e^{W_3W_4L_1}\right)\right]\sin W_4y_1 \tag{8.24}$$

Therefore, the strain energy of the ploy region can be expressed as

$$U_\mathrm{p} = \frac{1}{2}\int_0^{L_1}\int_{-\frac{R\theta}{2}}^{\frac{R\theta}{2}}\left[D_{22}\left(\kappa_{y_1} - \frac{1}{R}\right)^2 + D_{66}\kappa_{x_1y_1}^2\right]\mathrm{d}y_1\mathrm{d}x_1 \tag{8.25}$$

Substituting Eqs. (8.19) and (8.24) into (8.25), and applying the minimum energy principle, the length of the ploy region can be obtained:

$$\frac{\mathrm{d}U_\mathrm{p}}{\mathrm{d}L_1} = 0 \tag{8.26}$$

Obviously, Eq. (8.26) is an implicit function, which can be solved numerically by Newton iterative method to solve the unknown quantity L_1. The strain energy of the ploy region can be obtained by substituting L_1 into Eq. (8.25).

When the deployable composite boom is folded to a certain extent, the free region disappears and the folding process enters the second stage, which only includes coiled region and free region, as shown in Fig. 8.3c. The critical polar angle between the first and second stages can be expressed as

$$L_1 = L - \int_0^{\phi_\mathrm{c}}\sqrt{\rho^2 + (\rho')^2}\,\mathrm{d}\phi \tag{8.27}$$

Similarly, Eq. (8.27) is an implicit function, and the critical polar angle can be numerically solved using the Newton-Raphson method. Therefore, the strain energy stored in the deployable composite boom in the first stage can be expressed as

$$U_1(\phi) = \int_0^{\phi}\rho u(\phi)\mathrm{d}\phi = \int_0^{\phi}(a\phi + r_0)u(\phi)\mathrm{d}\phi, \phi \in [0, \phi_\mathrm{c}] \tag{8.28}$$

Using the Newton-Leibniz equation, the folding moment of the deployable composite boom during the first stage can be derived as

$$T_1(\phi)=\frac{\mathrm{d}U_1(\phi)}{\mathrm{d}\phi}=(a\phi+r_0)u(\phi),\phi\in[0,\phi_c] \tag{8.29}$$

In the second stage, the length of the ploy region decreases gradually with the folding process. The reduced length can be expressed as

$$L_2=\int_{\phi_c}^{\phi}\sqrt{\rho^2+(\rho')^2}\mathrm{d}\phi,\phi\in[\phi_c,\phi_1] \tag{8.30}$$

The strain energy stored in the reduced length of the ploy region can be derived as

$$U'(\phi)=\frac{l}{2}\int_{L_1-L_2}^{L_1}\int_{-\frac{R\theta}{2}}^{\frac{R\theta}{2}}\left[D_{22}\left(\kappa_{y_1}-\frac{1}{R}\right)^2+D_{66}\kappa_{x_1y_1}^2\right]\mathrm{d}y_1\mathrm{d}x_1,\phi\in[\phi_c,\phi_1] \tag{8.31}$$

Therefore, the incremental strain energy stored in the tubular deployable composite boom during the second stage can be expressed as

$$U_2(\phi)=\int_{\phi_c}^{\phi}(a\phi+r_0)u(\phi)\mathrm{d}\phi-U'(\phi),\phi\in[\phi_c,\phi_1] \tag{8.32}$$

Using the Newton-Leibniz formula, the folding moment of the tubular deployable composite boom during the second stage can be derived as

$$T_2(\phi)=\frac{\mathrm{d}U_2(\phi)}{\mathrm{d}\phi}=(a\phi+r_0)u(\phi)-\frac{\mathrm{d}U'(\phi)}{\mathrm{d}\phi},\phi\in[\phi_c,\phi_1] \tag{8.33}$$

In summary, the folding moment of the tubular deployable composite boom throughout the entire folding process can be expressed as

$$T(\phi)=\begin{cases}(a\phi+r_0)u(\phi),\phi\in[0,\phi_c]\\(a\phi+r_0)u(\phi)-\dfrac{\mathrm{d}U'(\phi)}{\mathrm{d}\phi},\phi\in[\phi_c,\phi_1]\end{cases} \tag{8.34}$$

According to assumption (2), the longitudinal curvature radius gradually increases and the curvature gradually decreases in the equal-sense and opposite-sense folding processes of the tubular deployable composite boom. Therefore, the tubular deployable composite boom has the maximum strain and stress in the x direction at the start point. The maximum value of the change of longitudinal curvature is

$$\Delta\kappa_x\Big|_{\phi=0}=\pm\frac{2a^2+r_0^2}{\left[a^2+r_0^2\right]^{\frac{3}{2}}} \tag{8.35}$$

The maximum value of the change of transverse curvature is

$$\Delta\kappa_y\Big|_{\phi=0} = -\frac{1}{R} \tag{8.36}$$

Based on the transformation equation for stress component in the kth layer (Eq. 3.21), substituting Eqs. (8.35) to (8.36) into (3.21) yields the maximum principal stresses. Then, by substituting the maximum principal stresses into the Tsai-Hill criterion and the maximum stress criterion from Sect. 3.2.2, the corresponding indices can be obtained.

8.2 Validation Analysis

8.2.1 Design of Folding Mechanism

To complete the equal-sense and opposite-sense folding experiments of the tubular deployable composite boom samples, a folding mechanism was designed and manufactured, as shown in Fig. 8.4. The mechanism consists of three parts: (1) folding system, (2) driving system, and (3) data acquisition system, as shown in Fig. 8.4a.

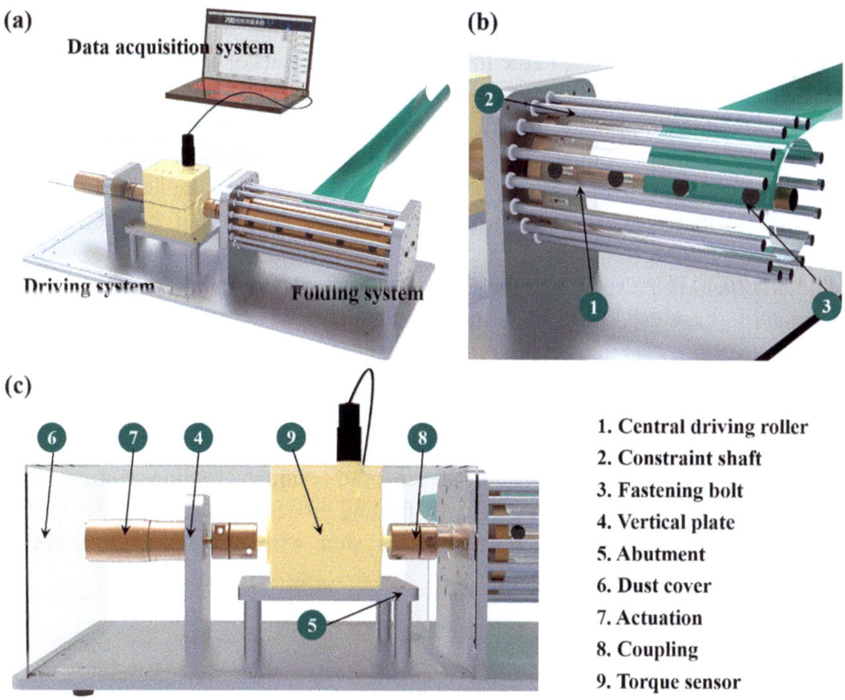

Fig. 8.4 Composition of the folding mechanism: (**a**) Folding mechanism. (**b**) Folding system. (**c**) Driving system

The folding system includes a central driving roller with 40 mm radius, five fastening bolts, and several constraint shafts. Considering the hollow and thin-walled characteristics of the tubular deployable composite boom, a central driving roller was chosen to facilitate its folding process while effectively saving storage space. The central driving roller is equipped with a groove and five fastening holes. One end of the tubular deployable composite boom is flattened and then inserted into the groove. Then, the tubular deployable composite boom with one end flat is fixed using fastening bolts to obtain boundary conditions where one end of the tubular deployable composite boom is tangent to the central driving roller, as shown in Fig. 8.4b. The clearance distance between the central driving roller and constraint shaft is 5 mm. The constraint shaft helps the deployable composite boom to be folded smoothly. In addition, it is worth noting that the constraint shaft can maintain smooth rotation during the folding process of the tubular deployable composite boom, which can reduce the friction between the tubular deployable composite boom and the constraint shafts. The folding system is a critical component of the overall mechanism, as it provides the necessary support and boundary conditions for the tubular deployable composite boom sample during the folding process.

The driving system is responsible for driving the folding mechanism and measuring the folding moment of the tubular deployable composite boom sample. The driving system mainly consists of a vertical plate, an abutment, a dust cover, an actuation, two couplings, and a torque sensor, as shown in Fig. 8.4c. The actuation is a key component of the driving system. The actuation with appropriate range should be selected according to the folding moment results of numerical simulation. The locked-rotor torque of the actuation should not be too small to avoid damage to the experimental equipment due to the required folding moment exceeding the locked-rotor torque in equal-sense and opposite-sense folding processes. At the same time, the locked-rotor torque of the actuation should not be too large, which will cause problems such as low measurement accuracy and high cost. Generally, the required folding moment is obtained through numerical simulation, and the folding moment is multiplied by the safety factor to select the appropriate actuation. The couplings are used to connect the actuation and the central driving roller, and the dust cover protects the driving system from external contamination.

The data acquisition system is responsible for collecting and analyzing the data generated during the folding process. It consists of a wire and a computer. The data acquisition system includes a wire and a computer. The two ends of the wire are connected to the computer and the torque sensor to obtain the folding moment of the tubular deployable composite boom in the folding process. In summary, the folding system, driving system, and data acquisition system work together to ensure accurate and reliable results.

8.2.2 Folding Experiments

The tubular deployable composite boom can achieve folding and deploying functions through elastic deformation during large-deformation processes. To ensure reversibility, the tubular deployable composite boom should be designed with appropriate mechanical properties to guarantee that the strain level remains within the elastic range and failure does not occur throughout the entire deformation process. Therefore, six samples were fabricated. Table 8.1 lists basic properties of T700/epoxy ply, and Table 8.2 presents the geometric parameters and stacking schemes of the six samples.

Under dry room temperature conditions, the folding tests of the tubular deployable composite boom samples were conducted with a uniform loading rate. Care was taken to ensure that the rotational speed was not too high to prevent significant fluctuations in the measured folding moment versus rotational displacement curves. Figure 8.5 shows the equal-sense and opposite-sense folding processes of Sample 1. First, one end of Sample 1 was flattened and fixed to the central driving roller in a tangential state. The folding deformation was then initiated, continuing until Sample 1 reached a fully folded state. The folding moment and rotational angle during the entire deformation process were recorded using the folding mechanism. Throughout the folding process, Sample 1 did not suffer any failure and could return to its deployed state after unloading, meeting the requirement for multiple cycles of use for the tubular deployable composite boom. Using the same method, the remaining five samples were subjected to folding, the folding moment and rotational angle throughout the process were also recorded. The results show that all six samples are capable of multiple circles between folding and deploying states.

Table 8.1 Specifications and properties of T700/epoxy composite ply

Specifications and properties	Values
E_1 (GPa)	128.61
E_2 (GPa)	7.52
G_{12} (GPa)	4.82
G_{13} (GPa)	4.50
G_{23} (GPa)	4.50
ν_{12}	0.314
X_t (MPa)	2103.44
Y_t (MPa)	75.97
X_c (MPa)	1233.65
Y_c (MPa)	181.46
S_{12} (MPa)	216.36
ρ_d (g/mm^3)	1.60×10^{-3}
t_0 (mm)	0.03

Table 8.2 The geometric parameters and stacking configurations of the six samples

No.	R (mm)	θ (°)	L (mm)	Layup schemes
1	20	270	800	[0°/45°/−45°/90°/−45°/45°/0°]
2	30	270	800	[0°/45°/−45°/90°/−45°/45°/0°]
3	20	270	800	[0/°45°/−45°/−45°/45°/0°]
4	30	270	800	[0°/45°/−45°/−45°/45°/0°]
5	20	270	800	[0°/90°/90°/0°]
6	30	270	800	[0°/90°/0°/90°/0°]

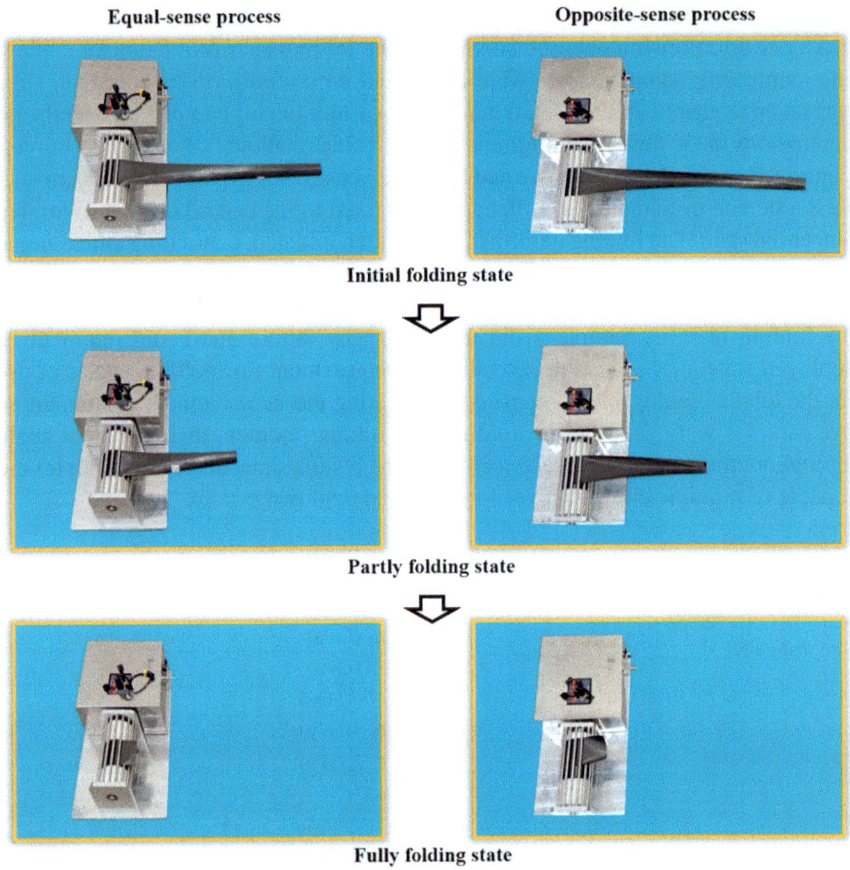

Fig. 8.5 The equal-sense and opposite-sense folding processes of Sample 1

8.2.3 Model Validation

Based on the analytical model established in Sect. 8.1 for predicting the folding behavior of the tubular deployable composite boom, the solution procedure was designed and conducted using the MATLAB code. Figure 8.6 illustrates the folding

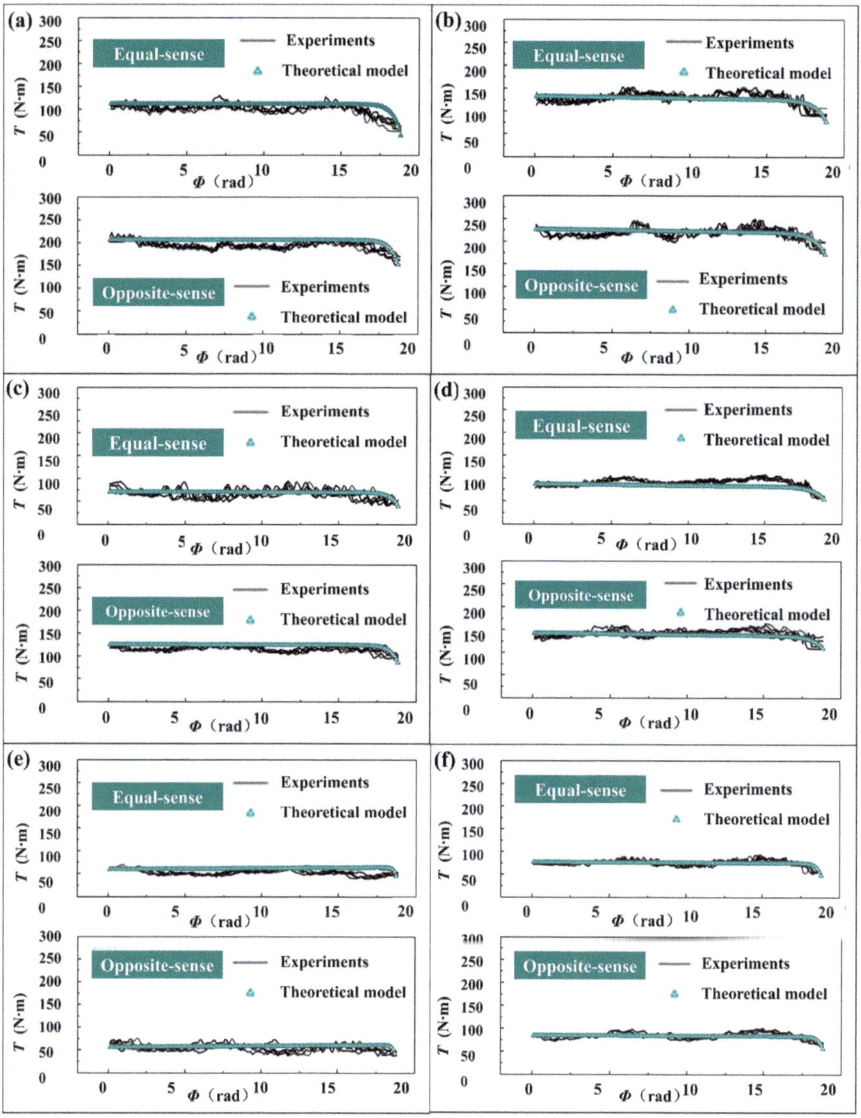

Fig. 8.6 Folding moment versus rotational displacement curves of six samples during the equal-sense and opposite-sense folding processes: (**a**) No. 1. (**b**) No. 2. (**c**) No. 3. (**d**) No. 4. (**e**) No. 5. (**f**) No. 6

moment versus rotational displacement curves of the analytical and experimental results.

From Fig. 8.6, it is obvious that folding moment versus rotational displacement curves predicted by the analytical model are in good agreement with experimental

Table 8.3 The external work required for the six samples during the folding processes

No.		Experimental results (mJ)						Analytical model (mJ)	
		1	2	3	4	5	Average	Values	Error
1	Equal-sense	1850.5	1964.5	1889.4	1856.1	1861.6	1884.4	2064.5	9.56%
	Opposite-sense	3631.6	3620.5	3601.0	3645.5	3653.8	3630.5	3826.1	5.39%
2	Equal-sense	2467.4	2417.4	2409.0	2400.7	2406.3	2420.1	2364.6	2.29%
	Opposite-sense	4106.7	4067.8	4009.5	4079.0	4098.4	4072.3	4123.4	1.25%
3	Equal-sense	1278.1	1280.9	1217.0	1242.0	1214.2	1246.5	1294.8	3.87%
	Opposite-sense	2161.7	2142.3	2172.9	2178.4	2170.1	2165.1	2320.1	7.16%
4	Equal-sense	1700.5	1708.8	1717.2	1711.6	1714.4	1710.5	1550.4	9.36%
	Opposite-sense	2684.1	2648.0	2670.2	2650.8	2661.9	2663.0	2575.7	3.28%
5	Equal-sense	1055.9	1000.3	989.2	1005.8	1005.8	1011.4	1089.2	7.69%
	Opposite-sense	1067.0	1044.7	1030.9	1011.4	1042.0	1039.2	1150.3	10.69%
6	Equal-sense	1417.1	1422.6	1433.7	1428.2	1400.4	1420.4	1408.7	0.82%
	Opposite-sense	1589.3	1589.3	1583.8	1583.8	1564.3	1582.1	1553.2	1.83%

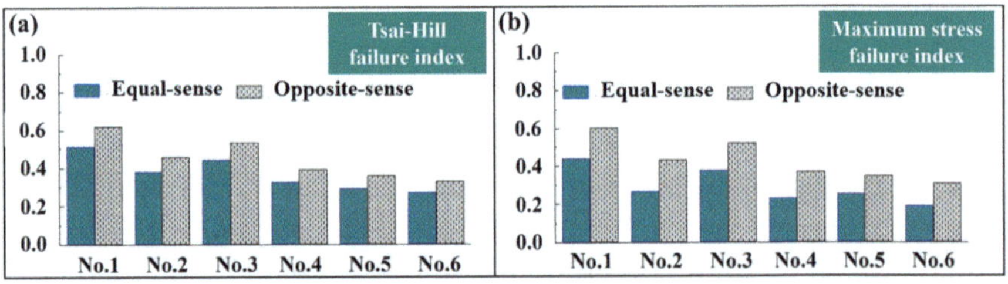

Fig. 8.7 Failure indices of the folded state predicted by the analytical model: (**a**) Tsai-Hill failure index. (**b**) Maximum stress failure index

results. By integrating the folding moment versus rotational displacement curves in Fig. 8.6, the external work required for the equal-sense and opposite-sense folding processes of the deployable composite boom was obtained, as shown in Table 8.3. According to Table 8.3, the maximum error between the experimental and analytical results is 10.69%, further indicating that the analytical model can effectively predict the folding behavior of the tubular deployable composite boom. Figure 8.7 and Table 8.4 provide the failure indices for the six samples during the equal-sense and opposite-sense folding processes. Among all the failure indices, the maximum value is only 0.6246, suggesting that all six samples can achieve multiple testing cycles without failure, which is consistent with the experimental results. In summary, the analytical model proposed demonstrates good prediction accuracy for the folding behavior of the tubular deployable composite boom.

Table 8.4 Failure indices of the six samples during the folding processes

	Tsai-Hill failure index		Maximum stress failure index	
No.	Equal-sense	Opposite-sense	Equal-sense	Opposite-sense
1	0.5166	0.6246	0.4406	0.6048
2	0.3803	0.4584	0.2664	0.4305
3	0.4429	0.5353	0.3777	0.5184
4	0.3259	0.3929	0.2283	0.3690
5	0.2952	0.3569	0.2518	0.3456
6	0.2716	0.3274	0.1903	0.3075

8.3 Results and Discussion

8.3.1 *Error Analysis*

The folding moment versus rotational displacement curves predicted by the analytical model are in good agreement with the experimental results, with the maximum error between the predicted and experimental required external work being 10.69%. A detailed analysis of the sources of this error reveals two main factors. First, the folding moment versus rotational displacement curve measured during the experiments exhibits some fluctuations throughout the entire coiling process. This is primarily due to the difficulty in achieving a completely quasi-static loading condition during the actual operation. To reduce such fluctuations, it is recommended to optimize experimental conditions, particularly by lowering the loading rate (i.e., the actuator speed in the folding mechanism), which will help capture and analyze the mechanical characteristics of the deployable composite cylindrical shell more accurately during the folding process. Second, the contact and friction between the tubular deployable composite boom and the folding mechanism are inevitable in the actual operation, and these factors are not sufficiently considered in the current analytical model. These factors may also have an impact on the predicted results. To further improve the predictive accuracy of the analytical model, future research should fully account for these physical phenomena and make necessary improvements to the model.

8.3.2 *The Value of Engineering Application*

As shown in Fig. 8.6, in the first stage, the folding moment of the tubular deployable composite boom remains relatively stable, with almost no significant changes. This is mainly because, during this stage, the free region gradually transforms into the coiled region, while the ploy region remains unchanged, thus maintaining the stability of the folding moment. However, in the second stage, as the free region completely disappears and the ploy region gradually transforms into the coiled region, the folding moment of the tubular deployable composite boom begins to show a gradual decrease. It is worth noting that existing linear models, due to their inherent limitations, are unable to effectively capture the nonlinear mechanical response of the tubular deployable composite boom during the folding process. Therefore, accurately predicting its folding behavior during both equal-sense and opposite-sense folding processes is critical. This not only aids in advancing practical deployment in space applications but also provides an analytical foundation for optimizing the design of the folding machine.

8.4 Parametric Study

Analytical model analysis indicates that the geometric parameters significantly affect the folding behavior of the tubular deployable composite boom. Therefore, this chapter uses the analytical model to investigate the influence of cross-sectional radius, cross-sectional angle, and thickness on the folding behavior of the tubular deployable composite boom. The material properties and other geometric parameters remain consistent with those of Sample 1 in Sect. 8.2.2. The specific values of the variables are as follows:

- The cross-sectional radius is varied as 15 mm, 17.5 mm, 20 mm, 22.5 mm, and 25 mm.
- The cross-sectional angle is varied as 180°, 225°, 270°, 315°, and 360°.
- The thickness is varied as 0.14 mm, 0.175 mm, 0.21 mm, 0.245 mm, and 0.28 mm.

From Fig. 8.8a, b, it can be seen that the folding moment versus rotational displacement curves of equal-sense and opposite-sense folding almost do not change with the increase of the cross-sectional radius. According to Eqs. (8.10) and (8.34), when the cross-sectional radius increases, the first term increases, the second term remains unchanged, and the third term decreases, which results in very small changes in Eq. (8.10), and then the folding moment versus rotational displacement curves almost do not change. In addition, when the five samples reach the fully folded state, the failure indices for both equal-sense and opposite-sense folding processes do not exceed 1, indicating that all five samples can be fully folded without failure. With the increase of the cross-sectional radius, the failure indices of equal-sense and opposite-sense folding decrease significantly, which is due to the reduction of the change of transverse curvature in Eq. (8.36), leading to a significant decrease in the failure indices for both equal-sense and opposite-sense folding processes.

From Fig. 8.8c, d, it can be observed that folding moments of equal-sense and opposite-sense folding increase with the increase of the cross-sectional angle, but the degrees of increase are not the same. According to Eqs. (8.10) and (8.34), when the cross-sectional angle increases, all three terms of Eq. (8.10) increase. However, the first and third terms of equal-sense folding are positive, the second term is negative, while the three terms of opposite-sense folding are positive, which results in a greater increase in the folding moment versus rotational displacement curve of

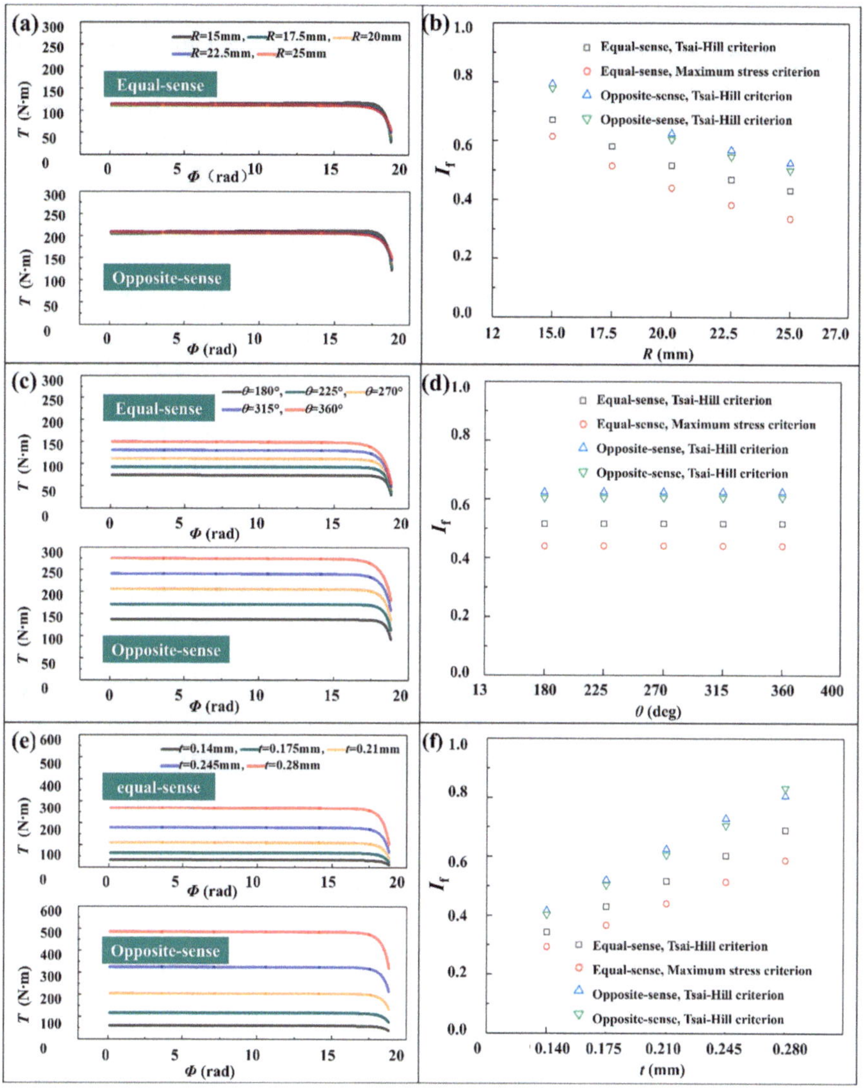

Fig. 8.8 Effect of geometric parameters on the folding behavior of the tubular deployable composite boom: (**a**) Folding moment versus rotational displacement curves with different cross-sectional radii. (**b**) Failure indices with different cross-sectional radii. (**c**) Folding moment versus rotational displacement curves with different cross-sectional angles. (**d**) Failure indices with different cross-sectional angles. (**e**) Folding moment versus rotational displacement curves with different thicknesses. (**f**) Failure indices with different thicknesses

equal-sense folding than that of opposite-sense folding. Furthermore, when the five samples reach the fully folded state, the failure indices for both equal-sense and opposite-sense folding do not exceed 1, indicating that all samples can be fully folded without failure. As the cross-sectional angle increases, the failure indices for both equal-sense and opposite-sense folding processes remain unchanged. The reason is that the calculation method of the failure index is not affected by the cross-sectional angle, but by the change of curvature.

From Fig. 8.8e, f, it can be seen that as the thickness increases, both the folding moments for equal-sense and opposite-sense folding increase, but the degree of increase is different. According to the calculation method of the bending stiffness matrix of laminates, when the layup angle remains unchanged, D_{11}, D_{12}, and D_{22} in the bending stiffness matrix increase with the increase of the thickness when the ply angle does not change. As shown in Eqs. (8.10) and (8.34), when the thickness increases, all three terms in Eq. (8.10) increase. However, the first and third terms of equal-sense folding are positive, the second term is negative, while the three terms of opposite-sense folding are positive, which results in a greater increase in the folding moment versus rotational displacement curve of equal-sense folding than that of opposite-sense folding. This pattern is consistent with the effect of the cross-sectional angle on the folding performance of the deployable composite boom. Additionally, when the five samples reach a fully folded state, the failure indices for both equal-sense and opposite-sense folding processes do not exceed 1, indicating that all samples can be fully folded without failure. Moreover, as the thickness increases, the failure indices for both equal-sense and opposite-sense folding increase

References

1. Liu TW, Bai JB, Fantuzzi N et al (2023) Analytical models for predicting folding behaviour of thin-walled tubular deployable composite boom for space applications. Acta Astronaut 208:167–178
2. Liu TW, Bai JB, Fantuzzi N et al (2023) Folding behaviour of thin-walled tubular deployable composite boom for space applications: experiments and numerical simulation. Acta Astronaut 209:159171
3. Bai JB, Xiong JJ, Gao JP et al (2013) Analytical solutions for predicting in-plane strain and interlaminar shear stress of ultra-thin-walled lenticular collapsible composite tube in fold deformation. Comp Struct 97:64–75
4. Liu TW, Bai JB, Fantuzzi N et al (2024) Thin-walled deployable composite structures: a review. Progress in aerospace. Sciences 146:100985
5. Bai JB, You FY, Wang ZZ et al (2024) An efficient multi-objective optimization framework for thin-walled tubular deployable composite boom. Comp Struct 327:117713
6. Bai JB, Chen D, Xiong JJ et al (2019) Folding analysis for thin-walled deployable composite boom. Acta Astronaut 159:622–636
7. Iqbal K, Pellegrino S, Daton-Lovett A (1998) Bistable composite slit tubes. In: IUTAM-IASS symposium on deployable structures. Cambridge, UK
8. Yang LY, Tan HF, Cao ZS (2018) Modeling and analysis of the ploy region of bistable composite cylindrical shells. Comp Struct 192:347–354

Chapter 9
Coiling Folding of Bistable Deployable Composite Booms

Abstract This chapter investigates the bistable behavior of tubular composite booms, where antisymmetric laminates enable stable folded/deployed states without external constraints, driven by Poisson effect-induced curvature-moment coupling. Model 1 assumes an Archimedean spiral, solving folded configurations via minimum energy; Model 2 uses spatial spiral equations and a two-parameter method, determining stability via stiffness matrix positive definiteness. Six samples validate the models, with theoretical polar radius/angle errors <9.42%/10.08% and failure indices <1. Parametric studies show increasing radius reduces failure risk, while thickness increases it, with length having minimal impact.

Deployable composite booms with symmetric laminates demonstrate significant advantages in space applications due to their distinctive folding and deploying functionalities. In the fully folded state, these structures occupy minimal storage space, facilitate transportation, and rapidly recover to their initial configuration by releasing stored elastic strain energy to achieve deployment. However, it is worth noting that this type typically remains in the fully folded state due to the high elastic strain energy, necessitating the use of additional restraining devices to maintain their configuration. In contrast, deployable composite booms with antisymmetric laminates, through precise ply design, exhibit unique bistable characteristics. This bistability allows to maintain stable configurations in both the folded and deployment states without requiring external loads or constrained devices. As illustrated in Fig. 9.1, this bistable property offers significant flexibility in the design of deployable structures for space applications and also enables optimization and lightweight of auxiliary devices. The stability of the bistable deployable composite boom primarily stems from the special distribution of their bending stiffness and also shows great potential and promising prospects in the field of deployable structures.

The two stable states of the bistable deployable composite boom correspond to curvatures on the same side of the structure. One stable state represents the initially deployed configuration, which provides superior load-bearing performance, while the other stable state corresponds to the folded configuration, where the boom is tightly and stably coiled. Previous studies have shown that this bistable behavior primarily arises from the Poisson effect [1]. When the boom is subjected to

J.-B. Bai et al., *Folding Mechanics of Deployable Composite Structures*, Solid Mechanics and Its Applications 283,
https://doi.org/10.1007/978-3-031-99254-4_9

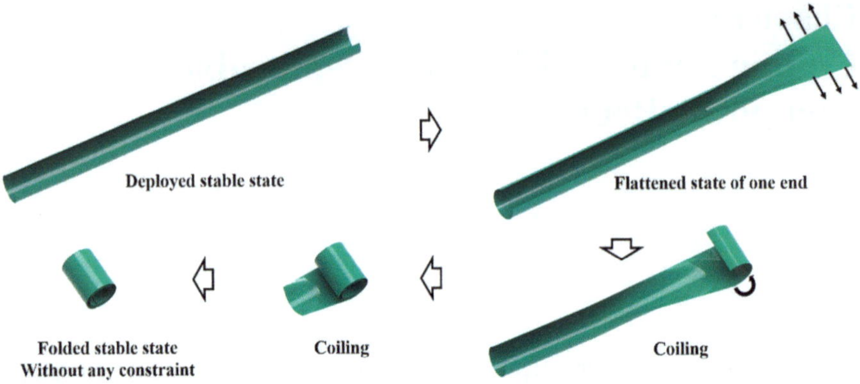

Fig. 9.1 Bistable deformation process of the deployable composite boom

transverse tension, the induced transverse bending moment causes the structure to flatten transversely, and the change of transverse curvature generates a substantial longitudinal bending moment. Due to the lack of coupling between bending and twisting, this process enables the shell to transition from the deployed stable state to the folded stable state [1–6]. The chapter primarily introduces the bistability mechanism and deformation characteristics of the bistable deployable composite boom.

9.1 Analytical Model 1

The bistable deployable composite boom can realize the conversion between folded and deployed stable states by storing and releasing elastic strain energy. To characterize the geometric behavior in the two stable states, the following basic assumptions are made:

(1) Central line of the longitudinal section of the bistable deployable composite boom in folded stable state is Archimedes' helix, and any two circles are closely attached, which models the contact behavior between surfaces.
(2) Thickness change of the bistable deployable composite boom is negligibly small in the bistable deformation process, so the overall deformation can be described with the changes of shape and curvature of the neutral surface without local extension.
(3) Curvature in the y direction of the bistable deployable composite boom is uniform in the bistable deformation process, and the curvature in the y direction is zero in folded stable state.

According to assumption (1), the bistable deployable composite boom shares a similar geometric model with the monostable deployable composite boom described in Sect. 8.1, where the longitudinal cross-sectional centerline is an Archimedean spiral. However, in Sect. 8.1, due to the constraints of the folding mechanism, r_0 and r_1 are known quantities, while they are unknown in the section. And if any two adjacent turns are in tight contact, the following relationship can be derived:

$$a = \frac{t}{2\pi} \tag{9.1}$$

Substituting Eq. (9.1) into (8.5), the quantitative relationship between r_0 and r_1 can be obtained, as follows:

$$\begin{aligned} &\frac{t}{2\pi}\left[\frac{\pi r_1}{t}\sqrt{\left(\frac{2\pi r_1}{t}\right)^2+1}+\frac{1}{2}\ln\left(\frac{\pi r_1}{t}+\sqrt{\left(\frac{2\pi r_1}{t}\right)^2+1}\right)\right] \\ &-\frac{t}{2\pi}\left[\frac{\pi r_0}{t}\sqrt{\left(\frac{2\pi r_0}{t}\right)^2+1}+\frac{1}{2}\ln\left(\frac{\pi r_0}{t}+\sqrt{\left(\frac{2\pi r_0}{t}\right)^2+1}\right)\right]=L \end{aligned} \tag{9.2}$$

Similarly, substituting Eq. (9.1) into (8.6) yields the x direction curvature of the bistable deployable composite cylindrical shell in its folded stable state, as shown below:

$$\kappa_x = \frac{2(\rho')^2+\rho^2}{\left[(\rho')^2+\rho^2\right]^{\frac{3}{2}}} = \frac{\frac{t^2}{2\pi^2}+\rho^2}{\left(\frac{t^2}{4\pi^2}+\rho^2\right)^{\frac{3}{2}}}, \phi \in \left[0, \frac{2\pi(r_1-r_0)}{t}\right] \tag{9.3}$$

Consistent with the analytical modeling of the coiling folding of the monostable deployable boom in Sect. 8.1, the energy method is employed to determine the geometric configuration of the bistable deployable composite boom in its folded stable state. Using the ABD matrix from classical laminate theory, the bistable deformation process of the bistable deployable boom with the antisymmetric laminates can be expressed, and the relationship between internal forces and deformations is given as follows:

$$\begin{bmatrix} N_x \\ N_y \\ N_{xy} \\ M_x \\ M_y \\ M_{xy} \end{bmatrix} = \begin{bmatrix} A_{11} & A_{12} & 0 & 0 & 0 & B_{16} \\ A_{12} & A_{22} & 0 & 0 & 0 & B_{26} \\ 0 & 0 & A_{66} & B_{16} & B_{26} & 0 \\ 0 & 0 & B_{16} & D_{11} & D_{12} & 0 \\ 0 & 0 & B_{26} & D_{12} & D_{22} & 0 \\ B_{16} & B_{26} & 0 & 0 & 0 & D_{66} \end{bmatrix} \begin{bmatrix} \Delta\varepsilon_x \\ \Delta\varepsilon_y \\ \Delta\gamma_{xy} \\ \Delta\kappa_x \\ \Delta\kappa_y \\ \Delta\kappa_{xy} \end{bmatrix} \tag{9.4}$$

where

$$\begin{cases} N_x = 0 \\ N_y = 0 \\ N_{xy} = 0 \\ M_{xy} = 0 \\ \Delta\kappa_{xy} = 0 \end{cases} \tag{9.5}$$

The bending strain energy per unit area in the mid-surface of a laminated plate is detailed below:

$$u' = \frac{1}{2}\left(M_x\Delta\kappa_x + M_y\Delta\kappa_y + M_{xy}\Delta\kappa_{xy}\right) \tag{9.6}$$

where

$$\begin{cases} \Delta\kappa_x = \kappa_x \\ \Delta\kappa_y = \kappa_y - \dfrac{1}{R} \end{cases} \tag{9.7}$$

Substituting Eqs. (9.4), (9.5), and (9.7) into (9.6), it is feasible to deduce as

$$\begin{aligned} u' &= \frac{1}{2}\kappa_x\left[B_{16}\gamma_{xy} + D_{11}\kappa_x + D_{12}\left(\kappa_y - \frac{1}{R}\right)\right] \\ &+ \frac{1}{2}\left(\kappa_y - \frac{1}{R}\right)\left[B_{26}\gamma_{xy} + D_{12}\kappa_x + D_{22}\left(\kappa_y - \frac{1}{R}\right)\right] \end{aligned} \tag{9.8}$$

Substituting Eq. (9.5) into (9.4) and solving Eq. (9.4), it is possible to determine as follows:

$$\gamma_{xy} = B_{61}^{*}\kappa_x + B_{62}^{*}\kappa_y \tag{9.9}$$

where

$$\boldsymbol{B}^{*} = -\boldsymbol{A}^{-1}\boldsymbol{B} \tag{9.10}$$

Based on assumption (3), it is known that u' is uniform. By substituting Eqs. (9.9) and (9.10) into (9.8), the strain energy stored per unit length in the x direction of the bistable deployable composite boom in its folded stable state can be expressed as follows:

$$
\begin{aligned}
u &= \frac{R\theta}{2}\left(D_{11}+B_{16}B_{61}^{*}\right)\kappa_x^2 \\
&+\frac{R\theta}{2}\left(2D_{12}+B_{16}B_{62}^{*}+B_{26}B_{61}^{*}\right)\kappa_x\left(\kappa_y-\frac{1}{R}\right) \\
&+\frac{R\theta}{2}\left(D_{22}+B_{26}B_{62}^{*}\right)\left(\kappa_y-\frac{1}{R}\right)^2
\end{aligned}
\tag{9.11}
$$

When the bistable deployable composite boom is in the folded stable state, the curvature in the y direction is zero. Thus, Eq. (9.11) can be simplified as

$$
\begin{aligned}
u &= \frac{R\theta}{2}\left(D_{11}+B_{16}B_{61}^{*}\right)\kappa_x^2 \\
&-\frac{R\theta}{2}\left(2D_{12}+B_{16}B_{62}^{*}+B_{26}B_{61}^{*}\right)\frac{\kappa_x}{R} \\
&+\frac{R\theta}{2}\left(D_{22}+B_{26}B_{62}^{*}\right)\frac{1}{R^2}
\end{aligned}
\tag{9.12}
$$

The total strain energy of the bistable deployable composite boom in the folded stable state can then be expressed as

$$
U=\int_{\phi_0}^{\phi_1} u(\phi)\rho(\phi)\mathrm{d}\phi \tag{9.13}
$$

Substituting Eqs. (8.3), (9.1), and (9.12) into (9.13), the following expression can be obtained:

$$
U=\frac{R\theta}{t}\int_{r_0}^{r_1}\left[\begin{array}{l}\left(D_{11}+B_{16}B_{61}^{*}\right)\kappa_x^2+\dfrac{\left(D_{22}+B_{26}B_{62}^{*}\right)}{R^2} \\ -\left(2D_{12}+B_{16}B_{62}^{*}+B_{26}B_{61}^{*}\right)\dfrac{\kappa_x}{R}\end{array}\right]\rho\mathrm{d}\rho \tag{9.14}
$$

It is widely recognized that folding stable state of Bi-DCB corresponds to an energy minimum. Consequently, the minimum energy principle can be effectively applied to Eq. (9.14). It is feasible to consider

$$
\begin{cases}
\dfrac{\mathrm{d}U}{\mathrm{d}r_0}=r_1O_1\dfrac{\mathrm{d}r_1}{\mathrm{d}r_0}-r_0O_2=0 \\
\dfrac{\mathrm{d}^2\Pi}{\mathrm{d}r_0^{\,2}}>0
\end{cases}
\tag{9.15}
$$

where

$$O_1 = \frac{\pi R\theta}{t}\left[\begin{array}{l}\left(D_{11}+B_{16}B_{61}^{*}\right)O_3^2 - \left(2D_{12}+B_{16}B_{62}^{*}+B_{26}B_{61}^{*}\right)\dfrac{O_3}{R} \\ +\left(D_{22}+B_{26}B_{62}^{*}\right)\dfrac{1}{R^2}\end{array}\right] \tag{9.16}$$

$$O_2 = \frac{\pi R\theta}{t}\left[\begin{array}{l}\left(D_{11}+B_{16}B_{61}^{*}\right)O_4^2 - \left(2D_{12}+B_{16}B_{62}^{*}+B_{26}B_{61}^{*}\right)\dfrac{O_4}{R} \\ +\left(D_{22}+B_{26}B_{62}^{*}\right)\dfrac{1}{R^2}\end{array}\right] \tag{9.17}$$

$$O_3 = \frac{\dfrac{t^2}{2\pi^2}+r_1^2}{\left(\dfrac{t^2}{4\pi^2}+r_1^2\right)^{\frac{3}{2}}} \tag{9.18}$$

$$O_4 = \frac{\dfrac{t^2}{2\pi^2}+r_0^2}{\left(\dfrac{t^2}{4\pi^2}+r_0^2\right)^{\frac{3}{2}}} \tag{9.19}$$

It should be noted that there is a differential term $\frac{\mathrm{d}r_1}{\mathrm{d}r_0}$ in Eq. (9.15), and central difference algorithm is used to achieve numerical differentiation. Clearly, Eqs. (9.2) and (9.15) each contain only two unknown variables. By solving these two equations simultaneously using numerical methods, the values of r_0 and r_1 can be obtained. Substituting the obtained r_0 and r_1 into Eq. (8.3) allows for the determination of the terminal polar angle. The method for calculating the stress level in the folded stable state in analytical model 1 is consistent with that described in Sect. 8.1. Figure 9.2 provides a clear flowchart illustrating the modeling process of analytical model 1 proposed in this section.

9.2 Analytical Model 2

This section presents an alternative analytical model [2, 3] for predicting the bistable behavior of deployable composite booms, based on the following fundamental assumptions:

(1) The projection of the centerline of the bistable deployable composite boom with arbitrary configurations onto the *uov* plane forms an Archimedes' helix, and any two turns are tightly adhered.
(2) The deformation of the bistable deployable composite boom can be defined by alterations in the curvature of its neutral surface, which bends without stretching.

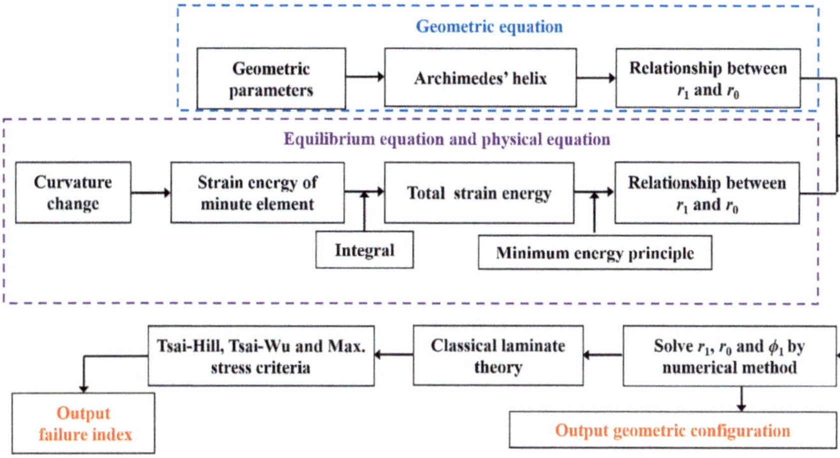

Fig. 9.2 The modeling process of analytical model 1

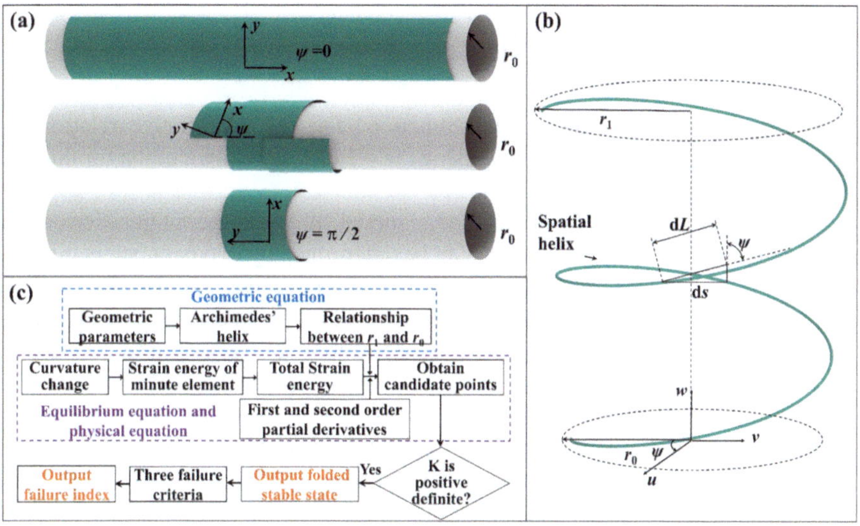

Fig. 9.3 Analytical model 2: (**a**) Geometric configuration of the bistable deployable composite boom in an arbitrary configuration. (**b**) Coordinate system defining the geometric configuration of the bistable deployable composite boom on a cylindrical surface. (**c**) Flowchart of the analytical modeling process

(3) For any arbitrary configuration, the Gaussian curvature of the bistable deployable composite boom remains constant, implying that all possible configurations can be accommodated on the surface of the cylinder, as visualized in Fig. 9.3a.

Based on assumption (1) and Fig. 9.3b, the projection of the centerline of the bistable deployable composite boom in any configuration can be described using cylindrical coordinates (ρ, ϕ, z):

$$\begin{cases} \rho = \dfrac{t}{2\pi}\phi + r_0 \\ z = \int_0^{\phi} \rho \cot\psi \, \mathrm{d}\phi \end{cases} \tag{9.20}$$

According to assumption (2), Fig. 9.3b, and geometric relationship, it can be calculated as

$$\int_0^{\phi_1} \frac{\sqrt{\rho^2 + (\rho')^2}}{\sin\psi} \mathrm{d}\phi = L \tag{9.21}$$

Substituting Eqs. (8.3), (9.1), and (9.20) into (9.21), Eq. (9.21) is integrated as

$$\left[\begin{array}{l} \dfrac{\pi r_1}{t}\sqrt{\left(\dfrac{2\pi r_1}{t}\right)^2 + 1} + \dfrac{1}{2}\ln\left(\dfrac{\pi r_1}{t} + \sqrt{\left(\dfrac{2\pi r_1}{t}\right)^2 + 1}\right) \\ -\dfrac{\pi r_0}{t}\sqrt{\left(\dfrac{2\pi r_0}{t}\right)^2 + 1} - \dfrac{1}{2}\ln\left(\dfrac{\pi r_0}{t} + \sqrt{\left(\dfrac{2\pi r_0}{t}\right)^2 + 1}\right) \end{array} \right] = \frac{2\pi L \sin\psi}{t} \tag{9.22}$$

Similarly, Eq. (9.22) is an implicit function, and the quantitative relationship between r_0 and r_1 can be determined using the Newton iteration method for numerical solving.

In the classical two-parameter analytical model, all possible configurations of the bistable deployable composite boom can be determined by two parameters, namely the cylinder curvature C and the orientation angle ψ in relation to the cylinder centerline [7]. This section proposes an improved two-parameter analytical model, which employs more precise equation for the spatial helix equation to describe all possible configurations of the bistable deployable composite boom. In the improved two-parameter analytical model, the cylinder curvature C is equal to the reciprocal of the radial distance of the shell centerline's projection starting point, i.e., $C = 1/r_0$. Therefore, the selected two parameters are r_0 and ψ.

The curvature at any position of the bistable deployable composite boom in any configuration can be inferred as

$$\begin{cases} \kappa_x = \dfrac{1}{2\rho}(1 - \cos 2\psi) \\ \kappa_y = \dfrac{1}{2\rho}(1 + \cos 2\psi) \\ \kappa_{xy} = \dfrac{\sin 2\psi}{\rho} \end{cases} \tag{9.23}$$

According to Fig. 9.3a, the initial curvatures of the bistable deployable composite boom are given by

$$\begin{cases} \kappa_{x,0} = 0 \\ \kappa_{y,0} = \dfrac{1}{R} \\ \kappa_{xy,0} = 0 \end{cases} \tag{9.24}$$

Therefore, the change in curvature of the bistable deployable composite boom at any configuration can be written as

$$\Delta\kappa = \begin{bmatrix} \dfrac{1}{2\rho}(1-\cos 2\psi) \\ \dfrac{1}{2\rho}(1+\cos 2\psi) - \dfrac{1}{R} \\ \dfrac{\sin 2\psi}{\rho} \end{bmatrix} \tag{9.25}$$

The strain energy per unit area in any configuration is given as

$$u' = \frac{1}{2}\Delta\boldsymbol{\kappa}^{\mathrm{T}}\boldsymbol{D}\Delta\boldsymbol{\kappa} \tag{9.26}$$

By substituting Eq. (9.25) into (9.26), it can be deduced as

$$\begin{aligned} u' &= \frac{D_{11}\sin^4\psi + D_{22}\cos^4\psi + (4D_{66} + 2D_{12})\sin^2\psi\cos^2\psi}{2\rho^2} \\ &\quad - \frac{D_{22}\cos^2\psi + D_{12}\sin^2\psi}{\rho R} + \frac{D_{22}}{R^2} \end{aligned} \tag{9.27}$$

The total strain energy of the bistable deployable composite boom in any configuration can be expressed as

$$U = R\theta\int_0^{\frac{2\pi(r_1-r_0)}{t}} \frac{\rho u'}{\sin\psi}\mathrm{d}\phi \tag{9.28}$$

It is explicit that Eq. (9.26) contains two variables, namely r_0 and ψ, which determine the total strain energy of the bistable deployable composite boom in any configuration. Therefore, the condition for minimizing the strain energy is given by

$$\delta U = \frac{\partial U}{\partial r_0}\delta r_0 + \frac{\partial U}{\partial \psi}\delta\psi = 0 \tag{9.29}$$

Enforcing the following conditions is crucial to satisfy the requirement of Eq. (9.29), and the solutions obtained from these conditions determine the equilibrium configurations, which are as follows:

$$\begin{cases} \dfrac{\partial U}{\partial r_0} = 0 \\ \dfrac{\partial U}{\partial \psi} = 0 \end{cases} \tag{9.30}$$

The stability of the solution can be determined using the stiffness matrix $\boldsymbol{K}$ of the bistable deployable composite boom, as defined in Eq. (9.31):

$$\boldsymbol{K} = \begin{vmatrix} \dfrac{\partial^2 U}{\partial {r_0}^2} & \dfrac{\partial^2 U}{\partial r_0 \partial \psi} \\ \dfrac{\partial^2 U}{\partial r_0 \partial \psi} & \dfrac{\partial^2 U}{\partial \psi^2} \end{vmatrix} \tag{9.31}$$

In particular, if $\boldsymbol{K}$ is positive definite, then the corresponding configuration is considered stable.

When the stiffness matrix satisfies the conditions given in Eq. (9.32), the stiffness matrix is positive definite:

$$\begin{cases} \dfrac{\partial^2 U}{\partial {r_0}^2} > 0, \\ \dfrac{\partial^2 U}{\partial \psi^2} > 0, \\ \dfrac{\partial^2 U}{\partial {r_0}^2} \dfrac{\partial^2 U}{\partial \psi^2} > \left(\dfrac{\partial^2 U}{\partial r_0 \partial \psi} \right)^2 \end{cases} \tag{9.32}$$

It is noteworthy that due to the corresponding relationship between r_0 and r_1 in Eq. (9.22), it is challenging to derive analytical solutions for the total energy and its first- and second-order partial derivatives in Eqs. (9.29)–(9.32). Therefore, numerical methods are employed to solve Eqs. (9.29)–(9.32). Figure 9.3c provides a clear representation of the modeling process for the proposed analytical model 2 using a flowchart.

9.3 Validation Analysis

9.3.1 Bistable Experiments

Similarly, six bistable deployable composite boom samples were fabricated using the vacuum bagging method. Table 8.1 lists the specifications and properties of the T700/epoxy composite materials and samples, while Table 9.1 lists the geometric

Table 9.1 Geometric parameters and stacking sequences of six bistable samples

No.	R (mm)	θ (°)	L (mm)	Stacking sequences
1	20	229	620	[45°/−45°/45°/−45°/45°/90°/−45°/45°/−45°/45°/−45°]
2	20	229	400	[45°/−45°/45°/−45°/0°/45°/−45°/45°/−45°]
3	20	229	200	[45°/−45°/45°/−45°/90°/45°/−45°/45°/−45°]
4	27.5	167	650	[45°/−45°/45°/−45°/0°/90°/45°/−45°/45°/−45°]
5	27.5	167	650	[45°/−45°/45°/−45°/45°/−45°/45°/−45°/45°/−45°]
6	30	191	620	[45°/−45°/45°/−45°/45°/90°/−45°/45°/−45°/45°/−45°]

parameters and stacking sequences of the six bistable samples. Under dry conditions at room temperature, the bistable samples were manually coiled into their folded stable states, as shown in Fig. 9.4a. A camera was used to capture images of the longitudinal cross section of the folded stable state of the bistable samples, obtaining the geometric configuration of the longitudinal cross section. The photos were imported into GetData software to extract the coordinates of 30 data points located on the longitudinal cross section. These data points were then linearly fitted using the Archimedes' helix, and the polar radii and polar angles corresponding to the starting and ending points of the spiral were recorded. The Archimedes' helix can be described by the polynomial shape function in Eq. (8.1).

Using the least squares method to perform linear fitting on these data points, the values a and b can be obtained as follows:

$$\begin{cases} a = \dfrac{\sum_{i=1}^{30}\left(\phi_{e,i} - \overline{\phi_e}\right)\left(\rho_{e,i} - \overline{\rho_e}\right)}{\sum_{i=1}^{30}\left(\phi_{e,i} - \overline{\phi_e}\right)^2} \\ b = \overline{\rho_e} - a\overline{\phi_e} \end{cases} \quad (9.33)$$

where $\overline{\alpha_e}$ and$\overline{\rho_e}$ are averages of 30 data points, respectively, and they can be expressed as

$$\begin{cases} \overline{\alpha_e} = \left(\sum_{i=1}^{30}\alpha_{e,i}\right)/30 \\ \overline{\rho_e} = \left(\sum_{i=1}^{30}\rho_{e,i}\right)/30 \end{cases} \quad (9.34)$$

According to the geometric relationships, the boundary conditions should be satisfied:

$$\begin{cases} r_0 = b \\ r_1 = a\phi_1 + b \end{cases} \quad (9.35)$$

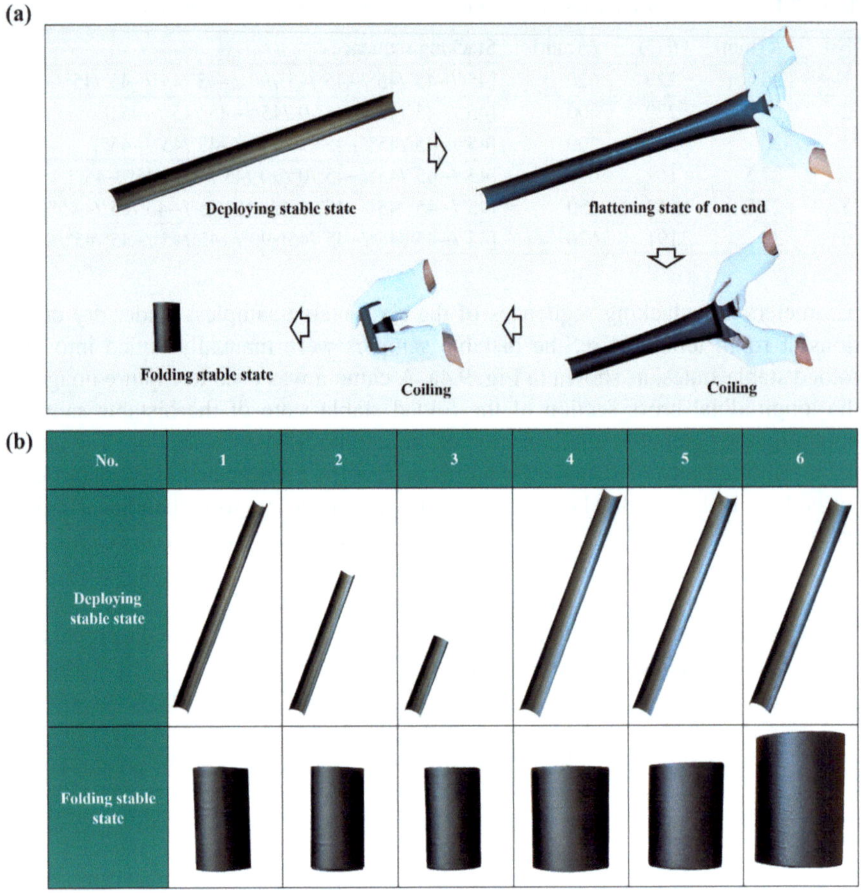

Fig. 9.4 Bistable experiments of the bistable deployable composite boom: (**a**) Bistable deformation process. (**b**) Deploying and folded states of the six samples

Under room temperature and dry conditions, the same method was used to carry out the bistable experiments on the six bistable samples. The deploying and folded stable states are shown in Fig. 9.4b; the results indicate high stability, and no failure occurred throughout the entire testing process.

9.3.2 *Analytical Solutions*

Based on the analytical models 1 and 2 established in Sects. 9.1 and 9.2 for predicting the bistable behavior of the bistable deployable composite boom, the solution procedure was designed with MATLAB software. To verify the effectiveness of the two analytical models proposed, the prediction results were compared with the experiments results from Sect. 9.3.1. Figure 9.5 shows the total strain energy versus polar radius at the start-point curve predicted by analytical model 1 for the six bistable samples. The results indicate that the six bistable shells exhibit different strain energy values for different geometric configurations. Their bistable behavior is closely related to the existence of a local minimum in their total strain energy, and all six bistable samples satisfy the bistability criterion defined by Eq. (9.15).

For analytical model 2, ψ is zero; it indicates that the bistable deployable composite boom is in the deployed stable state. The parameter intervals for r_0 and ψ for the six bistable samples are set as $r_0 \in [15, 45]$ and $\psi \in \left[\frac{\pi}{6}, \frac{2\pi}{3}\right]$. Based on Eq. (9.28), the strain energy contour maps for the six bistable samples in any configuration are calculated, as shown in Fig. 9.6. In each isothermal plot of the bistable samples, a candidate point satisfying Eq. (9.30) and with a positive definite stiffness matrix is found, marked as $V_i(i = 1, 2, 3, 4, 5, 6)$, where the corresponding values ψ are also provided. For example, for the Sample 1, when the first-order derivative in Eq. (9.30) equals zero, r_0 and ψ are found to be 24.78 and $\frac{\pi}{2}$, respectively. Substituting the obtained r_0 and ψ values into Eq. (9.31), the corresponding stiffness matrix $\boldsymbol{K}$ can be expressed as

$$\boldsymbol{K} = \begin{vmatrix} 13.14 & -9.56 \\ -9.56 & 20834.10 \end{vmatrix} \tag{9.36}$$

The stiffness matrix $\boldsymbol{K}$ satisfies the positive definiteness criterion in Eq. (9.32). Using the same method, the positive definiteness of the stiffness matrices corresponding to the other five candidate points was also verified. This indicates that the

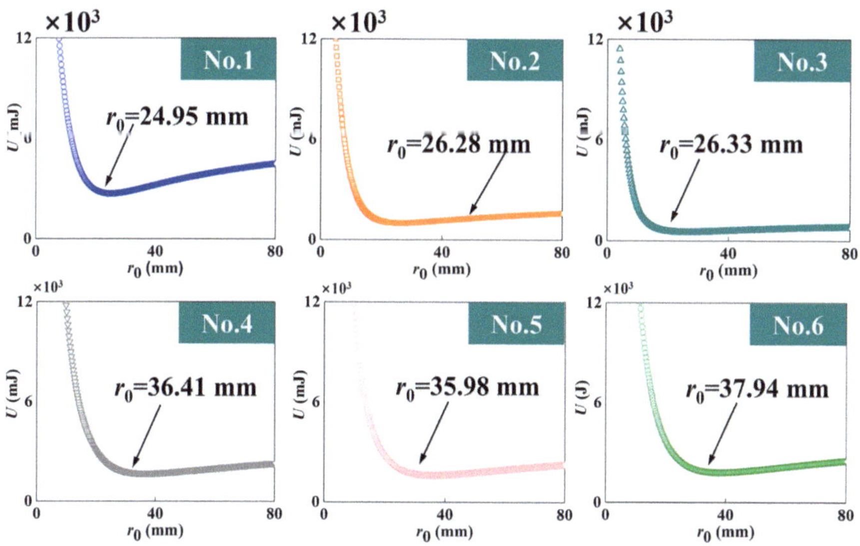

Fig. 9.5 Total strain energy versus polar radius at the starting point curve predicted by analytical model 1 for the six bistable samples

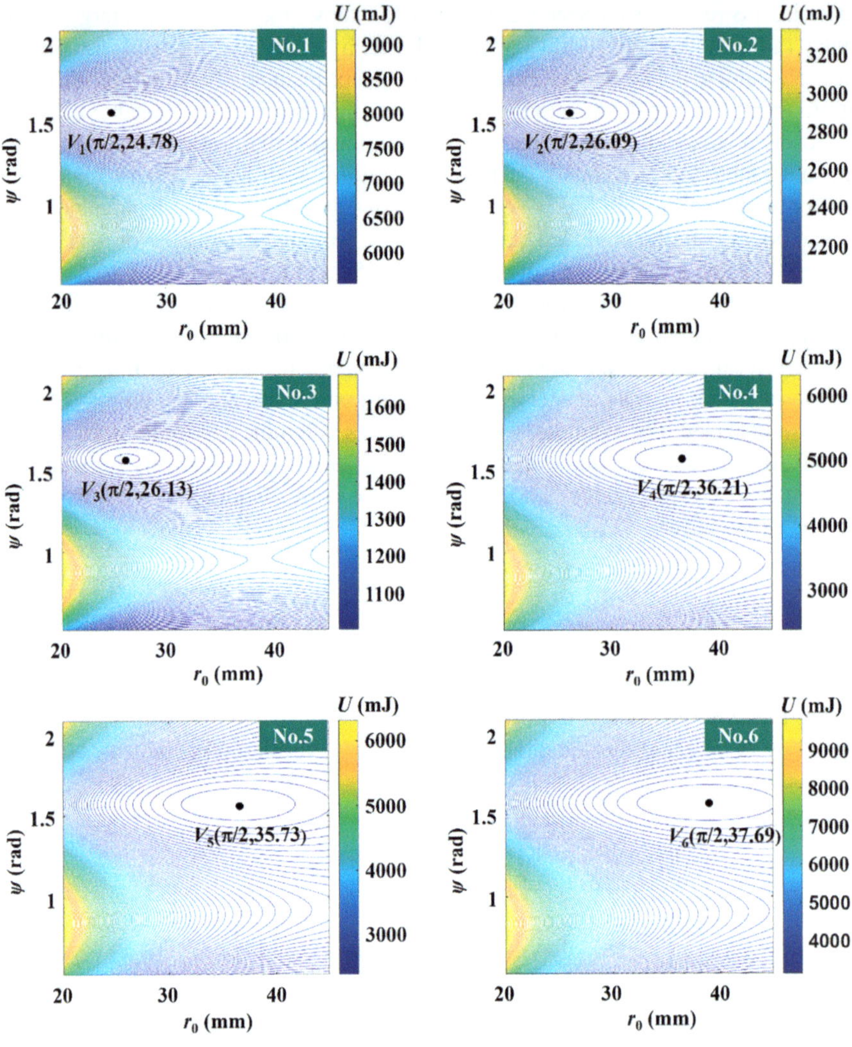

Fig. 9.6 The strain energy contour plots of the six bistable deployable composite cylindrical shell samples predicted by analytical model 2

geometric configurations represented by these candidate points, the folded stable configurations of the six bistable samples, possess sufficient stiffness to ensure their shape and stability under external loads. In conclusion, the geometric configurations corresponding to the V_i points represent the folded stable configurations of the six bistable samples.

9.3.3 Model Validation

The polar radii and polar angles corresponding to the start point and the end point of the Archimedes' helix in the folded stable state, determined by both the experiments and the two analytical models, are listed in Table. 9.2. As seen in Table 9.2, the values of the polar radii and angles at the start point and the end point of the Archimedes' helix, as determined by the three methods, align well. For the six bistable samples, the maximum errors between the experimental results and the predictions from analytical model 1 for the polar radii and angles at the start point and the end point of the Archimedes' helix were 9.42%, 8.82%, and 10.08%, respectively. The maximum errors between the experimental results and the predictions from analytical model 2 were 8.79%, 8.23%, and 9.29%, respectively. Based on the Tsai-Hill criterion and the maximum stress criterion, the maximum failure indices for the six bistable samples in the folded stable state were calculated using the two analytical models, as shown in Fig. 9.7 and Table 9.3. Both failure indices for the folded stable state were below 1, with low error, indicating that all six bistable samples could achieve bistable functionality without failure, which is consistent with the experimental findings.

In summary, the two models established are effective in predicting the folded stable configurations and stress levels of the bistable deployable composite boom, with high predictive accuracy. For a more intuitive comparison of the analytical and experimental results, the geometric and longitudinal cross-sectional configurations of the bistable deployable composite boom in the folded stable state, as determined by both the experiments and analytical model 1, are shown in Fig. 9.8a, b,

Table 9.2 Polar radius and polar angle corresponding to the start and end points of the Archimedes' helix for the six bistable samples in the folded stable state

Geometrical parameters			No. 1	No. 2	No. 3	No. 4	No. 5	No. 6
r_0 (mm)	Experiments		22.63	24.04	24.13	34.08	32.59	37.31
	Analytical model 1	Values	24.95	26.28	26.33	36.41	35.98	37.94
		Error	9.30%	8.52%	8.36%	6.40%	9.42%	1.66%
	Analytical model 2	Values	24.78	26.09	26.13	36.21	35.73	37.69
		Error	8.68%	7.86%	7.65%	5.65%	8.79%	1.01%
r_1 (mm)	Experiments		24.23	24.87	24.66	35.08	33.58	38.41
	Analytical model 1	Values	26.22	26.93	26.65	37.25	36.83	38.79
		Error	7.59%	7.65%	7.47%	5.82%	8.82%	0.98%
	Analytical model 2	Values	26.06	26.74	26.46	37.06	36.59	38.54
		Error	7.02%	6.99%	6.80%	5.34%	8.23%	0.34%
φ_1 (rad)	Experiments		26.46	16.36	8.20	18.80	19.65	16.38
	Analytical model 1	Values	24.23	15.04	7.55	17.65	17.85	16.16
		Error	9.20%	8.78%	8.61%	6.52%	10.08%	1.36%
	Analytical model 2	Values	24.39	15.14	7.59	17.74	17.98	16.26
		Error	8.49%	8.06%	8.04%	5.98%	9.29%	0.74%

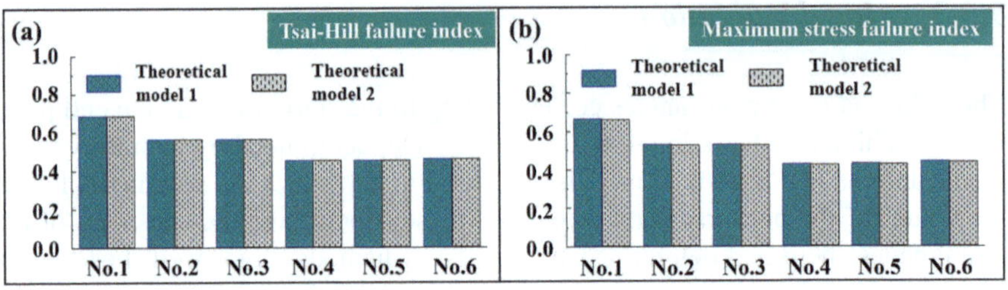

Fig. 9.7 Failure indices of the folded stable state predicted by two analytical models: (**a**) Tsai-Hill failure index. (**b**) Maximum stress failure index

Table 9.3 Failure indices of six bistable samples in the folding stable state

No.	Tsai-Hill failure index		Maximum stress failure index	
	Analytical model 1	Analytical model 2	Analytical model 1	Analytical model 2
1	0.3445	0.3448	0.3311	0.3321
2	0.2810	0.2810	0.2648	0.2656
3	0.2809	0.2810	0.2646	0.2655
4	0.2270	0.2270	0.2133	0.2138
5	0.2271	0.2271	0.2144	0.2150
6	0.2294	0.2295	0.2194	0.2201

respectively. The comparison further shows good prediction accuracy for evaluating folding stable state of the bistable deployable composite boom.

9.4 Results and Discussion

9.4.1 Error Analysis

A comparison between the experimental results and the predictions from the two analytical models for six bistable samples reveals that both models predict higher values for polar radii at the start and end points of the Archimedes' helix than the experimental results, while the predicted polar angles are lower than those observed experimentally. These discrepancies may result from two factors: (1) one of the fundamental assumptions of the analytical models is the uniform distribution of transverse deformations, which may not fully hold in practice, so leading to prediction errors; (2) another important factor is the unavoidable residual thermal stresses generated during the preparation manufacture. These thermal stresses have some impact on the folding stable state of the shells, but are not considered in the existed analytical models. To gain a deeper understanding of these discrepancies and improve the prediction accuracy of the models, further quantitative studies on these two factors will be conducted in future work.

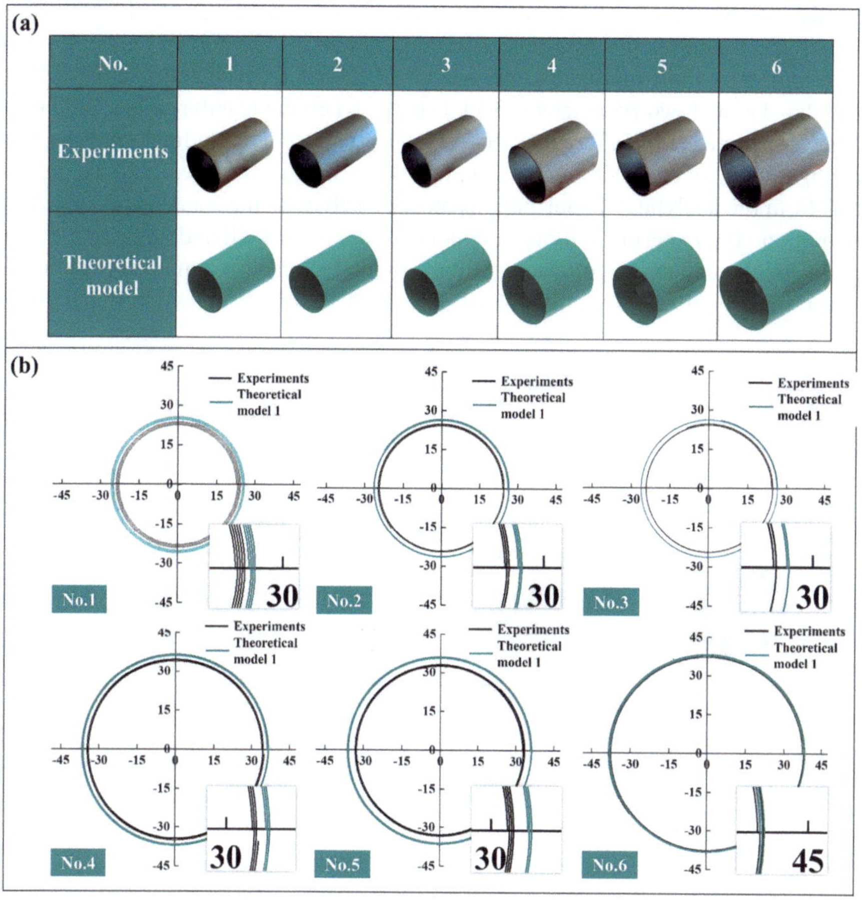

Fig. 9.8 Geometric configuration of the six bistable samples in the folded stable state: (**a**) Overall geometric configuration. (**b**) Longitudinal cross-section configuration

9.4.2 The Value of Engineering Application

As mentioned in Sect. 9.2, the classical two-parameter model [7] provides higher accuracy in predicting the folding stable state configuration for shorter bistable deployable composite booms. However, for longer bistable deployable composite cylindrical shells, the analytical model established in this section offers greater prediction accuracy. Among the six bistable samples, the longest length is only 650 mm, whereas the deployment arm length of solar sails for space applications can exceed 2000 mm, as seen in satellites such as NanoSail-D2, LightSail 1, LightSail 2, and NEA Scout, which carry solar sails. Therefore, a bistable deployable composite boom with a length of 2400 mm, a cross-sectional radius of 12.5 mm, and a cross-sectional angle of 300° was fabricated, and the stacking sequence is [45°/−45°/45° /−45°/0°/90°/45°/−45°/45°/−45°]. Figure 9.9a presents the geometric configurations of the sample in the deployed and folded stable states. Figure 9.9 b–d shows the longitudinal cross-section configurations of the folded stable state as determined by the experiments, the analytical model 1 developed in this section, and the classical two-parameter model. Clearly, when the length reaches 2400 mm, there is a noticeable difference between the predicted longitudinal cross-section configurations of the folded stable state from both analytical models, and the prediction from the analytical model 1 aligns more closely with the actual geometric configuration. The experimental measurements of the polar radii and angle at the start point and the end point, compared with the predicted results from analytical model 1, show errors of 6.87%, 7.12%, and 1.38%, respectively, as shown in Fig. 9.9e. In

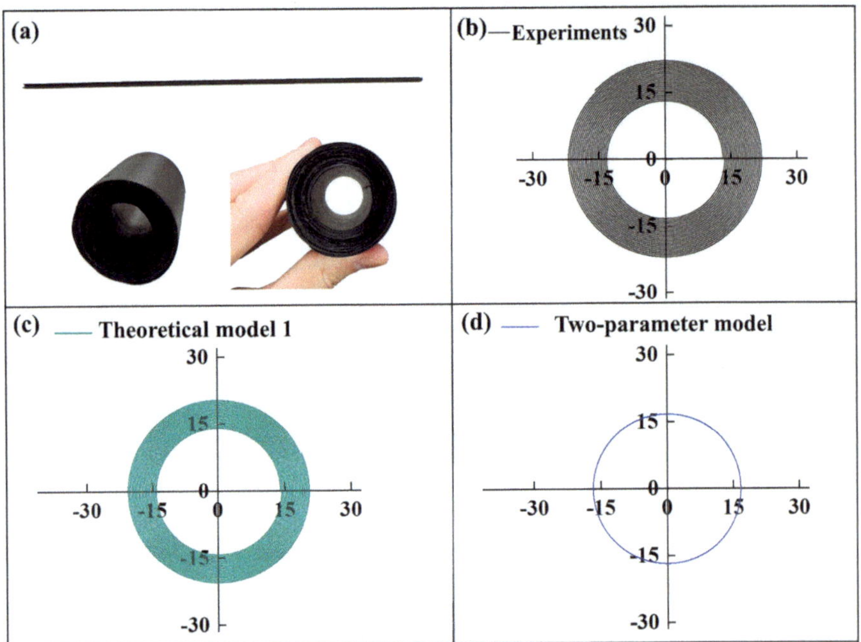

(e)

	r_0 (mm)	r_1 (mm)	ϕ_1 (rad)	$I_{f,1}$	$I_{f,2}$
Experiment	13.10	22.20	140.59	-	-
Theoretical model	14.00	20.62	138.65	0.5115	0.5060
Error	6.87%	7.12%	1.38%	-	-

Fig. 9.9 Effect of length on the folded stable state of bistable deployable composite boom samples: (**a**) Deployed and folded stable states. (**b**) Longitudinal cross-section configuration of the folded stable state determined by experiments. (**c**) Longitudinal cross-section configuration of the folded stable state determined by analytical model 1. (**d**) Longitudinal cross-section configuration of the folded stable state determined by classical two-parameter model. (**e**) Geometric configuration and failure indices determined by experiments and analytical model 1

summary, the proposed analytical model 1 has higher prediction accuracy for evaluating folding stable state of the bistable deployable composite boom, which proves the validity of the analytical model compared to the classical two-parameter model.

9.5 Parametric Study

According to the analytical model, it is known that geometric parameters significantly affect the bistable behavior of the bistable deployable composite boom. In this section, analytical model 1 is used to investigate the impact of cross-sectional radius, thickness, and length on the bistable behavior. The material properties and other geometric parameters are identical to those of No. 4. The specific variable values are as follows:

- The cross-sectional radius is varied as 22.5 mm, 25 mm, 27.5 mm, 30 mm, and 32.5 mm.
- The thickness is varied as 180°, 225°, 270°, 315°, and 360°.
- The length is varied as 250 mm, 450 mm, 650 mm, 850 mm, and 1050 mm.

As shown in Fig. 9.10 a–c, the local minimum of total strain energy exists in bistable deployable composite booms with different cross-section radii. With increase of cross-sectional radius, polar radii r_0 and r_1 at starting point and ending point of Archimedes' helix increase, while polar angle α_1 of Archimedes' helix decreases. Additionally, the failure indices for bistable deployable composite booms with different cross-sectional radii do not reach 1 when they are in the folded stable state, indicating that they can achieve bistable function without failure. As the cross-sectional radius increases, both the Tsai-Hill failure index and the maximum stress failure index decrease significantly. Reason is that change of curvature in the y direction decreases, resulting in decrease of Tsai-Hill failure index and maximum stress failure index.

As shown in Fig. 9.10 d–f, bistable deployable composite cylindrical shells with different thicknesses all exhibit local minima in total strain energy. As the thickness increases, there are no significant changes in the polar radius and polar angle at the start and end points. Additionally, the failure indices for bistable deployable composite booms with different thicknesses do not reach 1 in the folded stable state, indicating that the samples can achieve bistable functional without failure. With increase of thickness, Tsai-Hill failure index and maximum stress failure index increase. This is because the z value increases, resulting in Tsai-Hill failure index and maximum stress failure index increase.

As shown in Fig. 9.10 g–i, bistable deployable composite booms with different lengths all exhibit local minima in total strain energy. As the length increases, the polar radius at the start and end points decreases and increases slowly, respectively, while the polar angle increases significantly. Additionally, for bistable deployable composite booms with different lengths, the failure indices in the folded stable state do not reach 1, indicating that they can achieve bistable function without failure. As

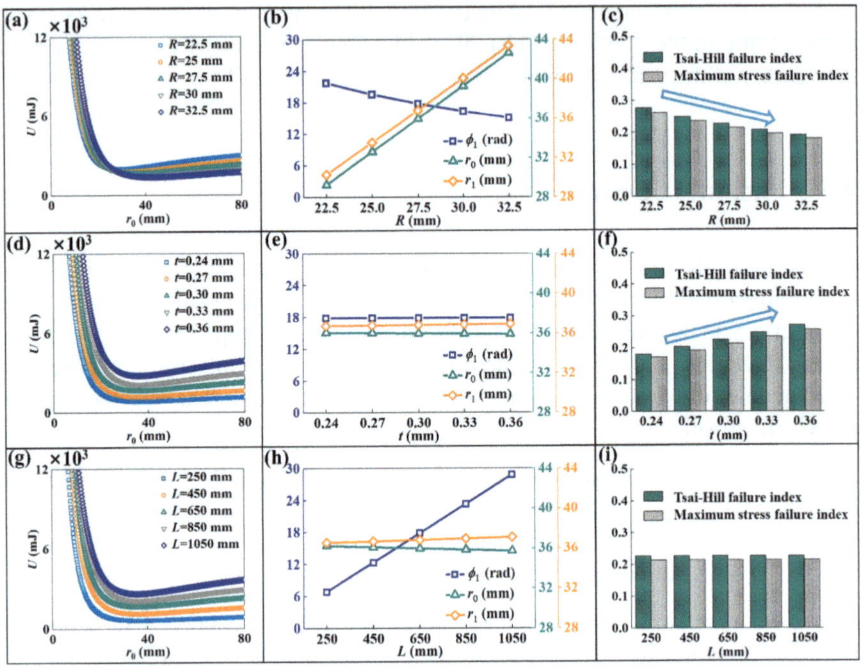

Fig. 9.10 Effect of geometric parameters on the bistable behavior of deployable composite booms: (**a**) Total strain energy versus polar radius at starting point curve with different cross-sectional radii. (**b**) Longitudinal configurations with different cross-sectional radii. (**c**) Failure indices in folded stable state with different cross-sectional radii. (**d**) Total strain energy versus polar radius at starting point curve with different thicknesses. (**e**) Longitudinal configurations in folded stable state with different thicknesses. (**f**) Failure indices in folded stable state with different thicknesses. (**g**) Total strain energy versus polar radius at starting point curve with different lengths. (**h**) Longitudinal configurations in folded stable state with different lengths. (**i**) Failure indices in folded stable state with different lengths

the length increases, Tsai-Hill failure index and maximum stress failure index almost do not change.

References

1. Liu TW, Bai JB, Fantuzzi N (2024) Analytical model for predicting folding stable state of bistable deployable composite boom. Chinese J Aeronaut 37(8):460–469
2. Liu TW, Bai JB, Xi HT et al (2023) Experimental and numerical investigation on folding stable state of bistable deployable composite boom. Comp Struct 320:117178
3. Liu TW, Bai JB, Wang ZZ et al (2023) Determining folding stable state of bistable deployable composite boom: an improved two-parameter analytical model. Acta Astronaut 213:507–515
4. Liu TW, Bai JB, Xi HT et al (2023) A curved bistable composite slit tube for deployable membrane structures. Comp Commun:101648

5. Liu TW, Bai JB, Xi HT et al (2023) Temperature effect on large deformation of bistable deployable composite boom. Comp Struct:117090
6. Liu TW, Bai JB, Fantuzzi N et al (2024) Thin-walled deployable composite structures: a review. Prog Aerosp Sci 146:100985
7. Guest SD, Pellegrino S (2006) Analytical models for bistable cylindrical shells. Proc Royal Soc A: Math Phys Eng Sci:839–854

Chapter 10
Numerical Simulations for Deployable Composite Structures

Abstract This chapter presents numerical simulations of large-deformation flexible composite structures, using geometrically nonlinear explicit FEM for deployable booms. For lenticular booms, explicit solvers handle contact in tension/compression/coiling, with long-step analyses matching experiments and negligible kinetic energy. Tubular boom models predict folding moment-displacement curves consistent with tests, with failure indices less than 1 confirming elastic folding. Bistable boom models match folded configurations, with max failure index 0.37, errors from material nonlinearity and residual thermal stress. Explicit FEM effectively addresses large-deformation contact problems.
It is highly complex for exploring the mechanisms of the folding and deployment of large-deformation flexible composite structures, involving significant nonlinearity characteristics, precision control strategies, intricate contact interactions, and complex boundary constraints. The authors of this book [1–16] have conducted extensive research on analytical modeling, experimental preparation, and numerical simulation of such structures, achieving notable effectiveness in practical engineering applications. This chapter focuses on the deployable composite booms to establish accurate finite element models, predict the folding behavior, and reveal the deformation mechanisms under large deformations, thereby providing a guidance for the deformation control strategies.

10.1 The Lenticular Deployable Composite Boom

In the folding deformation of the lenticular deployable composite boom, significant geometric configuration changes and complex interactions between different components often lead to the singularity of the stiffness matrix, resulting in convergence issues that are challenging to resolve in dynamic analyses. However, such convergence problems can be addressed using the explicit solver in the commercial finite element software ABAQUS. Accordingly, geometric nonlinear explicit dynamic

J.-B. Bai et al., *Folding Mechanics of Deployable Composite Structures*, Solid Mechanics and Its Applications 283,
https://doi.org/10.1007/978-3-031-99254-4_10

analysis was conducted to simulate the folding deformation process of the lenticular deployable composite boom.

10.1.1 Tensile/Compressive Finite Element Analysis

Based on the geometry and dimension of the lenticular deployable composite boom, a three-dimensional finite element model (FEM) was developed to simulate the deformation process under tensile and compression loading (as shown in Fig. 10.1a). In order to save computation time, the length of 10 mm, rather than the length of 65 mm, is chosen for the explicit analysis on tensile and compressive behavior considering structural symmetry, loading type, and boundary condition. Explicit analyses of the tensile and compressive behavior were performed using this simplified model. In both finite element models, the layered shell element S4R is utilized, and the total number of elements is 480. Under compressive loading, rigid plates were modeled using the same type of elements as the boom, with a total of 144 elements (as shown in Fig. 10.1b, c). The contact type between the lenticular deployable composite boom and the rigid plate was defined as general contact. The mechanical properties of the materials in the finite element model are listed in Table 7.1.

From Fig. 10.1d, e, it is evident that the lenticular deployable composite boom gradually stretches until completely flattening with an increase in the loading of tensile or compression. To validate the effectiveness of the finite element model, tensile and compressive experiments were conducted using an INSTRON servo-hydraulic testing machine. Detailed procedures and results of these experiments can be found in Chap. 7. In the tensile and compressive experiments, the lenticular deployable composite boom samples exhibited a flattening deformation state under increasing tensile loads and recovered their original shape upon unloading, consistent with the results observed from finite element analysis (FEA).

From Fig. 10.2, it can be observed that the load versus displacement curves predicted by standard analysis and geometrically nonlinear explicit analysis match well with experimental results when the given step time is greater than 0.05 s. Therefore, under sufficiently long step times, both standard analysis and explicit analysis can effectively simulate the quasi-static tensile and compressive deformation of the lenticular deployable composite boom: (1) The strain energy predicted by the standard analysis almost coincides with that predicted by explicit analysis (as shown in Fig. 10.3), whereas the kinetic energies predicted from nonlinear explicit analysis are much larger for step times of 0.01 s and 0.02 s than for other step times. This is because excessively small step times in explicit analysis result in pronounced dynamic effects, appreciably deviating from quasi-static tensile or compression deformation in tests. (2) All load versus displacement curves exhibit nonlinear behavior, the nonlinearity of these curves can be attributed to the fact that bending stress plays a dominant role within approximately linear range, whereas there are

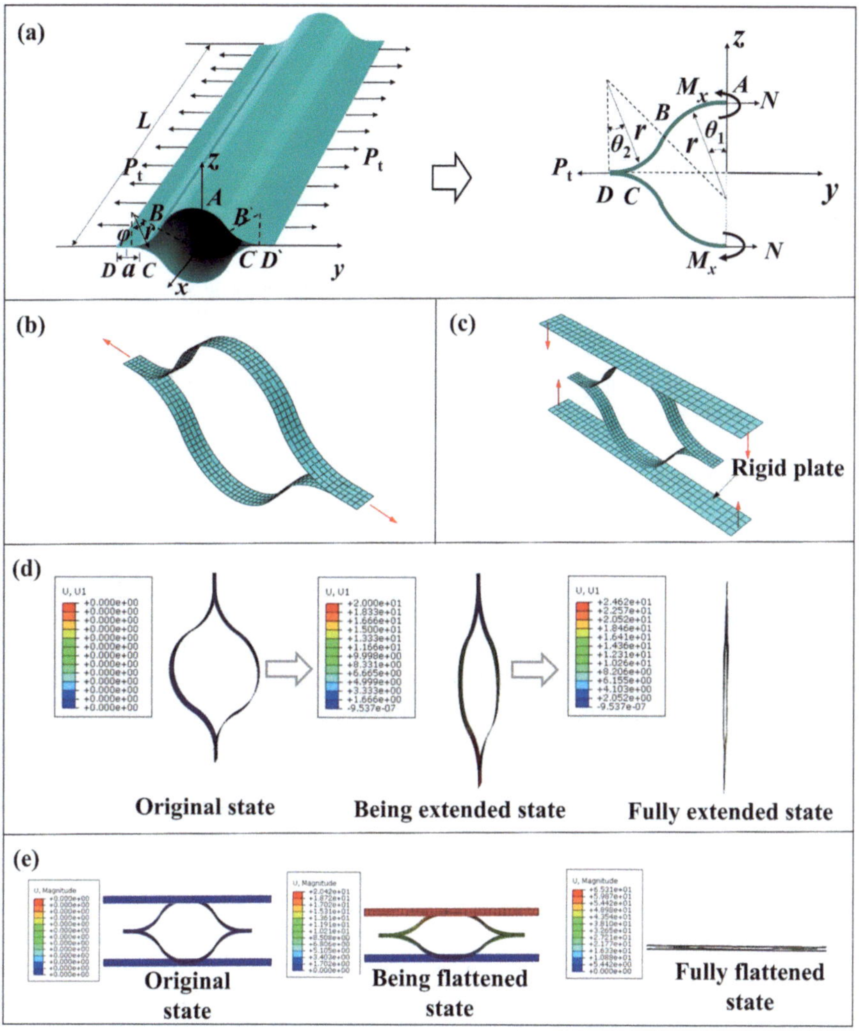

Fig. 10.1 Numerical simulations of tensile and compressive behavior of the lenticular deployable composite boom: (**a**) Geometric model. (**b**) Finite element model for tension. (**c**) Finite element model for compression. (**d**) Tensile process. (**e**) Compressive process

both bending and membrane stresses within nonlinear range. (3) The normal stress and shear stress along the cross-section path of the flattened booms are minimal, indicating that the lenticular deployable composite boom can stretch and recover elastically. (4) In geometric nonlinear explicit analysis, CPU time increases with an increase in the step time.

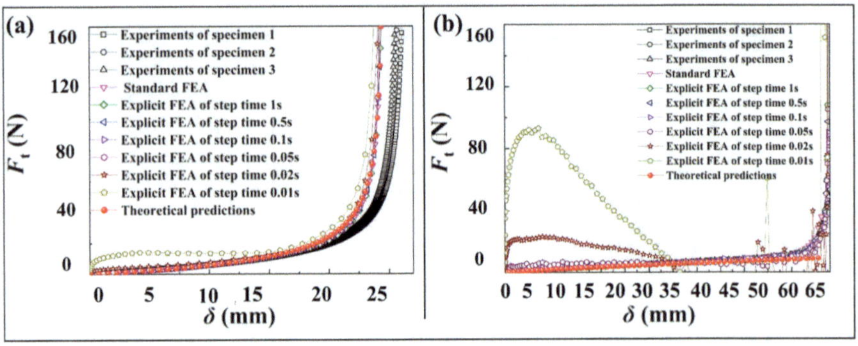

Fig. 10.2 Model validation of tensile and compressive deformation: (**a**) Tensile process. (**b**) Compressive process

10.1.2 Finite Element Model

Due to the complexity of contact modeling between the lenticular deployable composite boom and both the central driving roller and guide shaft during folding deformation, standard analysis is not appropriate and valid for folding deformation analysis. However, explicit analysis is still available to deal with the contact modeling. Considering the geometry and dimension of the lenticular deployable composite boom and the experimental device, a three-dimensional geometric nonlinear explicit finite element model was developed by using the ABAQUS code, as shown in Fig. 10.4a.

Due to the symmetry features, a half-model with symmetric boundary conditions was adopted to reduce the computational time. Shell element S4R was used to model the lenticular deployable composite boom, central driving roller, and guide shaft, with a total of 9750, 2200, and 550 elements, respectively, as well as the mechanical properties of material were consistent with those used previously. The contacts between the lenticular deployable composite boom and the central driving roller, and between the lenticular deployable composite boom and the guild shafts, are defined as the general ones in the nonlinear explicit finite element model, and the kinematic coupling constraint between the central driving roller and the lenticular deployable composite boom is defined as the hinged support constraint to make the lenticular deployable composite boom stretching and coiling along the central driving roller. The folding deformation of the lenticular deployable composite boom in the finite element model was divided into two steps: (1) a tensile deformation at one end until completely flattened and (2) a coiling deformation of the fully flattened boom. Based on the aforementioned analysis on step time, the step times for these two steps were set to 1 s and 2 s, respectively.

From Fig. 10.4, the following observations can be made: (1) The lenticular deployable composite boom is stably folded along the central driving roller, which is correlated well with the experiments. (2) The strain energy of the lenticular deployable composite boom increases with the tensile and rotational displacements.

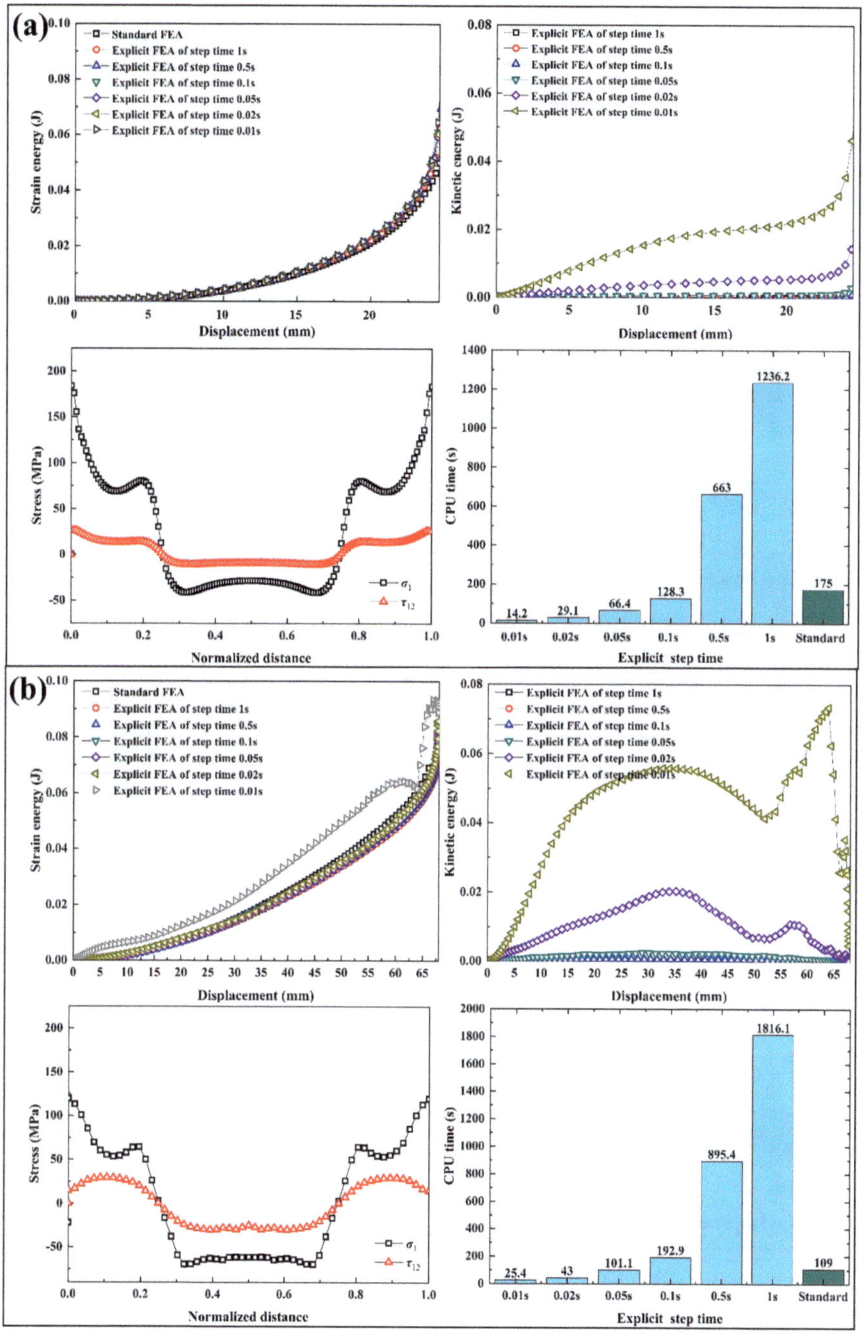

Fig. 10.3 Numerical simulation analysis of tensile and compression processes: (**a**) Tensile deformation. (**b**) Compression deformation

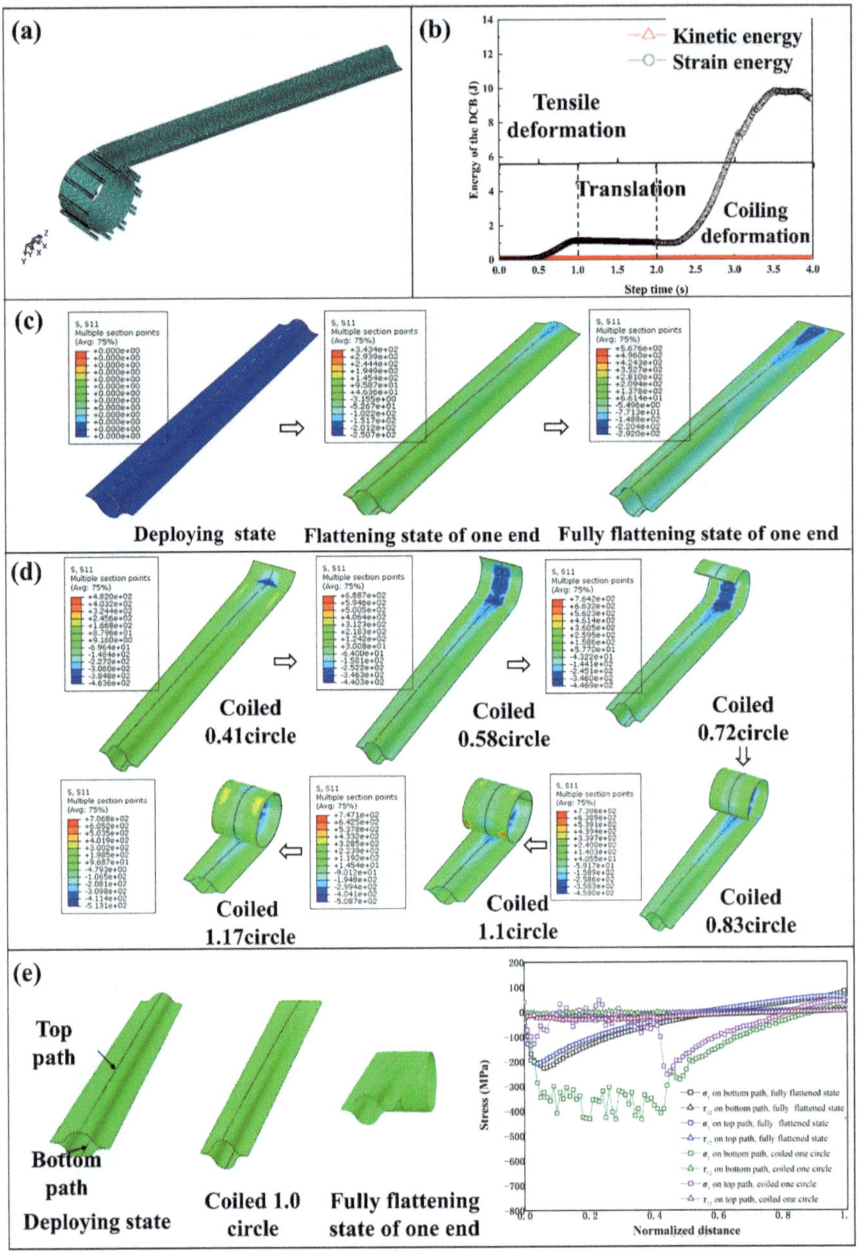

Fig. 10.4 Numerical simulation of folding deformation of the lenticular deployable composite boom: (**a**) Finite element model. (**b**) The change of energy variation. (**c**) Flattening process. (**d**) Coiling process. (**e**) Stress distribution during folding deformation

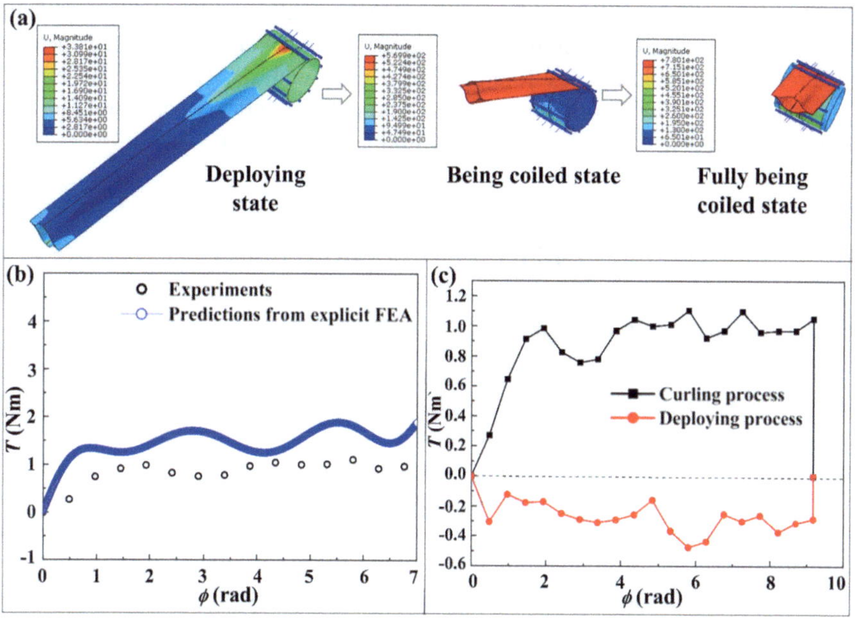

Fig. 10.5 Analysis of coiling deformation of the lenticular deployable composite boom: (**a**) Coiling process. (**b**) Comparison validation. (**c**) Folding moment and torsional displacement curve

Compared to the strain energy, the kinetic energy predicted by the finite element model is negligible as shown in Fig. 10.4b. (3) As shown in Fig. 10.4c, d, during the tensile deformation process, the maximum compressive stress and tensile stress separately appear on the internal surface and external surface (as shown in Fig. 10.4e). (4) Compared to the compressive and tensile stresses, the shear stress throughout the entire folding deformation is negligibly small. Furthermore, the maximum stress of the composite boom during the whole folding deformation process is much less than the ultimate strengths of material, enabling elastic folding and deployment. The finite element simulation results of the folding process, compared with experimental results as shown in Fig. 10.5, further confirm the accuracy of the finite element model.

10.1.3 Results and Discussion

The following conclusions can be drawn from the comparative analysis:

1. At the given short step time, the geometric nonlinear explicit method inevitably introduces some computational errors due to dynamic effects. However, the explicit finite element method appears to be more effective than standard analysis methods in addressing the instability issues in the solution during quasi-static large folding deformation analysis.

2. The geometric nonlinear explicit finite element analysis method and model are effective for tensile or compressive large-deformation analysis, but they are not directly applicable to large folding deformation analysis. This is because there are significant differences in geometric and dynamic effects between large tensile (or compressive) deformation and folding deformation.

10.2 The Tubular Deployable Composite Boom

10.2.1 Finite Element Model

In fact, in the equal-sense and opposite-sense folding processes of the tubular deployable composite boom, significant geometric changes and complicated contact interactions among various components were observed (i.e., the central driving roller, the constraint shaft); geometric nonlinear explicit finite element analysis was selected to address these complex contact modeling issues, as previously discussed. Utilizing the geometric configuration and dimension of the tubular deployable composite boom and the folding mechanism, a three-dimensional geometric nonlinear explicit finite element model of the tubular deployable composite boom was established based on its geometry and specifications, as well as those of the folding mechanism, as shown in Fig. 10.6.

Similarly, S4R shell elements were used to model the tubular deployable composite boom, the central driving roller, and the guide shafts. A mesh sensitivity analysis was conducted to investigate the influence of mesh size on the simulation of the folding behavior of the tubular deployable composite boom. In the finite element model, the mesh sizes for the tubular deployable composite boom, central driving roller, and constraint shafts were set to 5 mm, 7 mm, and 2 mm, respectively.

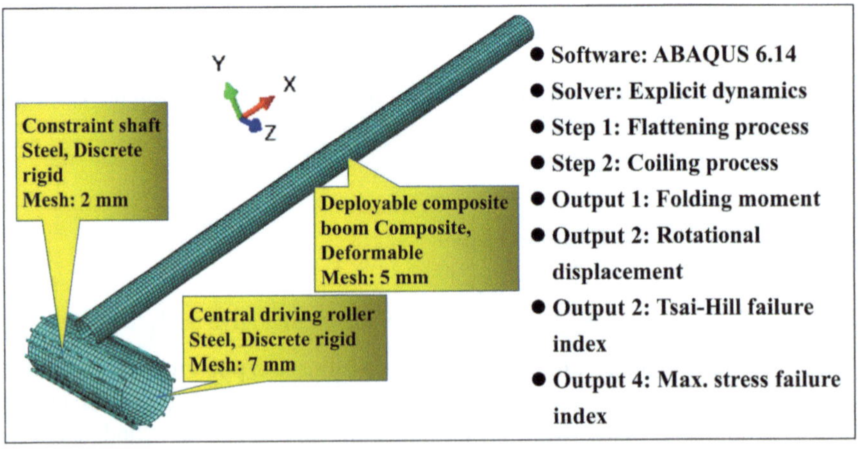

Fig. 10.6 Numerical simulation for the tubular deployable composite boom

The mechanical properties of the tubular deployable composite boom were adopted as specified in Chap. 8. The central driving roller and constraint shafts were assigned isotropic material properties with an elastic modulus of 20GPa, Poisson's ratio of 0.3, and density of 7.80×10^{-3} g/mm^3. To improve computational efficiency, the central driving roller and guide shafts were modeled as discrete rigid bodies. In the geometric nonlinear explicit finite element model, general contact was defined between the tubular deployable composite boom and both the central driving roller and several constraint shafts. The motion coupling constraint between the central driving roller and the tubular deployable composite boom was defined as a hinge support constraint, allowing the boom to stretch and fold along the central driving roller.

The folding deformation process of the tubular deployable composite boom was divided into two steps: (1) flattening process: tensile deformation of one end of the tubular deployable composite boom until it was completely flattened and (2) coiling process: coiling the completely flattened tubular deployable composite boom from one end. Based on the analysis of step time, the time for these two steps was set to 1 s and 15 s, respectively. The finite element model provided four outputs: the folding moment versus displacement curve, Tsai-Hill failure index, Tsai-Wu failure index, and maximum stress failure index, as shown in Fig. 10.6.

10.2.2 Coiling Analysis

The equal-sense and opposite-sense folding deformation processes of sample 1 are shown in Figs. 10.7 and 10.8. The analysis revealed the following: (1) Stable folding: Sample 1 folded equal-sense and opposite-sense along the central driving roller in a stable manner, consistent with the experimental results. (2) Energy analysis: The strain energy of sample 1 increased with tensile and rotational displacements during both equal-sense and opposite-sense folding. Compared to strain energy, the kinetic energy predicted by the finite element analysis was negligible, further validating the accuracy of the established finite element model. (3) Failure index: The maximums of Tsai-Hill failure index for sample 1 during the entire equal-sense and opposite-sense folding processes remained below 1, indicating that the sample was capable of elastic folding and deploying. The same numerical simulation method was applied to the folding deformation processes of the remaining five samples, and the results confirmed the reliability of the established model.

To further validate the numerical simulations, folding experiments were conducted on the tubular deployable composite boom samples under room temperature and dry conditions. During the coiling tests using the folding mechanism, the rotational speed was carefully controlled to prevent excessive fluctuations in the measured folding moment versus torsional displacement curves. Figures 10.9 and 10.10 illustrate the equal-sense and opposite-sense folding processes of sample 1 of the tubular deployable composite booms. For each test, one end of the sample was first flattened and fixed tangentially to the central driving roller. The coiling deformation

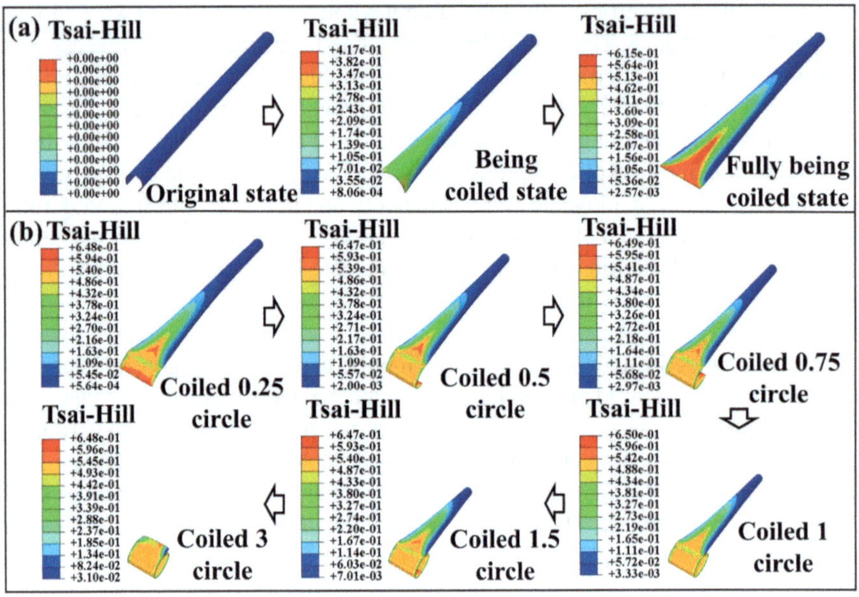

Fig. 10.7 Equal-sense folding process of the tubular deployable composite boom: (**a**) Flattening process. (**b**) Coiling process

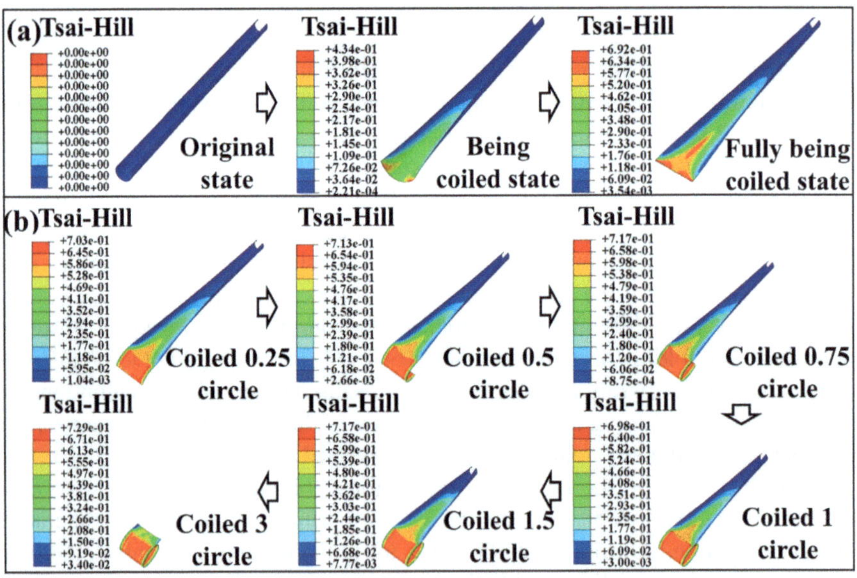

Fig. 10.8 Opposite-sense folding process of the tubular deployable composite boom: (**a**) Flattening process. (**b**) Coiling process

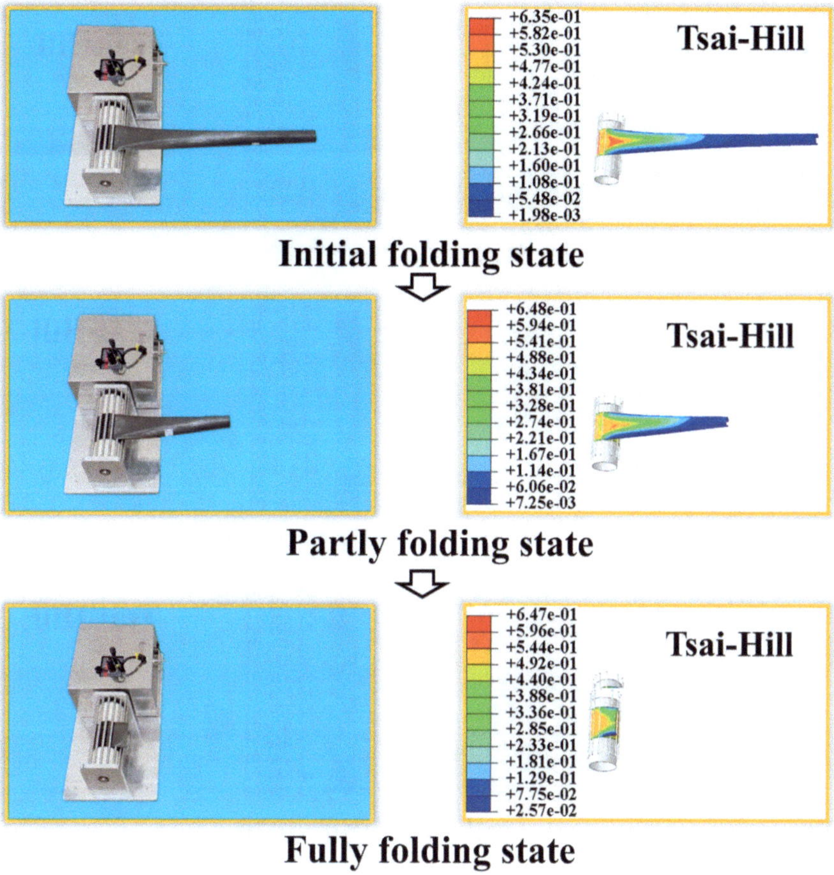

Fig. 10.9 Experimental and numerical simulation validation of equal-sense folding deformation of sample 1

was then initiated until the sample was fully folded. The folding mechanism recorded the folding moment and torsional displacement throughout the deformation process. Sample 1 of the tubular deployable composite booms experienced no damage during the entire folding process and was able to recover its original shape after unloading, satisfying the requirement for repeated use.

Three methods were used to confirm that no damage occurred to sample 1 during the loading and unloading processes: (1) Acoustic monitoring: no sound indicating damage or failure was detected during folding and deploying. (2) Visual inspection: surface smoothness was checked using a flashlight and magnifying glass to identify any cracks or edge delamination. (3) Tap test: internal delamination issues were assessed through tapping tests. The same procedure was applied to the remaining five samples of the tubular deployable composite booms. The folding mechanism recorded the folding moment and torsional displacement for each test, and the three assessment methods were employed to analyze potential damage or failure during equal-sense and opposite-sense folding processes. The results showed that all six

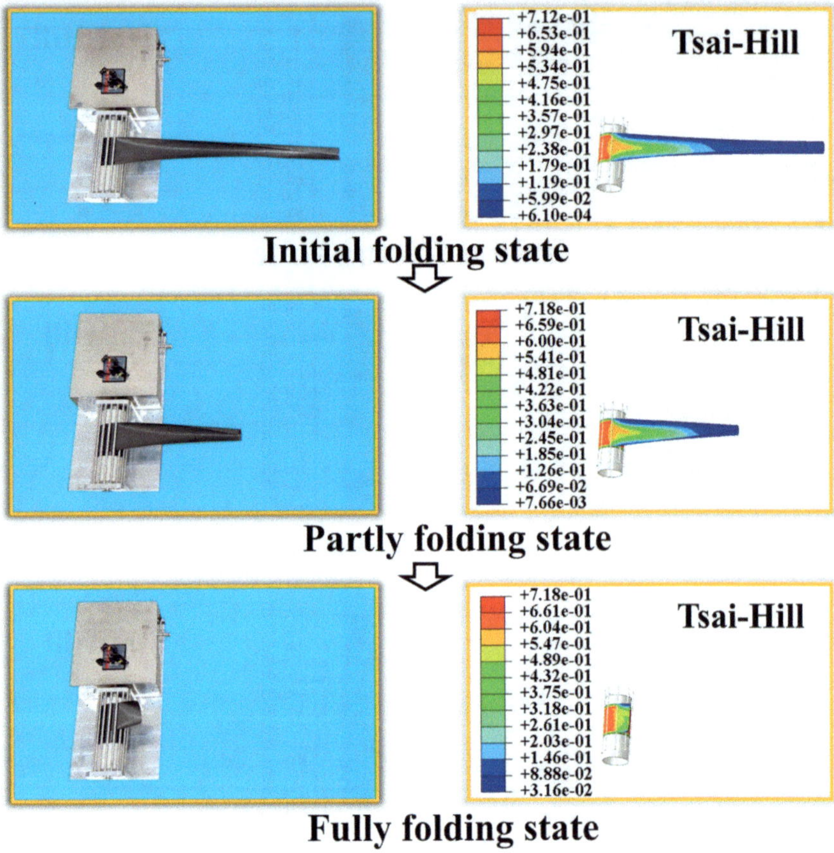

Fig. 10.10 Experimental and numerical simulation validation of opposite-sense folding deformation of sample 1

samples successfully achieved multiple folding and deploying cycles without any failure.

10.2.3 Results and Discussion

Figure 10.11 illustrates the folding moment versus displacement curves for the six tubular deployable composite boom samples during equal-sense and opposite-sense folding processes. It shows that the folding moments measured by the experiments keep a relatively stable level, which indicates that the equal-sense and opposite-sense folding processes are performed stably. Additionally, the folding deformation processes and folding moment versus rotational displacement curves predicted by numerical simulation have a good correlation with those of the experimental results.

Figure 10.12 and Table 10.1 present the maximum failure indices of each ply for the six tubular deployable composite boom samples during the folding deformation process, calculated using three strength criteria: the Tsai-Hill criterion, the Tsai-Wu criterion, and the maximum stress criterion. From Fig. 10.12 and Table 10.1, it can

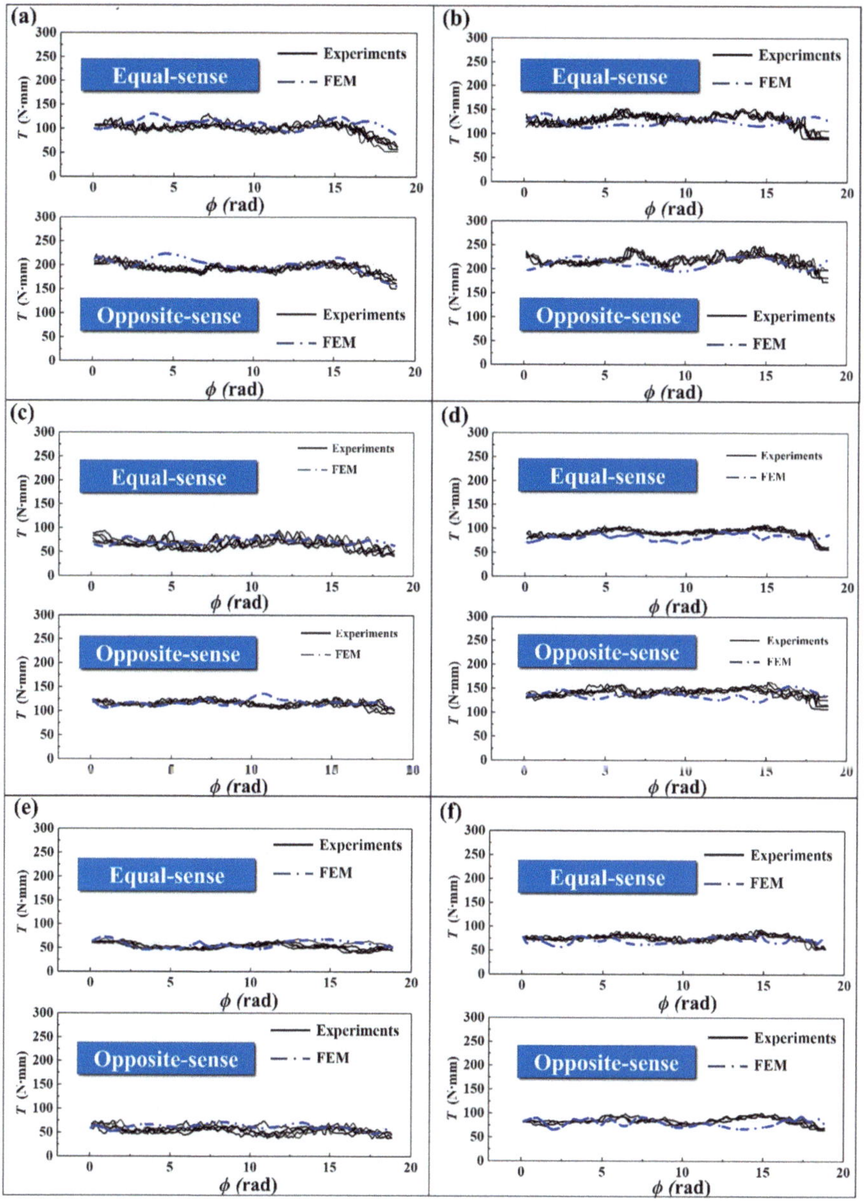

Fig. 10.11 Folding moment versus rotational displacement curves of six samples in the folding processes: (**a**) No. 1. (**b**) No. 2. (**c**) No. 3. (**d**) No. 4. (**e**) No. 5. (**f**) No. 6

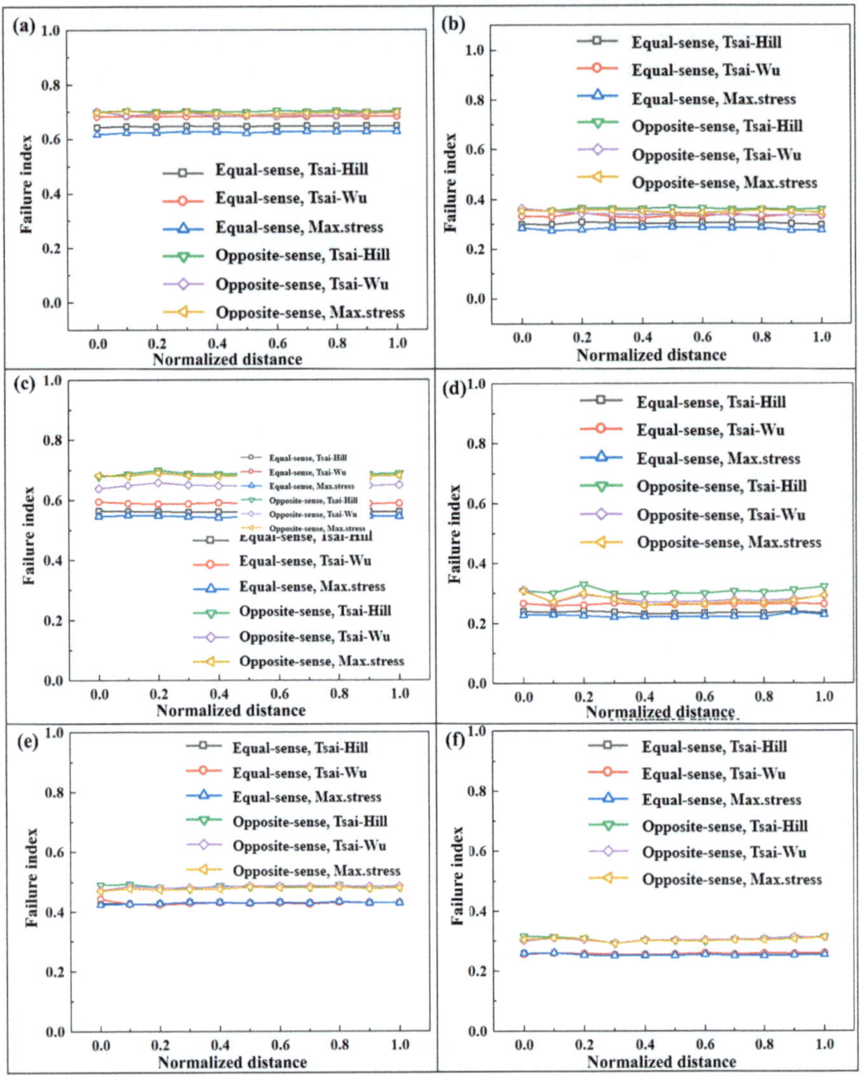

Fig. 10.12 The maximum failure indices during the folding process of six samples: (**a**) No. 1. (**b**) No. 2. (**c**) No. 3. (**d**) No. 4. (**e**) No. 5. (**f**) No. 6

be observed that the maximum failure index among all criteria is only 0.7. It indicates that all tubular deployable composite boom samples can realize the folding and deploying functions without failure, which is consistent with the experimental results. In summary, the finite element model established can effectively predict the folding behavior of the tubular deployable composite boom with high prediction accuracy.

Table 10.1 Maximum failure indices of six samples during the folding process

No.	Tsai-Hill failure index		Tsai-Wu failure index		The maximum stress failure index	
	Equal-sense	Opposite-sense	Equal-sense	Opposite-sense	Equal-sense	Opposite-sense
1	0.6488	0.7064	0.6844	0.7033	0.6292	0.7036
2	0.3068	0.3679	0.3478	0.3657	0.2894	0.3585
3	0.5653	0.6875	0.5959	0.6499	0.5498	0.6919
4	0.2422	0.3128	0.2681	0.3122	0.2299	0.3092
5	0.4327	0.4926	0.4441	0.4885	0.4346	0.4834
6	0.2611	0.3153	0.2601	0.3152	0.2609	0.3115

10.3 The Bistable Deployable Composite Boom

In this section, two nonlinear explicit finite element models were established to predict the bistable behavior of the bistable deployable composite boom. The bistable behavior, such as the cross-sectional geometric configuration and failure indices when in the folded stable state, was determined and analyzed using experimental methods and the two finite element models.

10.3.1 Finite Element Model 1

Based on the geometric configuration and dimension of the bistable deployable composite boom, a three-dimensional geometric nonlinear explicit finite element model 1 was created, as shown in Fig. 10.13a. Shell element S4R was used to model, and the effect of mesh size independence was investigated. Considering both computational accuracy and efficiency, the mesh sizes for the tubular bistable deployable composite boom and the roller were set to 5 mm and 5 mm, respectively, as shown in Fig. 10.13b.

The bistable deployable composite boom was fabricated using T700/epoxy composite ply, with its basic mechanical properties outlined in Chap. 9. The roller was set as the material properties of steel, with an elastic modulus of 200GPa, Poisson's ratio of 0.3, and a density of 7.80×10^{-3} g/mm^3. To improve computational efficiency, the roller was treated as a discrete rigid body. In the finite element model 1, the contact between the bistable deployable composite boom and the roller was defined as general contact. To allow the bistable deployable composite boom to quickly reach the folded stable state, the friction coefficient for self-contact was set to 0.05. The bistable deformation process of the tubular bistable deployable composite boom was divided into three steps, as follows:

In step 1, one end of the tubular bistable deployable composite boom is stretched until it is fully flattened. The time for step 1 is set to 0.5 s. Displacement is applied to both sides AB and CD (as shown in Fig. 10.13c), with directions along the

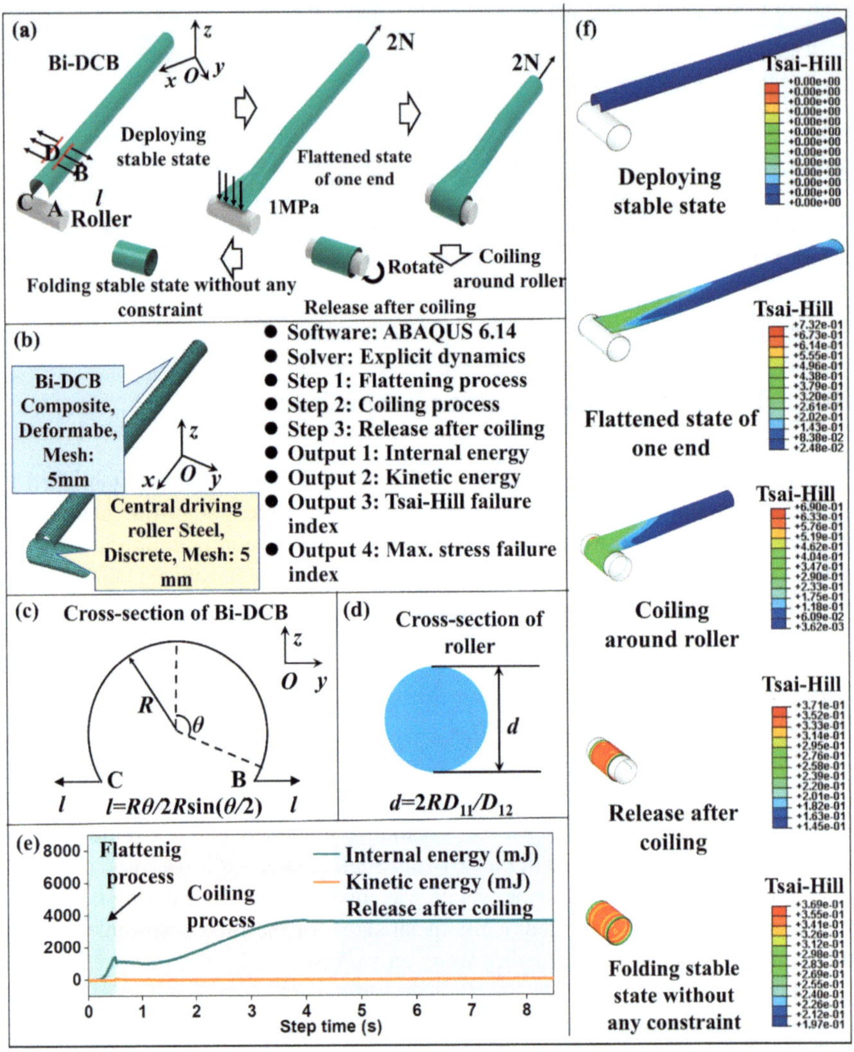

Fig. 10.13 Finite element model 1 of the bistable deployable composite boom sample 1: (**a**) Numerical simulation scheme. (**b**) Mesh size. (**c**) Applied displacement. (**d**) The roller diameter. (**e**) Energy changes in the three steps. (**f**) Numerical simulation of bistable deformation process

positive and negative directions of *y*-axis, respectively. The applied displacement l can be calculated using the initial geometry of the cross section of the bistable deployable composite boom, as follows:

$$l = \frac{R\theta}{2} - R\sin\frac{\theta}{2} \tag{10.1}$$

In step 2, the folding process of the bistable deployable composite boom along the roller is simulated. A pressure of 1 MPa was applied to the flattened region of the boom, while a torsional displacement is applied to the roller to drive the boom's folding motion. Additionally, an axial load of 2 N is applied to the other end of the bistable deployable composite boom. To minimize the kinetic energy during the folding process, the time for step 2 is set to 4 s.

In step 3, the folding deformation process of the bistable deployable composite boom is simulated after it has completely coiled. It is very important to select the roller diameter. A roller diameter that is too large or too small will significantly affect whether the boom reaches the folded stable state in step 3. Therefore, the closer the roller diameter is to the diameter of the folding stable state, the easier it is for the bistable deployable composite boom to reach the folding stable state. Iqbal et al. [17] established an analytical model to predict the diameter of the folding stable state, which is in good agreement with experimental results. In finite element model 1, the roller diameter is set to the folded stable diameter predicted by the above analytical model (as shown in Fig. 10.13d). The diameter of the folding stable state can be derived as

$$d = \frac{2RD_{11}}{D_{12}} \tag{10.2}$$

Step 3 is a stability analysis step where all constraints and boundary conditions are removed, allowing the bistable deployable composite boom to reach its folded stable state. The time for step 3 is set to 4 s. The self-contact of the bistable deployable composite boom is defined as general contact, in which the normal contact was set as hard contact and the tangential contact was set as smooth contact. The loading assignments of all three steps were set to table type. In this case, the Tsai-Hill criterion and the maximum stress criterion are used for failure assessment, with four output results set, including internal energy, kinetic energy, Tsai-Hill failure index, and maximum stress failure index.

As shown in Fig. 10.13e, the internal energy in the bistable deployable composite boom sample 1 increases with the increase in both the tensile and rotational displacements. In step 3, the internal energy does not change, indicating that the sample has reached the folded stable state. Compared to the internal energy, the kinetic energy predicted by the finite element analysis is very small and can be neglected. The bistable deformation process of the bistable deployable composite boom sample 1, as predicted by finite element model 1, is shown in Fig. 10.13f.

From Fig. 10.13f, it can be clearly observed that (1) sample 1 can fold stably along the roller and ultimately reach the folded stable state and (2) the Tsai-Hill failure index of sample 1 is always less than 1 throughout the bistable deformation process. Therefore, sample 1 can achieve bistable functionality. Using the same method, the bistable deformation processes of the other five samples were simulated and calculated, yielding analysis results similar to those of sample 1. Consistent with the experimental results, the coordinates of the data points from the numerical simulation were linearly fitted using an Archimedes' helix, and the corresponding radial distance and spiral angle at the start and end points of the Archimedean spiral were recorded.

10.3.2 Finite Element Model 2

Based on the geometric configuration and dimension of the bistable deployable composite boom sample, a three-dimensional geometric nonlinear explicit finite element model 2 was established (as shown in Fig. 10.14a), with the same procedure as above. The mesh sizes for the bistable deployable composite boom and the rigid plate were set to 5 mm and 10 mm, respectively, to achieve stable results (as shown in Fig. 10.14b). Similar to finite element model 1, the bistable deployable composite boom and the rigid plate in finite element model 2 were set as material properties of T700/epoxy composite ply and steel, respectively. The bistable deformation process of the bistable deployable composite boom in finite element model 2 was divided into three steps, as follows:

In step 1, one end of the bistable deployable composite boom was stretched until it was completely flattened, and the time of step 1 was set to 2 s. Displacements were applied to two side edges EF and GH, in the positive and negative directions of y-axis, respectively. Similar to finite element model 1, the applied displacement l can be determined based on the initial geometry y of the cross section of the bistable deployable composite boom (as shown in Fig. 10.14c).

Step 2 is used to simulate the folding process of the bistable deployable composite boom. To minimize the kinetic energy during the folding process, the time for analysis step 2 was set to 15 s. The contact between the bistable deployable composite boom and the rigid plate is defined as general contact, with normal contact set as hard contact and tangential contact set as smooth contact. A displacement loading method is employed, using the rigid plate to assist the folding of the boom (as shown in Fig. 10.14d). The displacement of the rigid plate in the x direction can be expressed as

$$L^* = L - \frac{RD_{11}}{D_{12}} \tag{10.3}$$

Step 3 is the stability step, where constraints and all boundary conditions were removed to allow the bistable deployable composite boom to reach the folded stable state. The time for step 3 is set to 2 s. The loading assignments for all three analysis steps are set as table type. Internal energy, kinetic energy, Tsai-Hill failure index, and maximum stress failure index are shown in Fig. 10.14b.

As shown in Fig. 10.14e, the internal energy in the bistable deployable composite boom sample 1 increases with the increase in the stretching displacement and the displacement of the rigid plate. In step 3, the internal energy no longer changes, indicating that the boom has reached the folded stable state. Compared to the internal energy, the kinetic energy predicted by the finite element analysis is very small

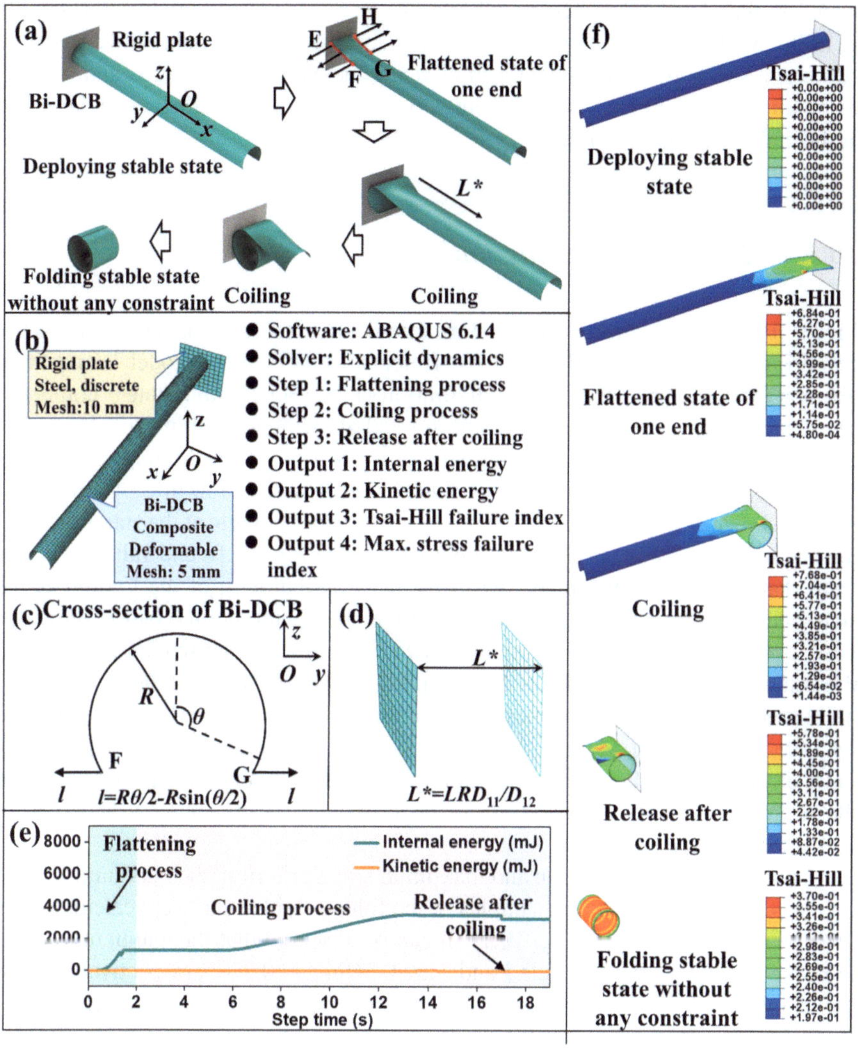

Fig. 10.14 Finite element model 2 of the bistable deployable composite boom sample 1: (**a**) Numerical simulation scheme. (**b**) Mesh size. (**c**) Applied displacement. (**d**) Rigid plate displacement. (**e**) Energy changes in the three steps. (**f**) Numerical simulation deformation process

and can be neglected. The bistable deformation process of sample 1, as predicted by the numerical simulation, is shown in Fig. 10.14f. From Fig. 10.14f, it is clearly observed that (1) sample 1 can fold stably with the movement of the rigid plate and eventually reach the folded stable state and (2) the Tsai-Hill failure index for the entire bistable deformation process of sample 1 is less than 1. Therefore, the finite element model 2 predicts that sample 1 can achieve bistable functionality. Using the same method, the bistable deformation processes of the remaining five samples were simulated and calculated, and the results were consistent with those of sample 1. In line with the experimental results, the coordinates of the data points from the numerical simulation were linearly fitted using the Archimedes' helix, and the corresponding polar radii at the start point and the end point as well as the polar angle at the end point were recorded.

10.3.3 Results and Discussion

Figure 10.15 shows the bistable deformation process of the bistable deployable composite boom as determined by experiments, finite element model 1, and finite element model 2. The geometric configuration of the folded stable state is compared between the three methods, and all are in good agreement. Archimedes' helix was used to linearly fit the data points of the cross section of the folding stable state determined by experiments and the two finite element models. The fitted Archimedes' helix is shown in Fig. 10.16, and the polar radii and polar angle at the start point and the end point are listed are listed in Table 10.2. From Fig. 10.16 and Table 10.2, it is clear that the numerically predicted geometric configuration of the folding stable state and the polar radii and polar angle at the start point and the end point are in good agreement with the experiment. For the six samples, the maximum errors in the polar radii and polar angles at the start point and the end point of the Archimedes' helix predicted by finite element model 1 compared to the experimental results are 18.21%, 16.51%, and 14.78%, respectively. For finite element model 2, the maximum errors in the polar radii and polar angles at the start point and the end point of the Archimedes' helix compared to the experimental results are 17.81%, 16.30%, and 14.55%, respectively.

Using the Tsai-Hill criterion and maximum stress criterion, the maximum failure indices for each ply of the six samples in the folded stable state were calculated, as shown in Fig. 10.17. From Fig. 10.17, it can be observed that the maximum failure indices for finite element models 1 and 2 are 0.3694 and 0.3705, respectively, which indicates that all six samples can achieve bistable function without failure, consistent with the experimental results. In conclusion, the two nonlinear finite element models established can show good prediction accuracy for evaluating the folded stable state of the bistable deployable composite booms.

For the six samples, the polar radii at the start point and the end point of Archimedes' helix predicted by both finite element models are higher than the experimental results, while the polar angle at the end point predicted by the finite element models is lower than the experimental results. The main reasons for the discrepancies between the predictions of the finite element models and the experimental results are the nonlinear behavior of the epoxy resin matrix material, as

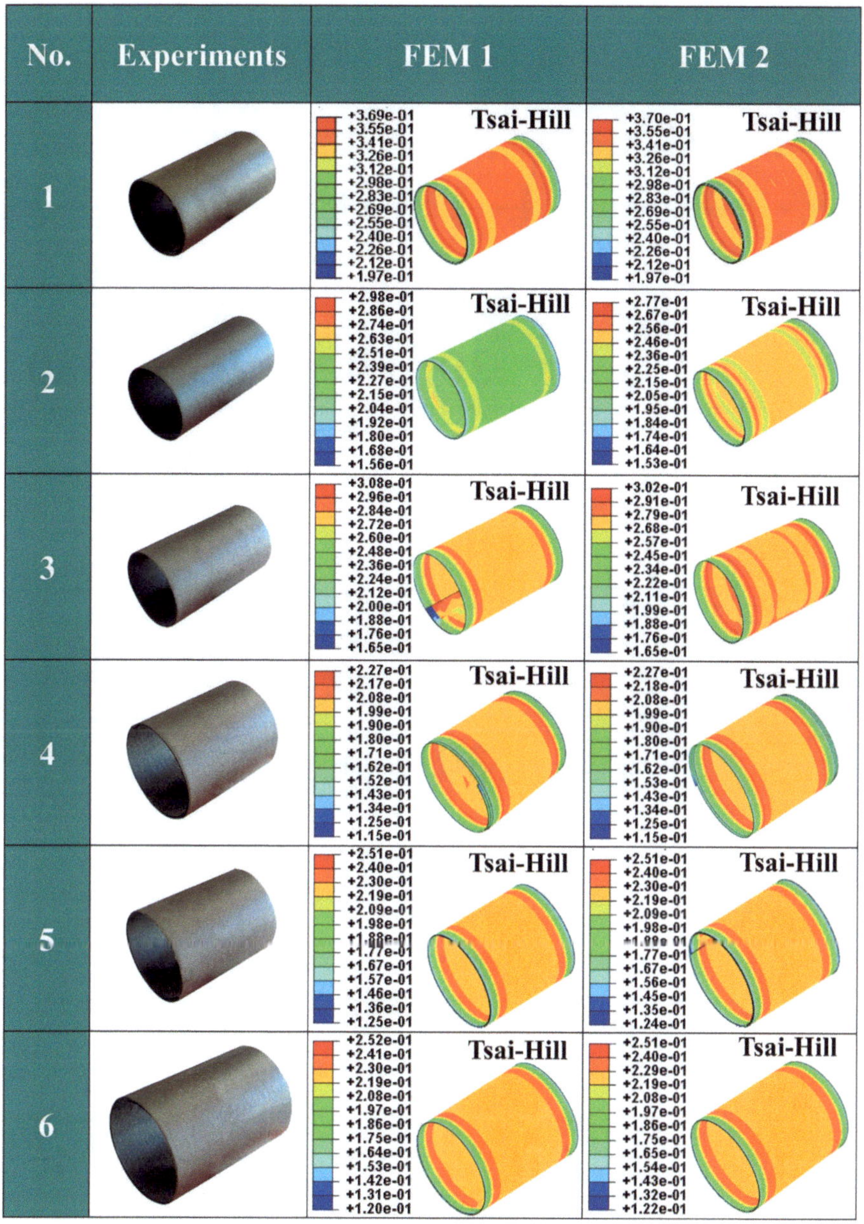

Fig. 10.15 Geometric configuration of the folded stable state for six samples

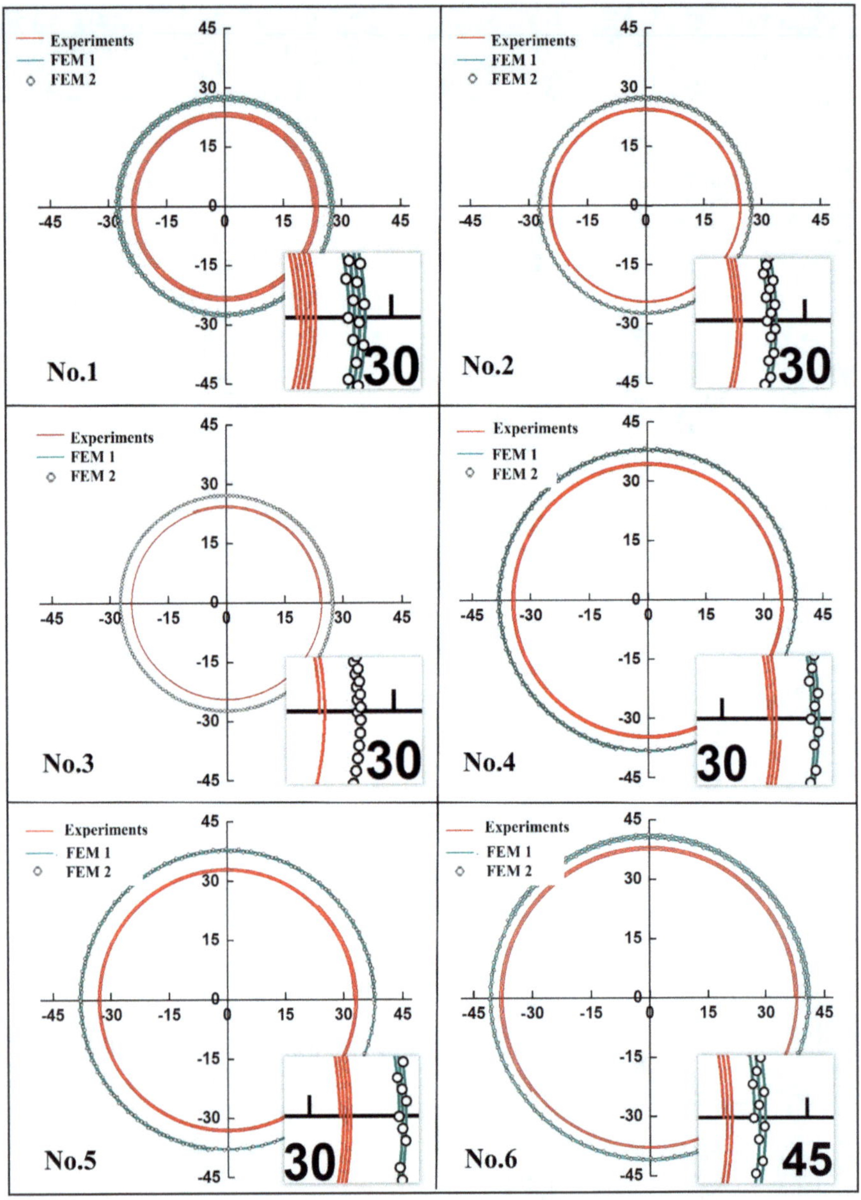

Fig. 10.16 Cross-sectional geometric configuration of six samples at the folded stable state

Table 10.2 Polar radii and polar angles at the start and end points of the folded stable state for six samples

Geometric parameters			No. 1	No. 2	No. 3	No. 4	No. 5	No. 6
r_0 (mm)	Experiments		22.63	24.04	24.13	34.08	32.59	37.31
	FEM 1	Values	26.75	26.89	27.16	37.75	37.49	40.21
		Error	18.21%	11.86%	12.56%	10.77%	15.04%	7.77%
	FEM 2	Values	26.66	26.85	27.15	37.68	37.47	40.15
		Error	17.81%	11.69%	12.52%	10.56%	14.97%	7.61%
r_1 (mm)	Experiments		24.23	24.87	24.66	35.08	33.58	38.41
	FEM 1	Values	28.23	27.75	27.52	38.57	38.32	41.47
		Error	16.51%	11.58%	11.60%	9.95%	14.11%	7.97%
	FEM 2	Values	28.18	27.67	27.44	38.55	38.31	41.35
		Error	16.30%	11.26%	11.27%	9.89%	14.09%	7.65%
α_1 (rad)	Experiments		26.46	16.36	8.20	18.80	19.65	16.38
	FEM 1	Values	22.55	14.64	7.32	17.03	17.15	15.07
		Error	14.78%	10.51%	10.73%	9.41%	12.72%	8.00%
	FEM 2	Values	22.61	14.67	7.33	17.05	17.15	15.11
		Error	14.55%	10.33%	10.61%	9.31%	12.72%	7.75%

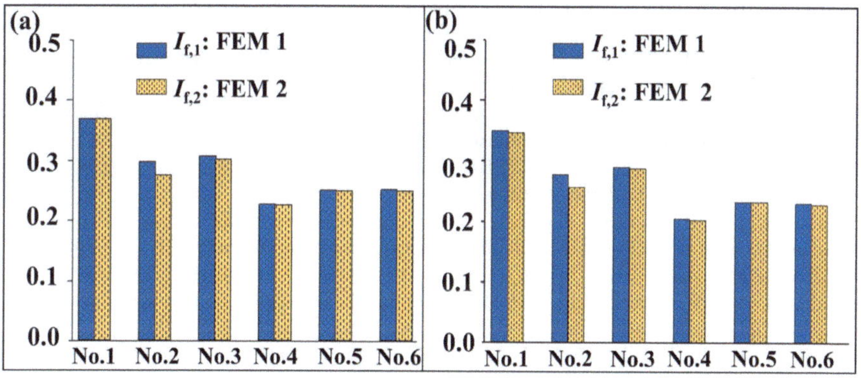

Fig. 10.17 Comparison of failure indices of six samples at the folded stable state: (**a**) Tsai-Hill failure index. (**b**) Maximum stress failure index

mentioned in [18]. However, both finite element models presented assume that the material is always linear elastic. Another factor contributing to the differences, as reported by Panesar et al. [19], is that residual thermal stresses inevitably arise during the manufacturing process of the samples, which causes noticeable differences in the deformation of laminates. A quantitative study of these two issues will be conducted in future work.

References

1. Bai JB, Chen D, Xiong JJ et al (2019) Folding analysis for thin-walled deployable composite boom. Acta Astronaut 159:622–636
2. Bai JB, Xiong JJ (2014) Temperature effect on buckling properties of ultra-thin-walled lenticular collapsible composite tube subjected to axial compression. Chin J Aeronaut 27(5):1312–1317
3. Liu TW, Bai JB, Fantuzzi N (2023) Folding behaviour of the thin-walled lenticular deployable composite boom: analytical analysis and many-objective optimization. Mech Adv Mater Struct 30(11):2221–2239
4. Liu TW, Bai JB (2022) Folding behaviour of a deployable composite cabin for space habitats-part 1: experimental and numerical investigation. Compos Struct 302:116244
5. Liu TW, Bai JB, Xi HT et al (2023) Experimental and numerical investigation on folding stable state of bistable deployable composite boom. Compos Struct 320:117178
6. Liu TW, Bai JB, Fantuzzi N (2023) Analytical models for predicting folding behaviour of thin-walled tubular deployable composite boom for space applications. Acta Astronaut 208:167–178
7. Liu TW, Bai JB, Fantuzzi N et al (2023) Folding behaviour of thin-walled tubular deployable composite boom for space applications: experiments and numerical simulation. Acta Astronaut 209:159–171
8. Liu TW, Bai JB, Fantuzzi N (2024) Analytical model for predicting folding stable state of bistable deployable composite boom. Chin J Aeronaut 37(8):460–469
9. Liu TW, Bai JB (2022) Folding behaviour of a deployable composite cabin for space habitats-part 2: analytical investigation. Compos Struct 297:115929
10. Liu TW, Bai JB, Xi HT et al (2023) Temperature effect on large deformation of bistable deployable composite boom. Compos Struct 318:117090
11. Bai JB, You FY, Wang ZZ et al (2024) An efficient multi-objective optimization framework for thin-walled tubular deployable composite boom. Compos Struct 327:117713
12. Liu TW, Bai JB, Xi HT et al (2023) Bistable behaviour of deployable composite boom with central flat region. Eng Struct 290:116357
13. Liu TW, Bai JB, Lin QH et al (2021) An analytical model for predicting compressive behaviour of composite helical structures: considering geometric nonlinearity effect. Compos Struct 255:112908
14. Bai JB, Liu TW, Wang ZZ et al (2021) Determining the best practice-optimal designs of composite helical structures using genetic algorithms. Compos Struct 268:113982
15. Liu TW, Bai JB, Xi HT et al (2023) A curved bistable composite slit tube for deployable membrane structures. Compos Commun 41:101648
16. Liu TW, Bai JB, Fantuzzi N et al (2022) Multi-objective optimization designs for thin-walled deployable composite hinges using surrogate models and genetic algorithms. Compos Struct 280:114757
17. Iqbal K, Pellegrino S, Daton-Lovett A (1998) Bistable composite slit tubes. In: IUTAM-IASS symposium on deployable structures, Cambridge, UK
18. Iqbal K, Pellegrino S (2000) Bi-stable composite shells. In: 41st structures, structural dynamics, and materials conference and exhibit, Atlanta, Georgia
19. Panesar AS, Hazra K, Weaver PM (2012) Investigation of thermally induced bistable behaviour for tow-steered laminates. Compos A: Appl Sci Manuf 43(6):926–934

Chapter 11
Optimization Designs for Deployable Composite Structures

Abstract This chapter performs multi-objective optimization of composite structures (lenticular booms, tubular booms, bistable booms) using hybrid methods combining surrogate models and evolutionary algorithms. For lenticular booms, NSGA-III minimizes ultimate coiling radius, folding moment, and linear density while maximizing moment of inertia, reducing radius by 0.92% and moment by 3.33%. Tubular boom optimization via MOEA/D maximizes bending stiffness and minimizes folding moment, with maximum radius increasing stiffness by 583.5%. Bistable boom optimization using NSGA-II balances light weight, small folded volume, high strain energy, and natural frequency, with NSGA-II outperforming other algorithms in benchmark tests, significantly improving multi-objective performance.

11.1 Introduction

Not only are engineers interested in designing different structural forms, but they also seek the optimal solution. The process of finding the best design is known as optimization and is widely applied in various engineering fields. The optimization design of deployable composite structures has been extensively reported. Design variables such as material type, geometric dimensions, number of layers, ply angles, and individual layer thickness can be considered as optimization parameters to achieve different objectives. These objectives may include maximizing natural frequency, maximizing stiffness, minimizing folding moment, minimizing stable state configuration, maximizing buckling load, minimizing total weight, and maximizing critical buckling temperature. The optimization problem of deployable composite structures can be categorized based on the quantitative relationship between design variables and optimization objectives, objective functions, and attributes of design variables and constraints. The following are common classification approaches:

J.-B. Bai et al., *Folding Mechanics of Deployable Composite Structures*, Solid Mechanics and Its Applications 283,
https://doi.org/10.1007/978-3-031-99254-4_11

- Univariate and multivariate optimization problems: This category is based on the number of design variables involved. For a single design variable, it is referred to as a univariate optimization problem, while multivariate problems involve multiple design variables.
- Discrete and continuous optimization problems: Based on the type of value range for design variables, problems can be classified as discrete or continuous optimization problems. Discrete optimization problems refer to cases where design variables take on discrete values, such as material type, number of layers in laminate, and ply angles. On the other hand, continuous optimization problems refer to cases where design variables take on continuous values.
- Constrained and unconstrained optimization problems: Some optimization methods require satisfying certain constraints, such as dimensions and mass of deployable composite structures during folding configuration and prevention of failure during large-deformation processes. These problems are known as constrained optimization problems, while unconstrained optimization refers to cases where design variables can freely vary without any constraints.
- Single-objective and multi-objective optimization problems: The classification of problems depends on the number of objective functions involved. Single-objective optimization problems evaluate the optimal solution based on a single objective function. Multi-objective optimization problems, based on the Pareto front, are often considered more interesting as engineers can better understand the design space and choose designs that match other considerations not included in the optimization.
- Static and dynamic optimization problems: If the optimization process depends on the time and the performance changes after a certain period, it is referred to as a dynamic optimization problem, which is different from static optimization problems. In static optimization, the optimal design does not depend on time.

The coupling of surrogate model and evolutionary algorithm has been widely applied in the design of deployable composite structures. Figure 11.1 presents the specific optimization process using this technology, as follows:

1. A set of initial samples are collected, which can be obtained through Design of Experiment (DoE) or random sampling methods. These samples serve as the basis for constructing the surrogate model.
2. The surrogate model is trained using the collected samples. The most common surrogate models include response surface method (RSM), Kriging model, elliptical basis function (EBF), radial basis function (RBF), artificial neural network (ANN), support vector regression (SVR), and random forest (RF) [1].
3. The predictive capability of the surrogate model is validated using statistical error functions. Table 11.1 summarizes nine high-performance statistical error functions, and specific evaluation methods can be found in the original references [2–10].
4. The surrogate model is embedded into the evolutionary algorithm. Each individual in the evolutionary algorithm consists of a set of decision variables, which are inputted into the surrogate model to obtain the corresponding objective func-

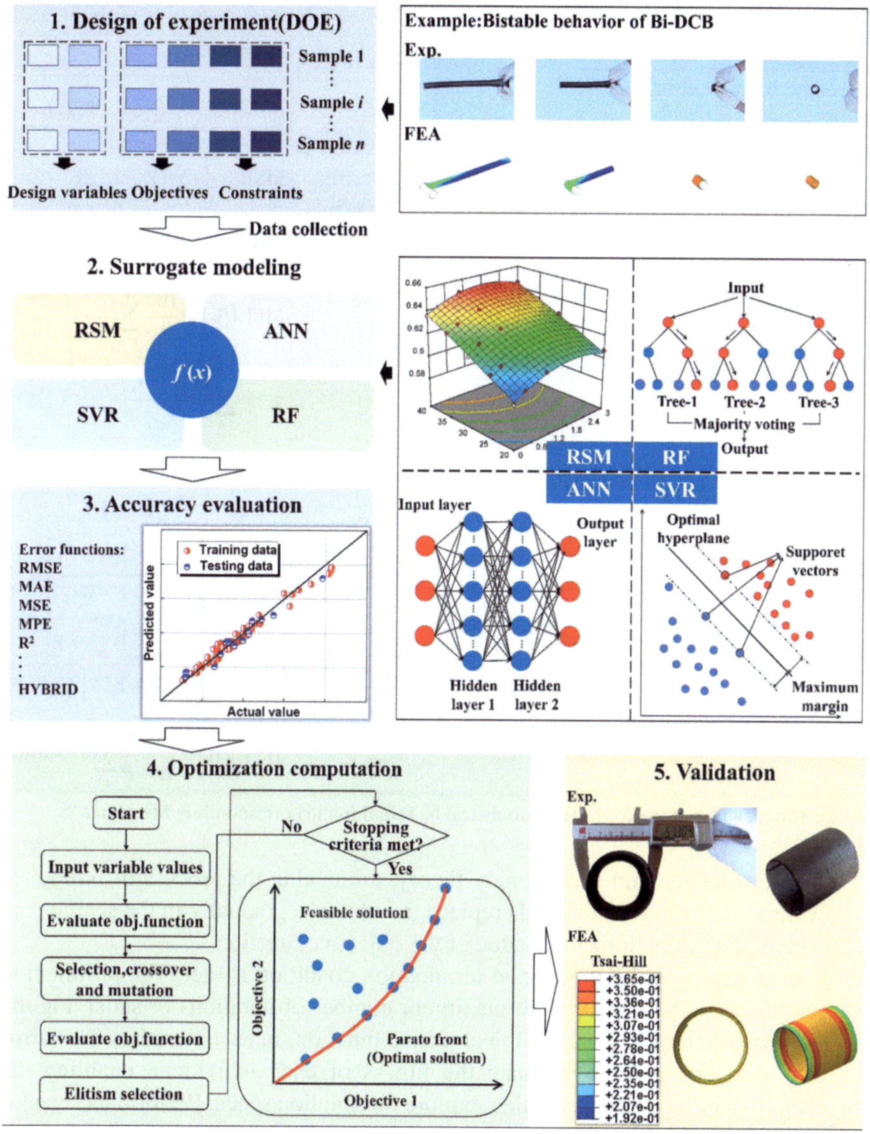

Fig. 11.1 Flowchart of coupling surrogate model and evolutionary algorithm technology for optimizing large-deformation flexible composite structures

tion value, serving as the fitness score of the individual. The most popular evolutionary algorithms include genetic algorithm (GA), simulated annealing (SA), particle swarm optimization (PSO), and ant colony optimization (ACO) [11, 12].

5. In evolutionary algorithms, mechanisms such as selection, crossover, and mutation are used to generate new individuals, which are then evaluated for their fitness using surrogate models. These mechanisms are based on the principles of

Table 11.1 Statistical error functions

Error function	Equation
Root mean square error [1]	$\mathrm{RMSE}=\sqrt{\frac{1}{N}\sum_{i=1}^{N}\left(Y_i-Y_{i,\mathrm{pre}}\right)^2}$
Mean square error [2]	$\mathrm{MSE}=\frac{1}{N}\sum_{i=1}^{N}\left(Y_i-Y_{i,\mathrm{pre}}\right)^2$
Mean absolute error [3]	$\mathrm{MAE}=\frac{1}{N}\sum_{i=1}^{N}\left\lvert Y_i-Y_{i,\mathrm{pre}}\right\rvert$
Mean predictive error (%) [4]	$\mathrm{MPE}(\%)=\frac{100}{N}\sum_{i=1}^{N}\left\lvert\frac{Y_i-Y_{i,\mathrm{pre}}}{Y_i}\right\rvert$
Chi square statistics (χ^2) [5]	$\chi^2=\sum_{i=1}^{N}\frac{\left(Y_{i,\mathrm{pre}}-Y_i\right)^2}{Y_{i,\mathrm{pre}}}$
Correlation coefficient (R^2) [6]	$R^2=1-\frac{\sum_{i=1}^{N}\left(Y_{i,\mathrm{pre}}-Y_i\right)^2}{\sum_{i=1}^{N}\left(Y_i-Y\right)^2}$
Adjusted correlation coefficient (R^2_{adj}) [7]	$R^2_{\mathrm{adj}}=1-\left[\frac{\left(1-R^2\right)\left(N-1\right)}{N-P-2}\right]$
Willmott's index of agreement ($d1$) [8]	$d1=1-\frac{\sum_{i=1}^{N}\left\lvert Y_{i,\mathrm{pre}}-Y_i\right\rvert}{\sum_{i=1}^{N}\left\lvert Y_i-Y_e\right\rvert+\left\lvert Y_{i,\mathrm{pre}}-Y_e\right\rvert}$
Hybrid fractional error function [9]	$\mathrm{HYBRID}=\frac{100}{N-P}\sum_{i=1}^{N}\frac{\left(Y_i-Y_{i,\mathrm{pre}}\right)^2}{Y_i}$

Note: The meaning of the formula symbols can be found in the corresponding literatures

natural selection and evolutionary theory, mimicking the process of biological evolution. The new individuals replace the old ones gradually in the search process, aiming to optimize the value of the objective function.

6. Repeat step 5 until a predefined termination condition is met. The termination condition can be reaching the maximum number of iterations or satisfying the convergence requirements of the objective function. In each iteration, the surrogate model is used to evaluate the fitness of each individual, enabling the evolutionary algorithm to further explore the solution space. Performance evaluation of different evolutionary algorithms and hyperparameters can be quantitatively assessed based on key indicators such as convergence, distribution diversity, and distribution range. Common quantitative evaluation methods for optimizers include hypervolume (HV) [13], maximum spread (MS) [14], generational distance (GD) [15], inverted generational distance (IGD) [16], and mimicked inverted generational distance (mIGD) [17].

It should be noted that although the computational process of coupling surrogate model and evolutionary algorithm is relatively complex, there are currently many integrated software options available, such as Insight, modeFRONTIER, OptiStruct,

MATLAB Optimization Toolbox, ANSYS DesignXplorer, and OpenMDAO. These software platforms integrate different models and algorithms into a unified platform for design and optimization workflows. By integrating different computational models and tools, engineers can find optimal structural design solutions faster and more accurately.

A brief review of existing optimization designs for deployable composite structures: Klimm et al. [18] investigated the influence of geometric parameters on the deployment behavior of viscoelastic deployable composite tape spring using Gaussian process regression. They applied non-dominated sorting genetic algorithm II (NSGA-II) for multi-objective optimization of residual deformation, strain, and mass. Su et al. [19] employed optimal Latin hypercube sampling method for DOE, established a surrogate model based on EBF, and then conducted sensitivity analysis. They achieved the optimal design of the integral deployable composite hinge in terms of lightweight, peak folding moment, and peak torsional moment using multi-objective particle swarm optimization algorithm (MOPSO). Jin et al. [20] focused on the slot shape optimization problem for the integral deployable composite hinge. They used four surrogate models and shape optimization methods to maximize the stored maximum strain energy during folding and the maximum bending moment during deployment. Li et al. [21] established an optimal model based on RSM to obtain the maximum strain energy and minimum peak moment for the integral deployable composite hinge, ensuring stable deployment. Additionally, the Hashin failure criterion was adopted to examine material damage during the folding process to prevent failures. Inspired by the work of Mallikarachchi et al. [22] and considering the requirements defined in ESA's 2016 work statement [23], Fernandes et al. [24] discussed the design concept incorporating damage tolerance, allowing for damage initiation during the folding process for the integral deployable composite hinge to meet stricter requirements. GAs were used to maximize the natural frequencies of two types of integral deployable composite hinge: one operating within the elastic range with restricted design, and another allowing for limited damage initiation during folding.

Yang et al. [25–28] performed the multi-objective optimization design of deployable composite booms, including tubular deployable composite boom, TRAC boom, N-shaped deployable composite boom, and lenticular deployable composite boom. Firstly, a full factorial design method was employed to create sampling designs for the four types of deployable composite booms mentioned above. Subsequently, efficient computational analyses were conducted on each design sample using the nonlinear finite element solver ABAQUS/Explicit, focusing on the deployment stiffness, folding behavior, and mass of the four types of deployable composite booms. RSM and Back Propagation Neural Network (BPNN) were utilized to establish surrogate models for the deployment stiffness and folding behavior. Lastly, NSGA-II and the Sequential Quadratic Programming (SQP) algorithm were employed for multi-objective optimization design. Similar to the optimization approach used by Yang et al., Shi et al. [29] optimized the M-shaped deployable composite boom using a coupling RSM and NSGA-II technique. Bessa and Pellegrino [30] proposed a data-driven computational framework combining Bayesian regression, uncertainty quantification, and multi-objective optimization for the design of the TRAC boom. The results show a significant increase in the ultimate buckling load and potential benefits of up to 100% compared to previously proposed design schemes. In addition to the single-stable deployable composite booms mentioned above, recent attention has been given to the multi-objective optimization design of bistable deployable composite booms. Zhang et al. [31, 32] proposed a combination of finite element analysis (FEA), multi-objective optimization techniques, and experimental methods to determine the optimal geometric parameters (including cross-sectional central angle, thickness, cross-sectional radius, and length) of the tubular bistable deployable composite boom. An optimal Latin hypercube sampling algorithm was employed to obtain sample points for DOE. RSM and NSGA-II were used to obtain the Pareto front of the tubular bistable deployable composite boom. Wu and Vinquerat [33] conducted natural frequency optimization design of the tubular bistable deployable composite boom using MATLAB Optimization Toolbox. To achieve this, they optimized the braid angles and stacking sequences with the constraint of folded stable radius while considering the maximum and minimum physically achievable braid angles as bounds.

At present, there is a lack of research on multi-objective optimization design of deployable composite reflector antennas using the coupling surrogate model and evolutionary algorithm. Given the wide applications of deployable composite reflector antennas in fields such as communication, radar, and satellites, future research can explore the application of multi-objective optimization design methods to the design of deployable composite reflector antennas. This would enable a balance and optimization between different performance indicators. It will be a key research direction in the design of deployable composite reflector antennas.

A brief review of literature on multi-objective optimization design of deployable composite structures reveals that most existing works utilize RSM to establish surrogate models between design variables and objective functions [34]. There are several reasons for this trend. Firstly, RSM is a simple and widely used mathematical regression model that establishes predictive equations by statistically fitting the relationship between design variables and objectives [35]. Secondly, RSM exhibits high prediction accuracy for fitting problems with small datasets. It is noted that most studies employ NSGA-II for multi-objective optimization design, which is likely due to its good performance as a general solver for many problems. However, it is important to consider that improving algorithm performance for a specific problem may inevitably lead to a decrease in performance for other types of problems, as stated by the "no free lunch" theorem. Hence, in specific cases, benchmarking multiple evolutionary algorithms is necessary to select the optimal algorithm and achieve the best optimization results. Lastly, other surrogate models such as ANN and RF possess certain limitations, such as the requirement for large datasets and susceptibility to parameter influence.

In addition, another optimization method, namely coupling analytical model and evolutionary algorithm, has also received some attention. For example, Liu and Bai used this technique to solve the multi-objective optimization problems of deployable composite structures such as lenticular deployable composite boom [36], tubular deployable composite boom [37], and deployable composite helical antennas [38]. Analytical models typically have clear expressions, simple forms, and high computational efficiency. However, compared to surrogate models, they have a limited range of applicability. It is difficult to establish analytical models for some key objectives of deployable composite structures, such as natural frequencies, buckling, and folding behavior in the presence of contacts. The pursuit of an efficient and highly adaptable optimization algorithm is one of the key directions for the future development of large-deformation flexible composite structures.

In this chapter, we take a typical large-deformation flexible composite structure, the deployable composite boom, used in space deployable structures as an example [36, 37]. Based on the analytical models established in the previous chapters, specifically the coupled analytical model and evolutionary algorithm, we conduct a multi-objective optimization design for different types of deployable composite booms.

11.2 The Lenticular Deployable Composite Boom

11.2.1 Problem Description

Optimal design of the lenticular deployable composite boom is a typical many-objective optimization problem. Unlike the single-objective optimization problem which only provides a single optimal solution, many-objective optimization is applied to identify a set of trade-off solutions to achieve the global optimization for objective functions. The set of trade-off solutions in the space of design variables is referred to as the Pareto non-inferior solution set. The map of Pareto non-inferior solution set in the objective space forms the Pareto front.

The optimization problem is formulated as a problem of minimization. The general many-objective optimization formula can be described as

$$\text{Opt.}\begin{cases} \mathit{Minimize}\ \ \phi(\boldsymbol{X}) = \left[\phi_1(\boldsymbol{X}),\phi_2(\boldsymbol{X})\ldots\phi_i(\boldsymbol{X})\right] \\ \text{subjected to:} \\ \boldsymbol{X} = \left(x_1,x_2\ldots x_m\right)^{\mathrm{T}} \\ \underline{x_g} \le x_g \le \overline{x_g} \end{cases} \tag{11.1}$$

where ϕ represents the objective function; $\boldsymbol{X}$ = ($x1$, $x2$, … xj)T are the design variables; i and m are, respectively, the number of objective functions and design variables; and x_g and x_g are the upper and lower limits of the variable space, respectively.

In practical engineering applications of the lenticular deployable composite boom, researchers focus on the folding behavior, lightweight, flexural stiffness, natural frequency, etc. To facilitate the design and preparation of the folding device, the ultimate coiling radius should be minimized as much as possible. The folding moment should be as small as possible for folding the lenticular deployable composite boom more conveniently and reliably. The linear density of the lenticular deployable composite boom should be minimized as possible to weight and launch cost. In addition, the flexural stiffness and natural frequency mainly depend on the inertia moments of the cross section, so the lenticular deployable composite boom needs to have enough inertia moment to obtain high flexural stiffness and natural frequency. Therefore, a total of five objectives are set in this research, as follows:

- Objective 1: minimizing the ultimate coiling radius (Eq. 7.56), that is,

$$\phi_1(\boldsymbol{X}) = R_{\text{ultimate}} \tag{11.2}$$

- Objective 2: minimizing the folding moment (Eq. 7.37), that is,

$$\phi_2(\boldsymbol{X}) = T \tag{11.3}$$

- Objective 3: minimizing the linear density (Eq. 7.13), that is,

$$\phi_3(\boldsymbol{X}) = \rho_d A \tag{11.4}$$

- Objective 4: maximizing the inertia moment of y-axis of the cross section (Eq. 7.16), that is,

$$\phi_4(\boldsymbol{X}) = 1/I_y \tag{11.5}$$

- Objective 5: maximizing the inertia moment of z-axis of the cross section (Eq. 7.16), that is,

$$\phi_5(\boldsymbol{X}) = 1/I_z \tag{11.6}$$

Chapter 7 has demonstrated that the geometric parameters of the lenticular deployable composite boom significantly influence the ultimate coiling radius, folding moment, linear density, and moment of inertia of the cross section. Therefore, the initial curvature radius of the concave and convex arcs, as well as the width of the adhesive interface, is selected as the three design variables, namely:

$$\boldsymbol{X} = \left(r_{1,0}\ r_{2,0}\ a\right)^{\mathrm{T}} \tag{11.7}$$

The optimization design of the lenticular deployable composite boom needs to satisfy some constraints. To obtain better stability, the inertia moments of y-axis and z-axis of the cross section should be as close as possible. The ultimate coiling radius must be greater than the practical coiling radius to avoid failure. In addition, the linear density of the lenticular deployable composite boom should not be too large to avoid increasing weight and launch cost. Therefore, the constraints are described as follows:

- Constraint 1: the geometric constraint of the inertia moment of the cross section is

$$\underline{h} \le h = \frac{I_z}{I_y} \le \overline{h} \tag{11.8}$$

- Constraint 2: the lenticular deployable composite boom does not fail in the folding process, that is,

$$R_{\text{ultimate}} \le R \tag{11.9}$$

- Constraint 3: the constraint of the linear density of the lenticular deployable composite boom is

$$S = \rho A \le \overline{S} \tag{11.10}$$

11.2.2 Optimization Algorithms

Unlike single-objective optimization problems that provide a single optimal solution, multi-objective optimization problems yield a set of solutions known as the Pareto optimal set, representing trade-offs between conflicting objectives. To obtain Pareto optimal solutions, many techniques have been compared in the literature [11, 35–38]. Among them, multi-objective evolutionary algorithms (MOEAs), such as NSGA-II [39], NSGA-III [40, 41], SPEA-II [42], RVEA [43], and MOEA/D [44], have garnered significant attention from researchers due to their effectiveness and ease of implementation.

Among these MOEAs, NSGA-III is considered one of the most powerful methods, serving as an improved version of NSGA-II. Over the past decade, NSGA-III has been widely applied to the design optimization of various engineering problems [45–52]. In this section, NSGA-III is utilized to solve the multi-objective optimization problem for composite lenticular rods. The hyperparameters for NSGA-III are defined in Table 11.2, consistent with those reported in the literature. For more detailed information about NSGA-III, please refer to relevant studies [40, 41].

11.2.3 Optimal Framework

The multi-objective optimization design framework for the lenticular deployable composite boom is illustrated in Fig. 11.2. The optimization process is implemented using a high-performance genetic algorithm toolbox coded in Python (Geatpy), which is available for download in the literatures [53, 54]. In addition, apart from the three design variables ($r_{1,0}$, $r_{2,0}$, a), all other material parameters of the lenticular deployable composite boom remain consistent with those defined in Chap. 7. In order to realize a

Table 11.2 NSGA-III algorithm parameter definition

	Type	Random
Initialization	Encoding type	Real value coded
Reproduction	Number of populations	1500
	Number of generations	200
	Selection type	Well-spread reference points
	Crossover type	Real variable SBX
	Crossover rate	0.7
	Mutation type	Polynomial mutation
	Mutation rate	0.08
	Elitism operator	Active
Termination	Stop criterion	Reach the maximum generation

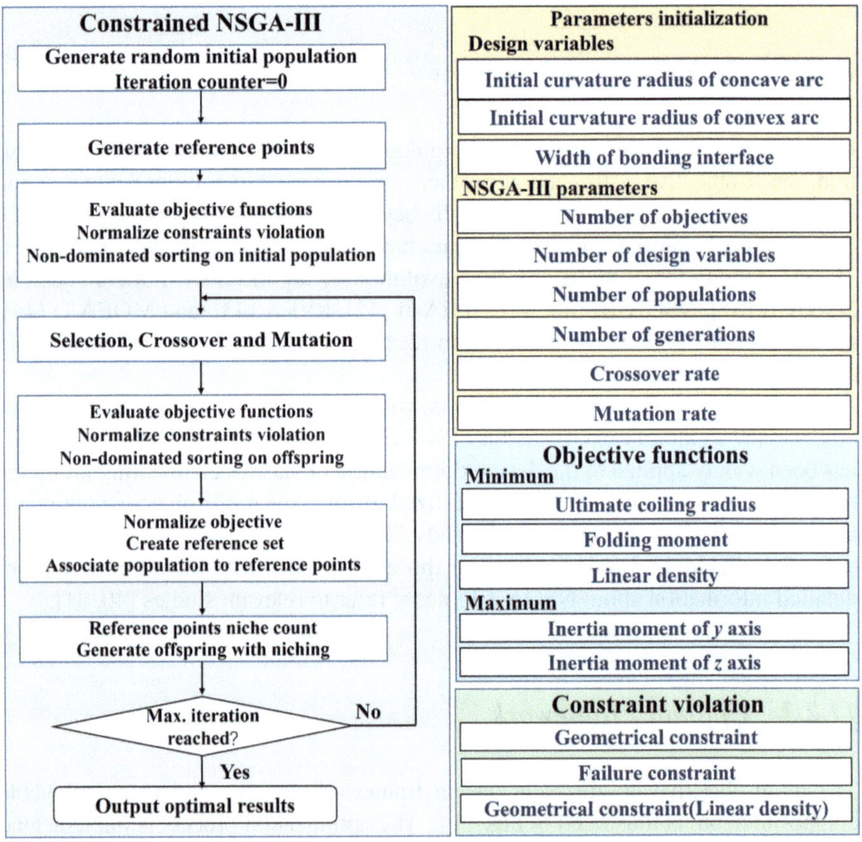

Fig. 11.2 Many-objective optimization design framework with NSGA-III

Table 11.3 Parameter setting of design variables and constraints

Parameter		Lower bound	Upper bound
Design variables	$r_{1,0}$ (mm)	20	50
	$r_{2,0}$ (mm)	20	50
	a (mm)	5	15
Constraints	h	1	5
	R_{ultimate} (mm)	–	80
	S (g/cm)	–	2.50

comprehensive Pareto front, the design variables were set as continuous variables. Table 11.3 shows the parameter setting of design variables and constraints.

11.2.4 Benchmarking for Algorithms

There is no "free lunch," meaning that no single algorithm can consistently perform optimally across all types of optimization problems. Although the multi-objective optimization framework proposed in this section adopts a well-known general optimization algorithm, NSGA-III, it is not guaranteed to exhibit the best performance for specific problems. Studies [35, 38] have demonstrated that MOEA/D, which is based on the decomposition approach, outperforms NSGA-II and its derivatives in certain optimization problems. However, in optimizing the performance of triaxially braided composites under tensile loading, MOEA/D performed the worst [17]. For the current optimization problem, it is evident that the minimum line density can be achieved when all three geometric design variables are at their smallest values, provided that all constraints are satisfied. However, the NSGA-III Pareto front did not include the design point corresponding to the minimum line density. Three primary reasons can explain this outcome: (1) The robustness and stochastic nature of the optimization algorithm may have prevented the discovery of the minimum line density design point, and repeated simulations could address this issue. (2) The choice of hyperparameters, particularly population size and generation count, significantly affects the Pareto front results. Generally, larger populations and more generations are required for problems with multiple objectives, albeit at the cost of computational efficiency. (3) NSGA-III may not be the optimal algorithm for this specific problem.

Therefore, four additional high-performance multi-objective genetic algorithms available in the Geatpy toolbox, NSGA-II, MOEA/D, awGA, and RVEA, were employed to optimize the lenticular deployable composite boom. To ensure a fair comparison, identical genetic operator types (selection, crossover, and mutation) were used across all algorithms. The Pareto front results for these algorithms are shown in Fig. 11.3. NSGA-III, NSGA-II, MOEA/D, and awGA discovered Pareto fronts with similar coverage, high quality, and diversity. This further demonstrates that the multi-objective optimization framework proposed in this section is effective for optimizing the lenticular deployable composite boom. However, RVEA,

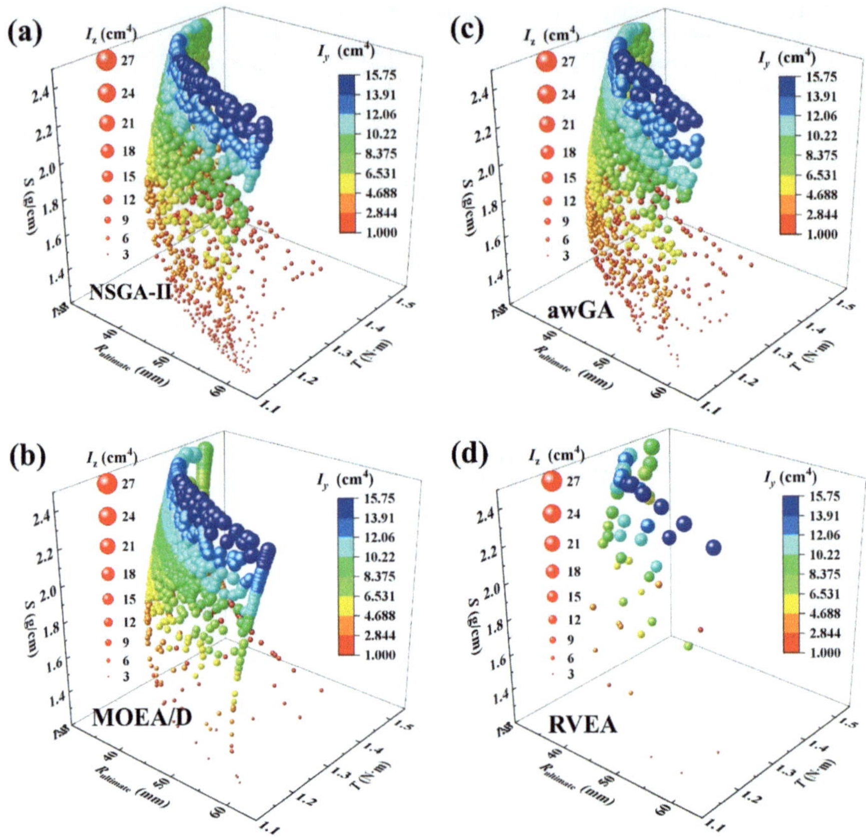

Fig. 11.3 Comparison of Pareto fronts for different state-of-the-art algorithms: (**a**) NSGA-II. (**b**) MOEA/D. (**c**) awGA. (**d**) RVEA

expected to perform well as a general optimization algorithm, exhibited poor performance in this optimization task, identifying only a few dozen design points on the Pareto front. Moreover, among the five benchmarked algorithms, only NSGA-II identified the design point with the minimum line density on its Pareto front.

This highlights the necessity of benchmarking the latest genetic algorithms for composite material optimization problems, as these engineering challenges possess characteristics distinct from algorithm benchmarks. For the evaluated genetic algorithms, their performance can be quantitatively assessed using key metrics such as convergence, diversity, and spread.

11.2.5 Design Schemes

Using the many-objective optimization design framework established in Sect. 11.2.3, Fig. 11.4 shows the optimization results of the Pareto front. There are a total of 1365 design points on the entire Pareto front, and the 1365 design points are

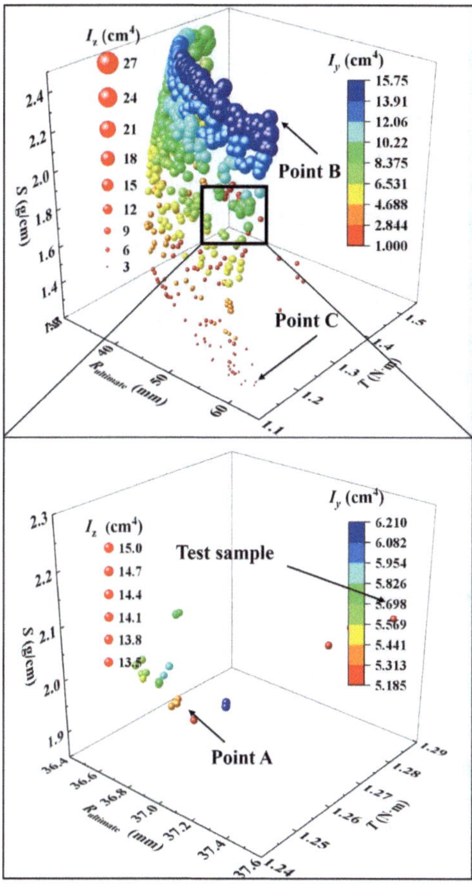

Samples	Design variables and Objectives
Point B	**Design variables:** $r_{1,0} = 46.02$mm $r_{2,0} = 37.79$mm $a = 5.08$mm **Objectives:** $R_{ultimate} = 35.485$mm $T = 1.341$Nm $S = 2.499$g/cm $I_y = 12.379$cm^4 $I_z = 27.124$cm^4
Point C	**Design variables:** $r_{1,0} = 20.02$mm $r_{2,0} = 22.55$mm $a = 5.01$mm **Objectives:** $R_{ultimate} = 60.835$mm $T = 1.153$Nm $S = 1.231$g/cm $I_y = 1.024$cm^4 $I_z = 2.878$cm^4
Point A	**Design variables:** $r_{1,0} = 34.58$mm $r_{2,0} = 34.79$mm $a = 5.53$mm **Objectives:** $R_{ultimate} = 37.093$mm $T = 1.246$Nm $S = 1.998$g/cm $I_y = 5.263$cm^4 $I_z = 13.458$cm^4
Test sample	**Design variables:** $r_{1,0} = 34.00$mm $r_{2,0} = 34.00$mm $a = 10.00$mm **Objectives:** $R_{ultimate} = 37.436$mm $T = 1.289$Nm $S = 2.079$g/cm $I_y = 5.002$cm^4 $I_z = 12.740$cm^4

Fig. 11.4 Optimal design points of the lenticular deployable composite boom

sorted in ascending order of the linear density. Appendix A summarizes 22 design points to exhibit how changes in design variables affect objectives, in which the 22 designs were picked from the first point of each 65 points interval. These 22 design points illustrate that the relationship between design variables and objectives is nonlinear.

The optimization results show that a total of 48 design points are found on the Pareto front, all of which are better than test sample for the five objectives (i.e., the minimum ultimate coiling radius, the minimum folding moment, the minimum linear density, and the maximum inertia moments of y-axis and z-axis of the cross section). These 48 design points are listed in Appendix B and Fig. 11.4. To further illustrate the optimization results, a design point was selected by hand from the Pareto front, namely Point A. The comparison results of test sample and Point A are shown in Fig. 11.4. According to Fig. 11.4, compared with test sample (R_{ultimate} = 37.436 mm, T = 1.289 N m, S = 2.079 g/cm, I_y = 5.002 cm^4, and I_z = 12.740 cm^4), Point A (R_{ultimate} = 37.093 mm, T = 1.246 N m, S = 1.998 g/cm,

I_y = 5.263 cm^4, I_z = 13.458 cm^4) is a better design scheme. The ultimate coiling radius is reduced by 0.92%, the folding moment decreases by 3.33%, the linear density reduces 3.90%, while the inertia moments of y-axis and z-axis of the cross section increase by 5.22% and 5.64%, respectively.

In order to further illustrate the effectiveness of the multi-objective optimization design framework proposed in this section, two points with special positions on the Pareto front were selected by hand, namely Point B and Point C. The comparison results of test sample, Points B and C, are shown in Fig. 11.4. Compared with test sample, the inertia moments of y-axis and z-axis of the cross section of Point B increase significantly, of which the inertia moment of y-axis of the cross section ascends by 147.48% and the inertia moment of z-axis of the cross section ascends by 112.90%. Compared with test sample, the linear density of Point C is greatly dropped by 40.79%. Therefore, the many-objective optimization design framework proposed in this section can effectively optimize the lenticular deployable composite boom.

11.3 The Tubular Deployable Composite Boom

11.3.1 Problem Description

In practical engineering applications, the load-bearing capacity and folding behavior of the deployable composite booms are of critical importance. To ensure better structural safety and stability, it is necessary to maximize the bending stiffness during the optimization process. Meanwhile, to facilitate more convenient and reliable folding, the folding moment should be minimized as much as possible. Consequently, the optimal design of a tubular deployable composite boom represents a typical multi-objective optimization problem. As shown in the results of Chap. 8, the opposite-sense folding moment is generally greater than the equal-sense folding moment, and there is a positive correlation between these two moments. Therefore, this section focuses on minimizing the equal-sense folding moment as the optimization objective.

In Sect. 8.1, an analytical model was established to predict the folding moment of tubular deployable composite booms during both equal-sense and opposite-sense folding processes. The geometric configuration of the tubular deployable composite boom during folding was divided into three regions: the coiled region, the transition region, and the free region. This section simplifies the analytical model by neglecting the influence of the transition region. As a result, the total strain energy generated by the tubular deployable composite boom throughout the folding process can be expressed as follows:

$$U = \int_0^{\frac{r_1 - r_0}{a}} \rho u(\phi) d\phi \tag{11.11}$$

The average folding moment of the tubular deployable composite boom during the folding process can be expressed as follows:

$$\overline{T} = \frac{Ua}{r_1 - r_0} \tag{11.12}$$

The moments of inertia of the cross section about the *y*-axis and *z*-axis are also positively correlated; therefore, the second objective is to maximize the smaller value between the two since the weaker bending stiffness better reflects the stability and rigidity of the tubular deployable composite boom. Additionally, based on the analytical model presented in Chap. 8, it is evident that the cross-sectional radius, central angle, and ply schemes have significant influences on the folding moment and bending stiffness of the tubular deployable composite boom. Subsequently, a comprehensive elaboration of the constraints is presented in the following content:

$$\text{Opt.}\begin{cases} Minimize:\left\{\overline{T}\left(R,\theta,\beta_k\right),1/E_xI(R,\theta,\beta_k)\right\} \\ I=\min\left\{I_y,I_z\right\} \\ \text{s.t}: 25\text{mm} \le R \le 40\text{mm} \\ \qquad 180^\circ \le \theta \le 270^\circ \\ \qquad \beta_k \in \left\{-45^\circ,0^\circ,45^\circ,90^\circ\right\} \quad k=1,2,\cdots,n \end{cases} \tag{11.13}$$

where E_xI represents the bending stiffness of the tubular deployable composite boom. Additionally, the length of the tubular deployable composite boom is set to 800 mm, and the number of plies is specified as 8. Other material parameters and specifications remain consistent with those defined in Chap. 8.

In the optimization design of the tubular deployable composite boom, several constraints need to be satisfied in order to guarantee the usability of obtained optimal designs. First, to prevent failures during the folding deformation process, two failure criteria are introduced to constrain the search space. Second, the laminate design of composite materials is one of the key design tasks and involves selecting appropriate ply angles and ply sequences. Third, the limitations of the folding mechanism range also need to be taken into consideration. Therefore, seven design constraints are involved in the optimization process:

- Constraint 1: the tubular deployable composite boom should not be a failure during the folding process, i.e., all two failure indices (Tsai-Hill and maximum stress failure indices) should be less than 1):

$$\max\left\{I_{f,1},I_{f,2}\right\} < 1 \tag{11.14}$$

- Constraint 2: the layup of laminate should be symmetric with respect to the neutral axis to minimize the post-curing deformation caused by coupling effects.

- Constraint 3: the lamina should be uniformly distributed along the thickness to avoid concentrated stacking of the same layer angle. Specifically, no more than four consecutive layers of the same ply angle are allowed.
- Constraint 4: the principle of equilibrium should be followed, where ±45° ply angles occur in pairs.
- Constraint 5: the outermost layer should be stacked at ±45° to reduce delamination damage caused by impact.
- Constraint 6: according to the actuator range of the folding mechanism, the folding moment should be less than 300 N mm.

This section addresses the constrained optimization problem using adaptive penalty techniques, wherein penalty coefficients are introduced into the fitness function to implement a dynamically efficient adjustment strategy [55–57]. For more detailed information about this method, please refer to [56].

11.3.2 Optimal Framework

The core idea of the multi-objective evolutionary algorithm based on decomposition (MOEA/D) [58, 59] is to decompose a multi-objective optimization problem into a series of single-objective optimization subproblems and optimize them simultaneously using a collaborative evolutionary algorithm. By employing a decomposition strategy, this algorithm effectively reduces the complexity of the problem and improves the efficiency of the optimization process. Additionally, MOEA/D has demonstrated significant advantages in preserving the diversity and distribution of solutions, making it well suited for addressing multi-objective optimization problems. Figure 11.5 depicts the flowchart of the framework for the multi-objective optimization of the tubular deployable composite boom. It is worth noting that the total number of function evaluations in this section was determined based on the implementation requirements of the algorithm. When selecting the number of function evaluations, factors such as computational accuracy and runtime were primarily considered. The goal was to minimize the number of function evaluations while ensuring computational accuracy, thereby improving overall efficiency. Table 11.4 presents the definitions of the hyperparameters for this algorithm.

11.3.3 Design Schemes

By utilizing the optimization framework established in Sect. 11.3.2, a discontinuous Pareto front consisting of two main Pareto fronts is obtained, which is presented in Fig. 11.6a. The entire Pareto front comprises 1000 design points, which have been sorted in descending order of folding moment. Starting from the first sample, a record is made every 20 samples and summarized in Appendices C and D. Appendix

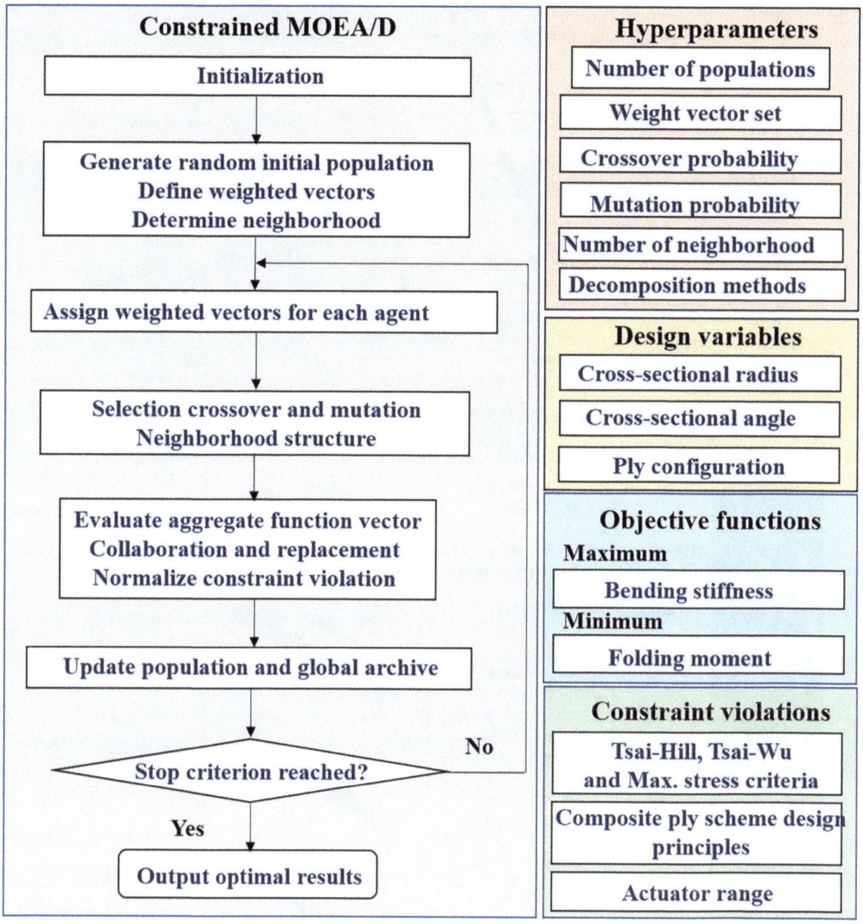

Fig. 11.5 Multi-objective optimization framework for the tubular deployable composite boom based on MOEA/D

Table 11.4 The hyperparameters of MOEA/D

	Initialization type	Random
Initialization	Encoding type	Real value coded
Reproduction	Number of populations	1000
	Function calls	100,000
	Selector operator	MOEA/D
	Crossover type	Real variable SBX
	Crossover rate	1
	Mutation type	Polynomial mutation
	Mutation rate	0.08
Termination	Stop criterion	Reach the total number of fitness calls

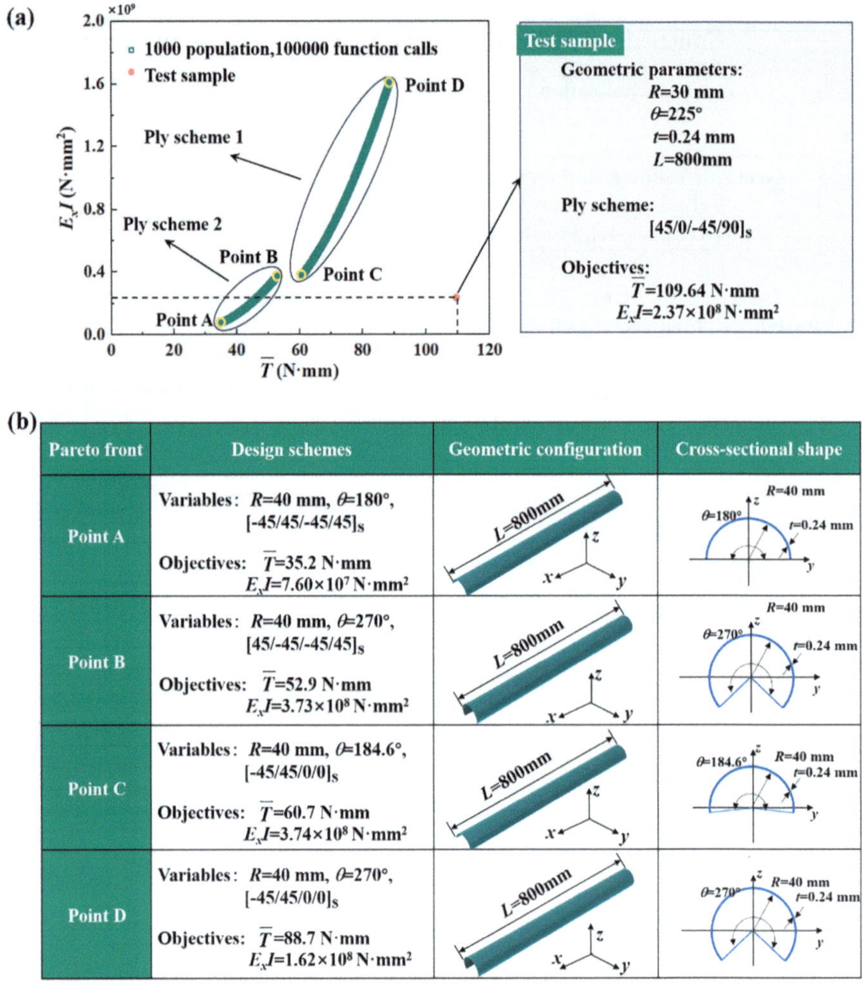

Fig. 11.6 Optimal designs of the tubular deployable composite boom: (**a**) Pareto front. (**b**) Topological structures of the four design points

C summarizes the variations in the variable space and objective space of the Pareto front on the right-hand side across 42 examples, while Appendix D provides a summary of 9 samples from the left-hand-side Pareto front. These design points illustrate the existence of nonlinear relationships between design variables and objectives, thereby increasing the complexity of design decision-making.

To further analyze the relationship between various variables and objective functions, four design points were manually selected on the Pareto front. These four points are the left end point (Point A: $T = 35.2\,\text{N mm}$ and EI = $7.6 \times 10^7\text{N mm}^2$) and right end point (Point B: $T = 52.9\,\text{N mm}$ and EI = $3.73 \times 10^8\text{N mm}^2$) on the left-hand-side Pareto front as well as the left end point (Point C: $T = 60.7\,\text{N mm}$ and

EI = 3.74 × 10^8N mm^2) and right end point (Point D: T = 88.7N mm and EI = 1.62 × 10^9N mm^2) on the right-hand-side Pareto front. The cross-sectional radius of these design points is equal to the maximum value within the variable range, i.e., 40 mm. For Point A and Point B, the stacking sequence is composed of only ±45° plies. On the other hand, for Point C and Point D, the stacking sequence includes ±45° plies for the outermost two layers and 0° ply angles for the middle four plies. Additionally, Point A exhibits the minimum folding moment, while Point D exhibits the maximum bending stiffness among all the Pareto optimal solutions. Figure 11.6b shows the topological structure of these four design points, along with the values of each design variable and optimization objective.

To validate the optimal designs from the Pareto front, this study compared the pre-optimized case (i.e., initial sample) to the optimal designs. The geometric parameters and stacking sequences of the pre-optimized case are shown in Fig. 11.6a, with a folding moment of 109.64 N mm and bending stiffness of 2.37 × 10^8 N mm^2. A total of 924 design points (i.e., maximize bending stiffness and minimize folding moment) outperform the initial sample. Among them, Point D (T = 88.7 N mm and EI = 1.62 × 10^9 N mm^2) exhibited a 19.1% reduction in folding moment but a 583.5% increase in bending stiffness when compared to the initial sample.

11.3.4 Effects of Ply Schemes and Cross-Sectional Radius

Figure 11.6a depicts the Pareto front, which exhibits an overall discontinuous shape. The primary reason for this issue is the discreteness of ply schemes, which encompass two distinct sets of stacking sequences. The stacking sequence corresponding to the right-side Pareto front is ply scheme 1 (combination of ±45° and 0°), while the stacking sequence corresponding to the left-side Pareto front is ply scheme 2 (solely combination of ±45°); the detailed specification of different ply schemes is shown in Table 11.5. Among all these design points, 90° ply angle is never selected in the stacking sequences. This is because a 90° ply angle significantly reduces the axial elastic modulus of the tubular deployable composite booms when compared to the 0° and

Table 11.5 Specifications of ply schemes

Type	Description
Ply scheme 1	$[45, -45, 0, 0]_s$
	$[-45, 45, 0, 0]_s$
Ply scheme 2	$[45, -45, -45, 45]_s$
	$[45, -45, 45, -45]_s$
	$[45, 45, -45, -45]_s$
	$[-45, 45, -45, 45]_s$
	$[-45, -45, 45, 45]_s$
	$[-45, 45, 45, -45]_s$

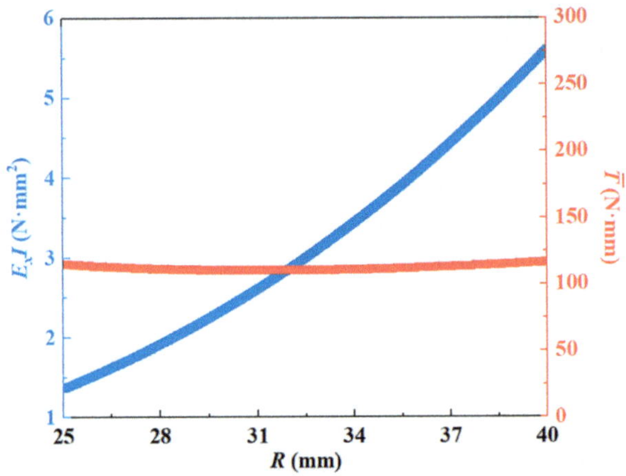

Fig. 11.7 The trends of changes in folding moment and bending stiffness with increased cross-sectional radius

±45° ply angles, leading to a decrease in bending stiffness. Furthermore, when the ply angles and the number of plies are determined, the stacking sequence has negligible influence on the bending stiffness but can affect the folding moment. When given specific geometric parameters (cross-sectional radius, central angle, thickness, and length), as the stacking sequence changes, the axial elastic modulus of the tubular deployable composite booms has no change, but the bending stiffness matrix changes. Consequently, the stacking sequence has negligible influence on bending stiffness of the tubular deployable composite booms, while it has an influence on folding moment.

Based on Appendices C and D, it is evident that all design points on the Pareto front have the same cross-sectional radius of 40 mm. To investigate the reasons for these optimal designs, curves depicting the folding moment and bending stiffness variation with cross-sectional radius are plotted in Fig. 11.7. Other geometric parameters and stacking sequences remained consistent with the initial sample. According to Fig. 11.7, the change of folding moment is subtle with increasing cross-sectional radius, while the bending stiffness is rapidly increased. Therefore, setting the cross-sectional radius to the maximum value within the variable space allows for obtaining maximum bending stiffness, as the influence of cross-sectional radius on the folding moment is negligible. Based on this finding, it is recommended that researchers should directly set the cross-sectional radius to the maximum value within the variable space when investigating similar optimization problems.

11.3.5 Benchmarking for Hyperparameters

In practical engineering applications, long optimization time caused by excessive function calls and population size can lead to decreased computational efficiency. Therefore, in order to improve the computational efficiency while obtaining a reasonable and convergent Pareto front, it is advisable to reduce the number of function calls and the population size appropriately. This section investigates the influence of hyperparameters (i.e., function calls and population size) on the optimization results. To determine the optimal number of function calls, the optimization results with 15,000, 35,000, 55,000, and 75,000 function calls are compared while maintaining the same population size. Figure 11.8a illustrates the Pareto front obtained with these four different numbers of function calls. From Fig. 11.8a, it is observed that the Pareto fronts obtained from different numbers of function calls have similar coverage ranges and exhibit high quality.

Similarly, the influence of population size on the optimization results is analyzed. Considering the special mechanism of MOEA/D, the number of updates on the best solution found so far for each subproblem, the neighboring solutions, and the external population for storing non-dominated solutions must be kept consistent. In general, function calls are represented as the product of population size and the number of weight vector updates. As shown in Fig. 11.8b, given the fixed number of 25 weight vector updates (i.e., function calls change), it can be seen that as the population size increases, the density and diversity of the Pareto front gradually improve. A larger population size provides a larger candidate solution space and encourages the algorithm to discover more and better solutions during the search process. However, when the population size reaches a certain threshold (e.g., 200), the Pareto front tends to converge. It reflects that there is a certain population size threshold for this type of optimization problem. The computational resources will be wasted by increasing the population size after passing the threshold.

11.3.6 Benchmarking for Algorithms

Although this study employed a well-known multi-objective optimization algorithm to design the deployable composite booms, a general solver that maintains optimal performance across all optimization problems has not been found. Therefore, in this section, four other high-performance multi-objective genetic algorithms from the Geatpy toolbox were benchmarked; to make a fair comparison among these algorithms, the same genetic operators and function calls were applied. Figure 11.9 illustrates the Pareto fronts obtained by these four algorithms. Among them, NSGA-III, NSGA-II, and MOEA/D found similar coverage ranges and high-quality Pareto fronts, demonstrating excellent performance and diversity. This further validates the proposed multi-objective optimization framework in this chapter. However, awGA and RVEA, as general optimization algorithms, performed poorly in this optimization process. This highlights once again the need of performing rigorous benchmarking on the state-of-the-art genetic algorithms when addressing an optimization problem, as these engineering problems have specific dominant characteristics that may differ from those of the mathematical optimization problems.

In summary, although MOEA/D demonstrated good performance in this study, different genetic algorithms may exhibit different performances in solving the

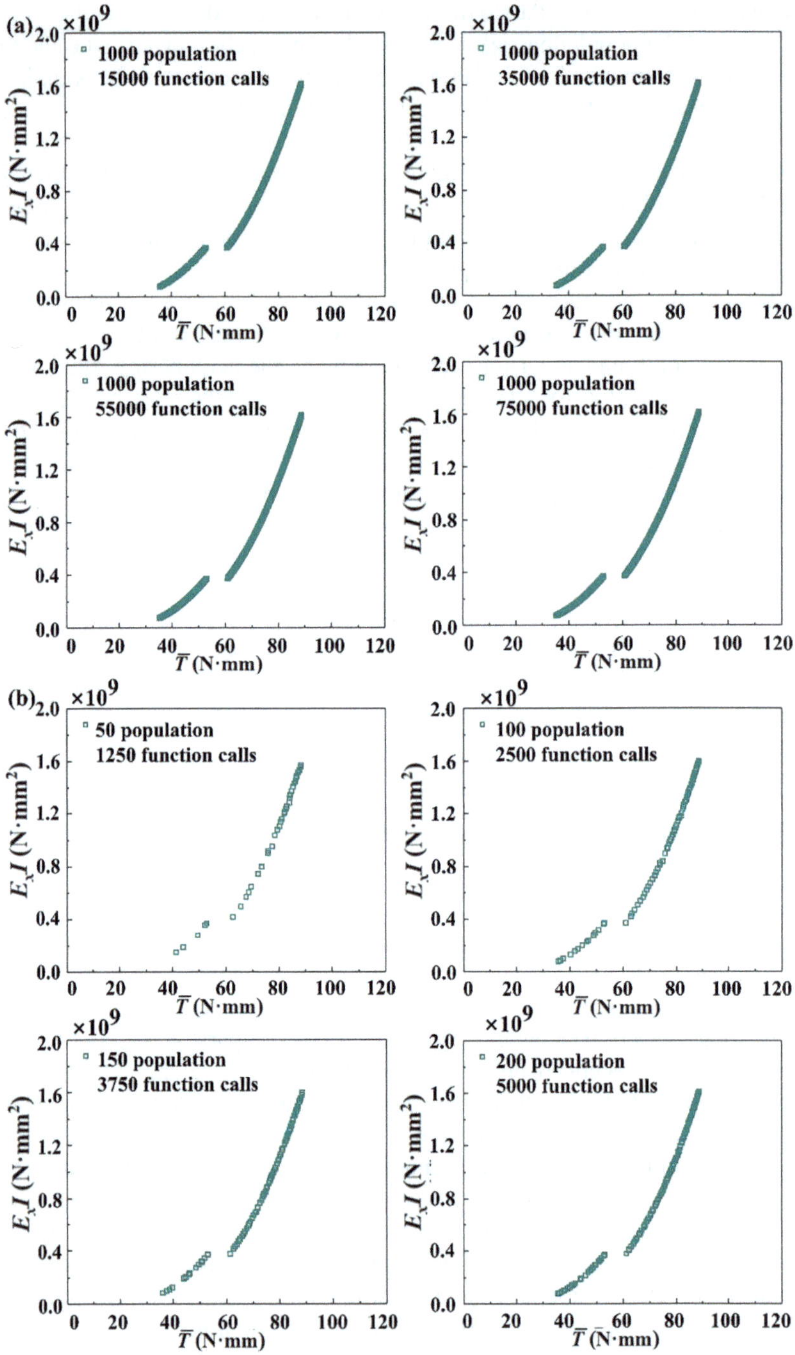

Fig. 11.8 The effect of hyperparameters on the Pareto front: (**a**) Function calls. (**b**) Population size

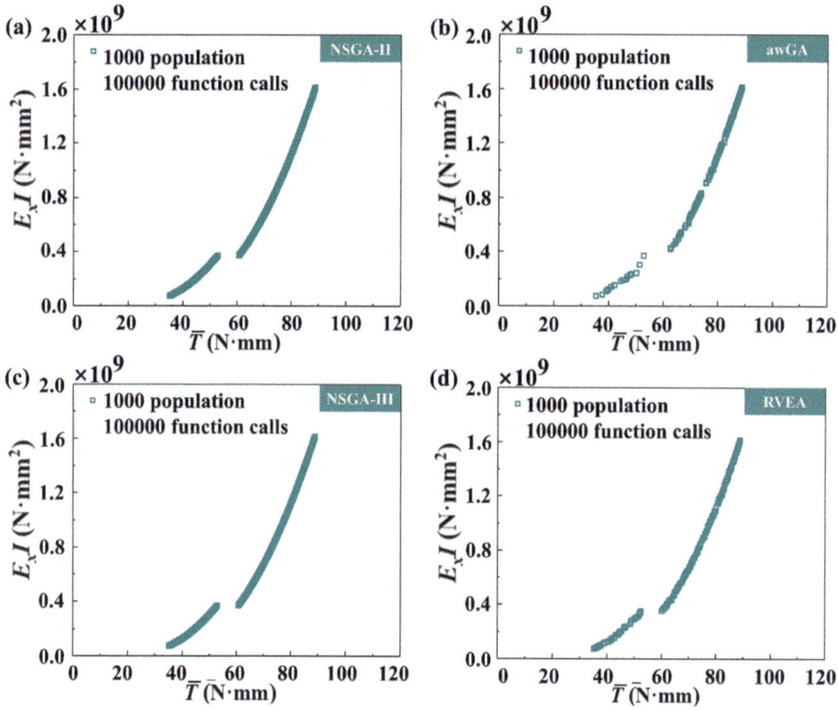

Fig. 11.9 The effect of genetic algorithm types on the Pareto front: (**a**) NSGA-II. (**b**) awGA. (**c**) NSGA-III. (**d**) REVA

optimization problem of the tubular deployable composite boom. Therefore, when selecting a genetic algorithm, the specific requirements of the problem should be considered, and benchmarking the state-of-the-art genetic algorithms is a task that requires further investigation.

11.4 The Bistable Deployable Composite Boom

11.4.1 Problem Description

To carry out complex space missions and adapt to harsh operational environments, the bistable deployable composite boom should achieve efficient stowage, lightweight, stability, reliability, and excellent mechanical properties. In space missions, due to cost constraints, the mass of the bistable deployable composite boom should be minimized to reduce launch costs as much as possible. Simultaneously, a compact geometric configuration in the folding stable state is crucial for efficient stowage within limited space. To ensure the deployment of the bistable deployable composite boom, strain energy should be stored as much as possible in the folding stable state. Furthermore, the bistable deployable composite boom should have a high natural frequency in the deploying stable state to ensure stability. In summary, to achieve the objectives of lightweight, efficient folding, and reliable use of the bistable deployable composite boom, the efforts should be devoted to reducing the linear density, decreasing the volume of the folding stable state, increasing the average strain energy, and enhancing the natural frequency. These measures will facilitate the optimization of the bistable deployable composite boom's performance and satisfy the diverse requirements of space missions.

Therefore, the cross-sectional radius, cross-sectional angle, and length are selected as design variables, and the Tsai-Hill criterion and maximum stress criterion as failure constraints; an optimization model is established with objectives including minimizing the linear density m and the longitude-sectional polar radius at the end point r_1, maximizing the average strain energy $\bar{u}$ and the natural frequency f. Through many-objective optimization design, design variables and objective functions are matched to obtain an optimal scheme.

Correspondingly, the aforementioned many-objective problem can be described as

$$\text{Opt.}\begin{cases} \mathit{Minimize}:\left\{m\left(R,\theta\right),r_1(R,\theta,L),1/\bar{u}(R,\theta,L),1/f(R,\theta,L)\right\} \\ \text{s.t}:\quad 25\ \text{mm} \le R \le 30\ \text{mm} \\ \qquad 150^\circ \le \theta \le 270^\circ \\ \qquad 500\ \text{mm} \le L \le 1000\ \text{mm} \\ \qquad \max\left(I_{\text{f1}},I_{\text{f2}}\right) < 1 \end{cases} \tag{11.15}$$

In addition, an antisymmetric layup configuration $[45°/-45°/45°/-45°/45°]_s$ was selected as the layup scheme for the bistable deployable composite boom. Other material properties and parameters remain consistent with those specified in Chap. 9.

11.4.2 Optimal Framework

Based on the aforementioned optimization problem, a multi-objective optimization framework for the bistable deployable composite boom was established by leveraging the advantages of analytical and surrogate models. The corresponding optimization process consists of four steps, as illustrated in Fig. 11.10.

Drawing upon the aforementioned optimization problem, a many-objective optimization framework for the bistable deployable composite boom is formulated, leveraging the synergies of analytical and surrogate models. The optimization process is delineated in four distinct steps, as depicted in Fig. 11.10. In the first stage, finite element analysis (FEA) is designed to gather a comprehensive dataset grounded in the design space. To guarantee the credibility and efficacy of the data,

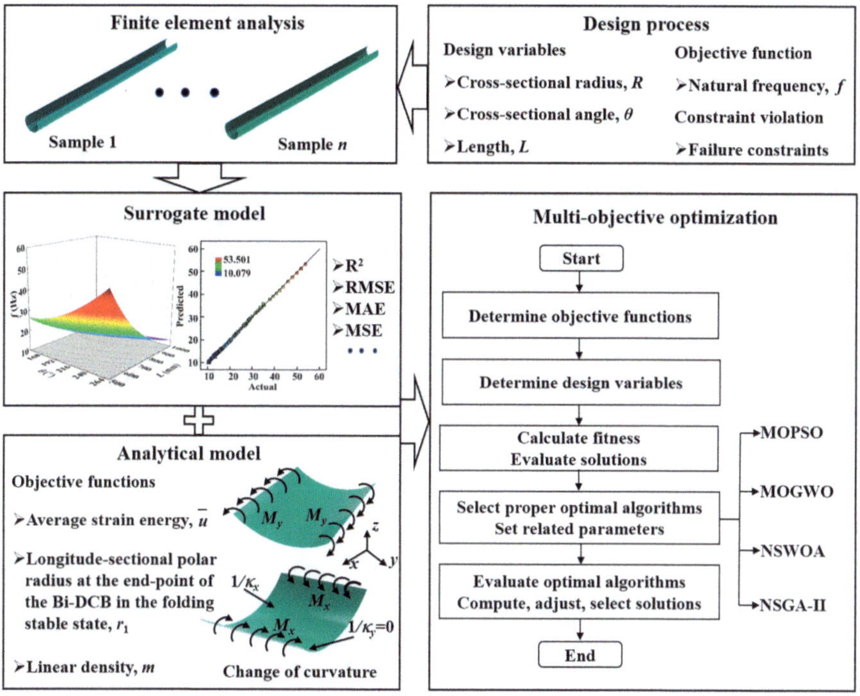

Fig. 11.10 Flowchart of modeling and optimization processes

a three-factor, five-level experimental design is employed, and the requisite dataset for the surrogate model is procured through FEA software. The second step involves using the dataset obtained from FEA as sample inputs to generate a surrogate model capable of predicting the natural frequency in the Design-Expert software. The accuracy of this surrogate model is evaluated through variance analysis. Subsequently, the analytical and surrogate models are integrated to ascertain the objective functions of the bistable deployable composite boom, namely the regression equation for the natural frequency, and analytical expressions for the average strain energy, the longitude-sectional polar radius at the end point, and the linear density. Finally, based on the objective functions, design variables, and constraint criteria, a many-objective optimization design framework is determined. To address the practical engineering challenge at the core of this study, four different optimization algorithms are employed for benchmark testing and performance assessment: multi-objective particle swarm optimization (MOPSO) algorithm, multi-objective grey wolf optimization (MOGWO) algorithm, non-dominated sorting whale optimization algorithm (NSWOA), and NSGA-II. Ultimately, pertinent verifications will be conducted to validate optimization results.

11.4.3 Determining Objective Functions

In this section, analytical and surrogate models are utilized to explicitly define the functional expressions for the four objectives. Based on analytical model 1 described in Sect. 9.1 and the line density analytical model, the functional expressions for the mean strain energy, the radial position of the longitude-sectional polar radius at the end point in the folded stable state, and the line density can be derived. However, it is challenging to establish an analytical model to predict the natural frequency of the tubular bistable deployable composite boom in its deployed stable state. To address this challenge, a surrogate model is employed to construct the functional expression for the natural frequency.

Analytical model 1 from Sect. 9.1 provides the functional expressions for the total strain energy and the radial position of the longitudinal section end point in the folded stable state of the tubular bistable deployable composite boom. Therefore, the mean strain energy in the folded stable state can be expressed as

$$\overline{u} = \frac{U}{L} \tag{11.16}$$

Based on the geometric model, the linear density of the bistable deployable composite boom can be expressed as

$$m = Rt\theta\rho_{\mathrm{d}} \tag{11.17}$$

Natural frequency is an important inherent property of a structure, which is usually closely related to mass, geometric configuration, and stiffness. It serves as a crucial indicator for structural stability assessment and vibration characteristic analysis. The bistable deployable composite boom belongs to long, thin-walled structures with a significant difference in size between the cross-sectional radius and thickness, making it prone to instability. The analytical derivation for stability analysis of such a structure is complex and cumbersome, while FEA is a relatively simple and efficient technique. Therefore, considering the limitations of analytical methods, this section adopts FEA and RSM to establish a computational model for the natural frequency of the bistable deployable composite boom.

To establish a surrogate model for predicting the natural frequency of the bistable deployable composite boom, it is first necessary to obtain a required dataset through FEA. In this process, a commercial software is employed as an FEA tool, with the cross-sectional radius, cross-sectional angle, and length serving as design variables for the experimental design. Based on variable ranges and experimental requirements, a three-factor five-level experiment design is devised, comprising a total of 125 combinations, as shown in Appendix E. The global mesh size is set at 2.5 mm, and the boundary conditions are set as one end fixed and the other end free, allowing for free vibration. Through 125 FEA iterations, the corresponding dataset is obtained, and Fig. 11.11 presents some of the FEA results.

Subsequently, based on the aforementioned dataset, a surrogate model for the natural frequency of the bistable deployable composite boom is established using

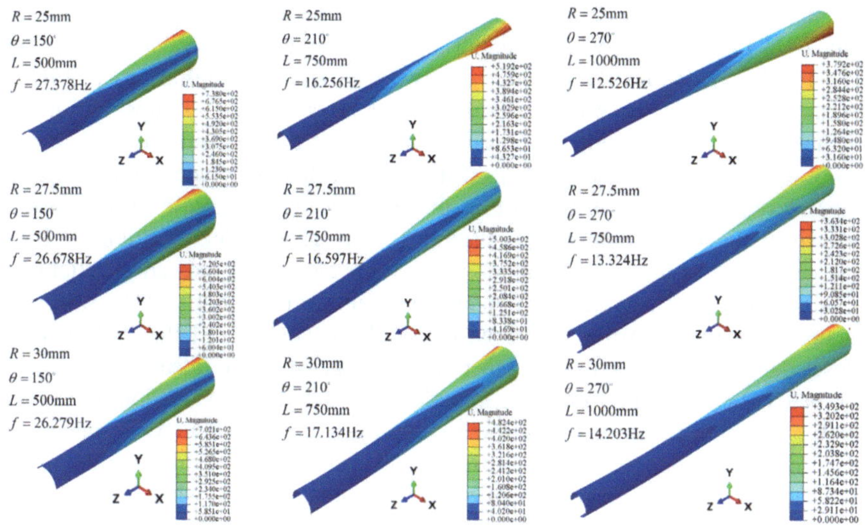

Fig. 11.11 Partial FEA results of the natural frequency

RSM. RSM is a typical approach for establishing surrogate models and is commonly used to study the mathematical relationships between multiple design variables and objectives. It integrates the advantages of experimental analysis and mathematical modeling, offering benefits such as low computational cost and fast convergence speed, making it widely applied in engineering. Among the various experimental design methods available in RSM, including Plackett-Burman design, central-composite design, and Box-Behnken design, this study uses Box-Behnken design as the experimental design method for establishing the surrogate model. Considering the fitting accuracy and computational efficiency of polynomials, a cubic polynomial regression equation is adopted to describe the variation pattern of the response value. Accordingly, the natural frequency of the tubular bistable deployable composite boom in the deploying stable state can be represented as

$$f = f\left(R,\theta,L\right) \tag{11.18}$$

The cubic polynomial obtained through the Box-Behnken design fitting for the natural frequency can be expressed as

$$\begin{aligned} f \quad &= 160.554 + 0.566R - 0.936\theta - 0.222L + 0.024R\theta \\ &+6.808\times10^{-4} RL - 6.832\times10^{-5}\theta L - 0.154R^2 \\ &+2.558\times10^{-3}\theta^2 + 2.221\times10^{-4} L^2 - 2.288\times10^{-5} R\theta L \\ &-5.141\times10^{-5} R^2\theta - 2.420\times10^{-5} R^2 L + 9.375\times10^{-6} R\theta^2 \\ &+2.887\times10^{-6} RL^2 - 1.870\times10^{-6}\theta^2 L + 7.690\times10^{-7}\theta L^2 \\ &+2.261\times10^{-3} R^3 - 1.006\times10^{-6}\theta^3 - 1.653\times10^{-6} L^3 \end{aligned} \tag{11.19}$$

Finally, the accuracy of the surrogate model is assessed through analysis of variance (ANOVA). In ANOVA, two important indicators are the F-value and P-value. The F-value quantifies the difference between groups and within groups, and a larger F-value indicates a more significant difference between groups, not due to chance factors. The P-value, as a probability value in the test, indicates the significance of differences between groups if it is smaller than the predetermined significance level. Therefore, an increase in the F-value and a decrease in the P-value effectively demonstrate the significance of the surrogate model. Additionally, R^2 is a key indicator of model correlation, with a value closer to 1 indicating better correlation and higher data fitting. Meanwhile, to ensure the validity of the surrogate model, the errors of Adj-R^2 and Pred-R^2 should be controlled below 0.2, and the closer the two values are to each other and the larger they are (ideally closer to 1), the better the model performance. In addition to the above evaluation indicators, coefficient of variation (CV) and the average relative prediction (ARP) values are also important parameters for measuring model accuracy. A CV value less than 10% indicates a small coefficient of variation and good data stability, while an ARP value greater than 4 further verifies the accuracy of data fitting. Table 11.6 provides specific results of the analysis of variance, from which the following conclusions can be drawn: the model fits the cubic regression equation with large F-values and small P-values, R^2 is 0.9993, the errors of Adj-R^2 and Pred-R^2 are less than 0.2, the CV value is 1.39%, and the ARP value is 374.862. All these parameter values are within reasonable range, demonstrating that the established surrogate model has high predictive accuracy and wide adaptability.

Figure 11.12 provides detailed graphical analyses, including the normal plot of residuals, the residuals versus predicted plot, the predicted versus actual plot, the residual distribution plot versus the test steps, the Box-Cox chart plot, and the perturbation chart plot. These plots intuitively demonstrate the results of the analysis of variance. Specifically, the residual normal probability plot in Fig. 11.12a shows an approximately linear distribution trend of data points, indicating good normality of residuals and thereby reflecting the high fitting degree and adaptability. The residual versus predicted plot in Fig. 11.12b exhibits an irregular spot distribution, indicating good dispersion of data and further confirming the feasibility. Figure 11.12c illustrates the correspondence between predicted and actual values, with data points showing an approximately linear distribution, demonstrating the consistency between predicted and actual values. Figure 11.12d shows the residual distribution between predicted values and FEA values, indicating a uniform distribution of residuals between FEA values and predicted values, with errors controlled within 4%, confirming the accuracy of the surrogate model. The Box-Cox plot in Fig. 11.12e demonstrates the process of transforming data into an approximate normal distribution through Box-Cox transformation. This transformation helps reduce the impact of noise and outliers on the model, improving the stability of the data. Through this transformation, the optimal parameter lambda is found to correspond to the

Table 11.6 ANOVA analyses for the dataset of Box-Behnken design

Source	Degree of freedom	Sum of square	F-value	P-value
Model	19	12,350	7829.57	<0.0001
A-R	1	1.01	12.16	0.0007
B-θ	1	75.55	909.98	<0.0001
C-L	1	411.44	4955.36	<0.0001
AB	1	47.84	576.18	<0.0001
AC	1	15.45	186.11	<0.0001
BC	1	761.81	9175.15	<0.0001
A^2	1	0.14	1.63	0.2039
B^2	1	172.38	2076.15	<0.0001
C^2	1	708.93	8538.34	<0.0001
ABC	1	11.50	138.54	<0.0001
A^2B	1	4.066 E-03	0.049	0.8253
A^2C	1	0.016	0.19	0.6652
AB^2	1	0.078	0.94	0.3351
AC^2	1	2.22	26.80	<0.0001
B^2C	1	30.96	372.86	<0.0001
BC^2	1	90.96	1095.48	<0.0001
A^3	1	7.022 E-03	0.85	0.7718
B^3	1	0.27	3.20	0.0766
C^3	1	37.51	451.81	<0.0001
Residual	105	8.72		
Cor total	124	1.236E+04		
Std. Dev		0.29		
Mean		20.68		
CV(%)		1.39		
Press		12.86		
Adeq. Prec		374.862		
R^2		0.9993		
Adj-R^2		0.9992		
Pred-R^2		0.9990		

minimum sum of squares and the response quantity, making the data closer to a normal distribution. Finally, Fig. 11.12f shows the influence of the cross-sectional radius, cross-sectional angle, and length on the objective function. The analysis indicates that the length has the most significant impact on the natural frequency, with a negative correlation; the cross-sectional angle follows, showing a positive correlation with natural frequency, while the influence of the cross-sectional radius is relatively small. These results provide another perspective to validate the accuracy and effectiveness of the surrogate model and also provide a reliable computational model for subsequent many-objective optimization.

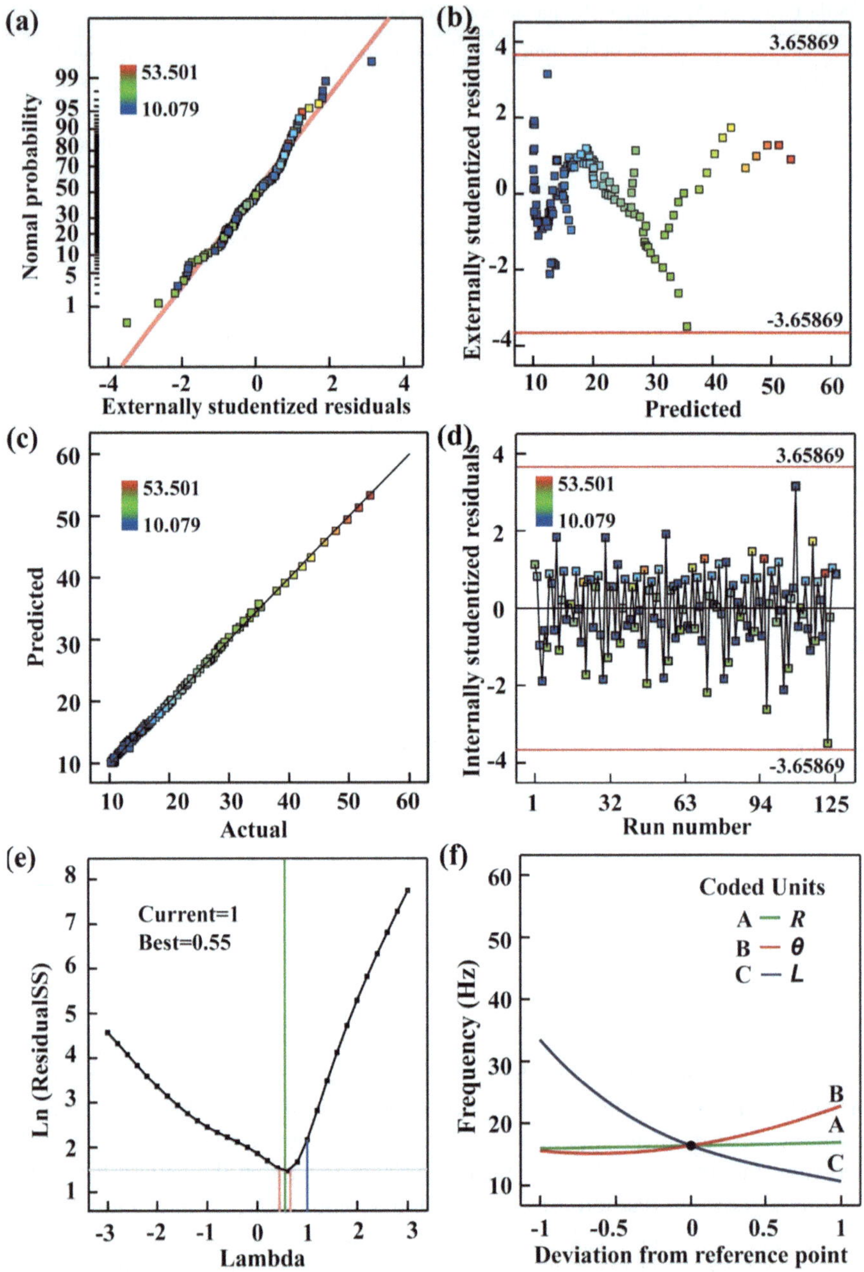

Fig. 11.12 Box-Behnken design analysis: (**a**) Normal plot of residual plot. (**b**) Residuals versus predicted plot. (**c**) Predicted versus actual plot. (**d**) Residual distribution versus the test step plot. (**e**) Box-Cox chart plot. (**f**) Perturbation chart plot

11.4.4 Optimization Algorithms

In this section, benchmark testing was conducted on four optimization algorithms, including MOPSO, MOGWO, NSWOA, and NSGA-II. Additionally, this section also introduced the basic principles, unique characteristics, and implementation processes of these four optimization algorithms, aiming to provide readers with a better understanding and evaluation of the performance and applicability of these algorithms.

MOPSO is a heuristic algorithm. It simulates the natural behavior of bird flock foraging by representing solutions in the problem space as "particles" and searching for the optimal solution based on the collaboration and competition among particles, as shown in Fig. 11.13a. Combining the advantages of single-objective particle swarm optimization with multi-objective optimization techniques, MOPSO exhibits strong global search capability and fast convergence speed. It is specialized in solving multi-objective optimization problems, providing effective solution strategies for complex problems. Figure 11.13b shows optimal program flowchart of MOPSO. The description of this algorithm refers to the detailed introduction of MOPSO by Coello et al. [60], and the hyperparameters are listed in Table 11.7.

MOGWO is an advanced version of the single-objective grey wolf optimization algorithm [61], which seeks solutions by simulating the unique social structure and hunting behavior of grey wolves, as illustrated in Fig. 11.14a. Based on the original grey wolf optimization algorithm, MOGWO further extends and expands its functionality by introducing social hierarchy-based optimization and archive

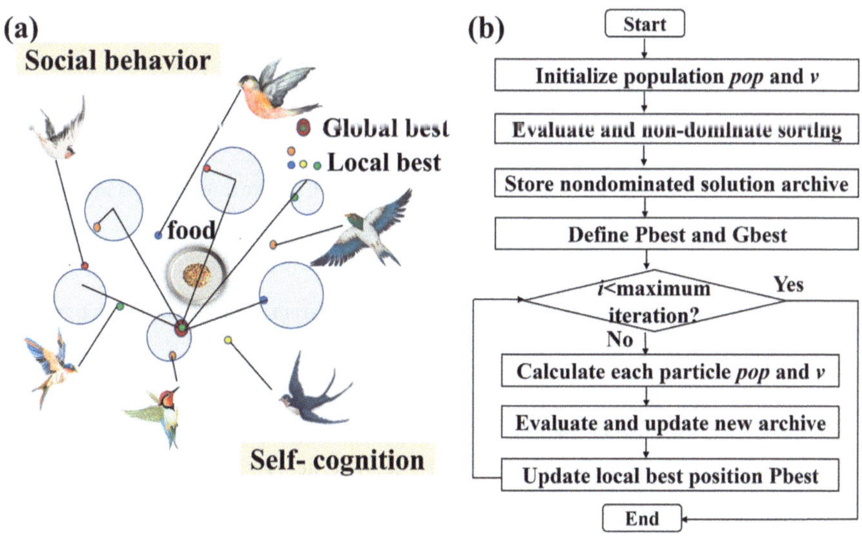

Fig. 11.13 The description of MOPSO: (**a**) Schematic diagram of principle. (**b**) Optimal program flowchart

Table 11.7 Parameter definition of MOPSO

Algorithm parameters	Values
Number of populations	200, 500, 1000
Number of generations	100
Number of archives	200, 500, 1000
Inertia weight	0.9
Inertia weight damping rate	0.9973
Personal learning coefficient	0.5
Global learning coefficient	2.5
Number of grids per dimension	12
Inflation rate	0.1
Leader selection pressure	2
Deletion selection pressure	4
Mutation rate	0.2

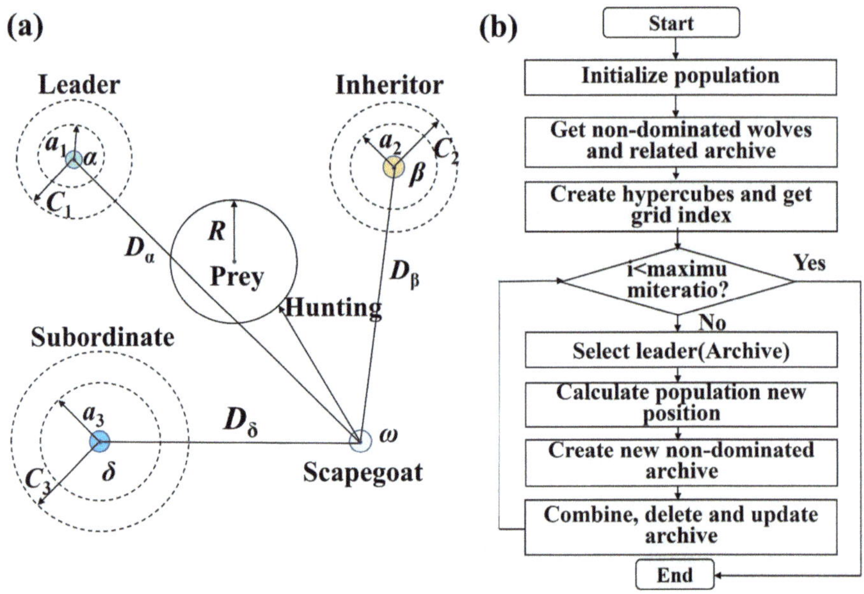

Fig. 11.14 The description of MOGWO: (**a**) Schematic diagram of principle. (**b**) Optimal program flowchart

management. The algorithm utilizes a pyramid-shaped social hierarchy system for non-dominated rank selection and leadership, while storing the best solutions in an external repository, as depicted in Fig. 11.14b. The description of this algorithm is based on the detailed introduction of MOGWO by Mirjalili et al. [62], and the hyperparameters are listed in Table 11.8.

Whale optimization algorithm (WOA) is a population-based intelligent optimization algorithm proposed by Mirjalili et al. [63] in 2016. Its design is inspired by

Table 11.8 Parameter definition of MOGWO

Algorithm parameters	Values
Number of populations	200, 500, 1000
Number of generations	100
Inflation rate	0.1
Number of grids per dimension	10
Number of archives	200, 500, 1000
Leader selection pressure	4
Extra repository of selection pressure	2

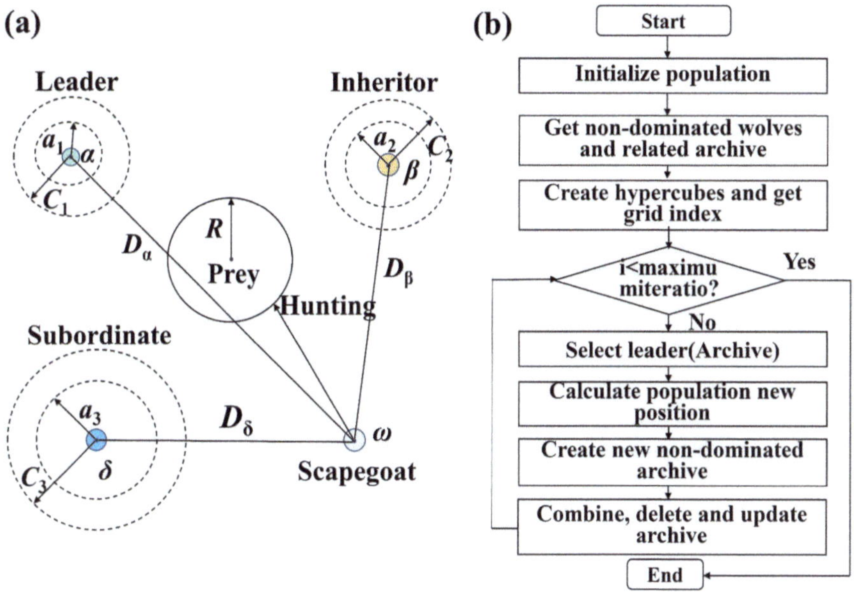

Fig. 11.15 The description of NSWOA: (**a**) Schematic diagram of principle. (**b**) Optimal program flowchart

the unique hunting behavior of whales, specifically the bubble-net feeding method used by humpback whales, as illustrated in Fig. 11.15a. This algorithm simulates the attack and foraging strategies of whales by mimicking their spiral encircling mechanism. However, WOA has difficulties effectively managing the relationships between multiple objective functions, thus limiting its application in the field of multi-objective optimization. To overcome this challenge, researchers have borrowed the handling strategy from NSGA-II and improved the whale optimization algorithm, developing the NSWOA. The algorithmic flowchart is illustrated in Fig. 11.15b. The description of this algorithm is based on the detailed introduction of NSWOA by Jangir et al. [64], and the hyperparameters are listed in Table 11.9.

NSGA is a genetic algorithm proposed by Srinivas and Deb [65] in the early 1990s, which adopts the Pareto dominance strategy. Subsequently, Deb et al. [66]

Table 11.9 Parameter definition of NSWOA

Algorithm parameters	Values
Number of populations	200, 500, 1000
Number of generations	100
Number of grids per dimension	10
Parameter, r	[0, 1]
Parameter, a	Decrease linearly from 2 to 0
Parameter, b	1
Parameter, l	[−1, 1]
Parameter, p	[0, 1]

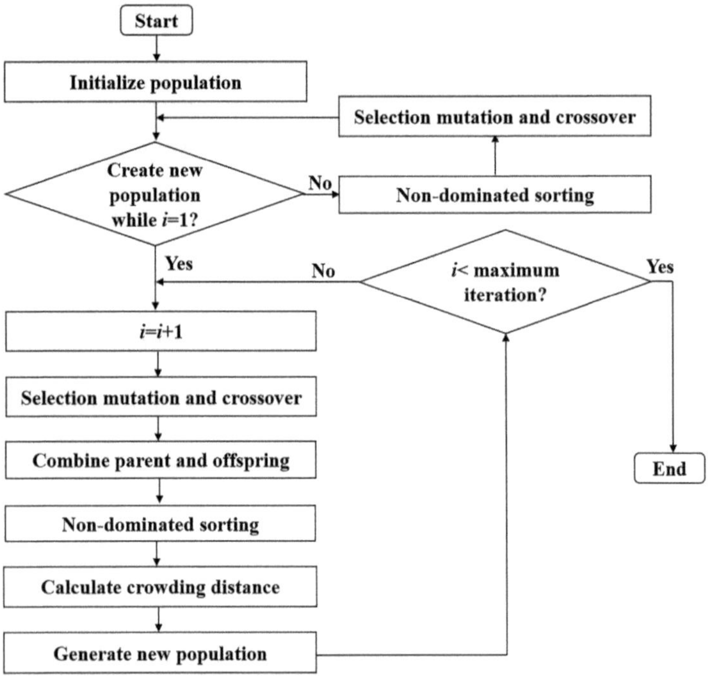

Fig. 11.16 Flowchart of NSGA-II

made significant extensions and upgrades to the original algorithm in 2000, introducing NSGA-II with fast non-dominated sorting characteristics. This new version of the algorithm incorporates concepts such as fast non-dominated sorting, crowding distance, and elitist strategy, significantly improving the computational efficiency and optimization capability of the algorithm. The algorithmic flowchart is illustrated in Fig. 11.16. The description of this algorithm is based on the introduction of NSGA-II from reference [66], and the hyperparameters are listed in Table 11.10.

Table 11.10 Parameter definition of NSGA-II

Algorithm parameters	Values
Number of populations	200, 500, 1000
Number of generations	100
Selection type	Crowding distance
Crossover type	SBX operator
Crossover rate	0.9
Mutation type	Polynomial operator
Mutation rate	0.33

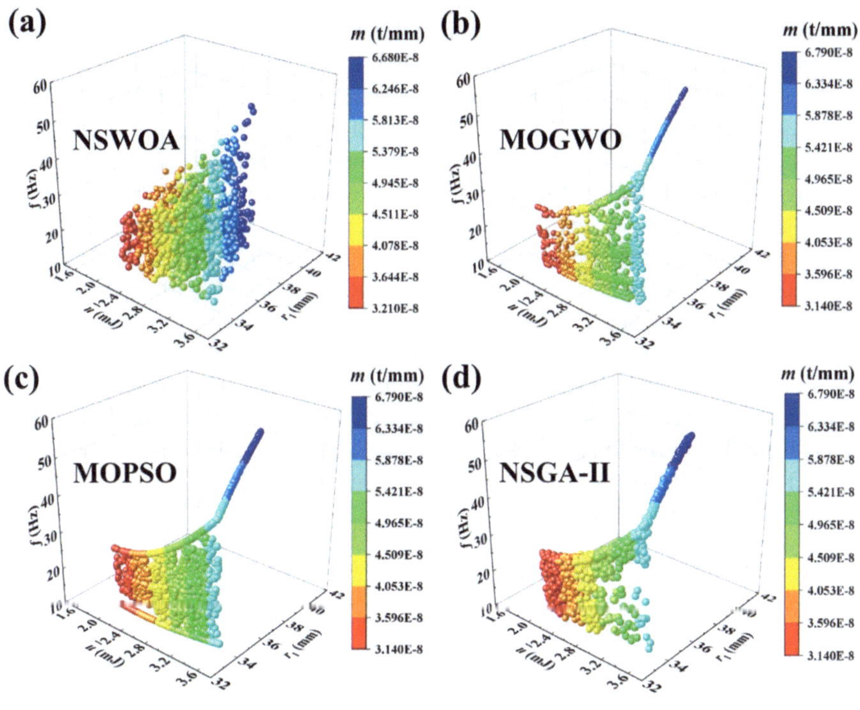

Fig. 11.17 Comparison of Pareto fronts with 1000 populations and 100 generations: (**a**) MOPSO. (**b**) MOGWO. (**c**) NSWOA. (**d**) NSGA-II

11.4.5 Benchmarking for Algorithms

Under the hyperparameter setting of 1000 populations and 100 generations, a comparison was made among four algorithms (i.e., MOPSO, MOGWO, NSWOA, and NSGA-II) in terms of the performance in generating the Pareto front, as shown in Fig. 11.17. It can be observed from Fig. 11.17 that the Pareto front generated by NSWOA exhibits a scattered distribution in space, with fuzzy boundaries and a lack of clear delineation. This reflects the significant randomness in NSWOA's search for the Pareto front, with the solution space displaying multidirectional and multilevel distribution characteristics, without clear regularity. Therefore, it fails to approximate the true Pareto front. This is mainly because NSWOA fails to balance and coordinate the conflicting relationships between multiple objectives, resulting in limited ability to search for reasonable Pareto front. In contrast, MOPSO, MOGWO, and NSGA-II perform better. The Pareto fronts generated by the three algorithms have clear contours and a uniform distribution, demonstrating high search efficiency and optimization capability. It indicates that they can better balance and coordinate the relationships between different objectives when dealing with multi-objective optimization problems. NSGA-II is renowned for its superior convergence and distribution properties, efficiently locating a Pareto front that closely resembles the true Pareto front and demonstrating a high degree of uniformity in the distribution of solutions. MOPSO possesses strong global search capability and rapid convergence, enabling it to discover a high-quality Pareto front with limited computational resources. Furthermore, MOGWO achieves an effective approximation of the Pareto front through its distinctive search strategy and information-sharing mechanism.

In practical engineering optimization problems, the accuracy and diversity of the Pareto front are critical metrics for evaluating algorithm performance. An ideal Pareto front should comprehensively cover the true Pareto solution set. However, due to the unknown nature of the true Pareto front, assessing algorithm performance becomes more challenging. To evaluate the quality of the Pareto front, this section conducts a convergence study of four optimization algorithms under population sizes of 200, 500, and 1000. This section employs the mimicked inverted generational distance (mIGD) method, introduced by Wang and Sobey [11]. Inspired by IGD [17], mIGD effectively assesses both the accuracy and diversity of the Pareto front. The computation of mIGD can be expressed as

$$\mathrm{mIGD}\left(O,M^{*}\right)=\frac{\sum_{v\in M} d\left(v,O\right)}{\left|M^{*}\right|} \tag{11.20}$$

where M^* represents a set of points along the mimicked Pareto front, O represents a set of points from the currently obtained Pareto front, ν represents each point in the set M^*, and $d(\nu, O)$ calculates the minimum Euclidean distance between v and the points in O. The lower mIGD value reflects better convergence, accuracy, and diversity of the Pareto front (Fig. 11.18).

Moreover, as the population size increases, the mIGD values of the three algorithms, excluding NSWOA, exhibit varying degrees of decline. This trend indicates that with larger populations, these algorithms improve their ability to explore the Pareto front space, bringing them closer to the ideal Pareto solution set. However, even at a population size of 1000, the algorithms have not yet reached convergence. Based on the current optimization results, it can be speculated that further increasing the population size might enhance the algorithms' performance, leading to the generation of superior Pareto fronts. Nevertheless, with the increase in population size, the computational complexity of the algorithms also rises, resulting in a

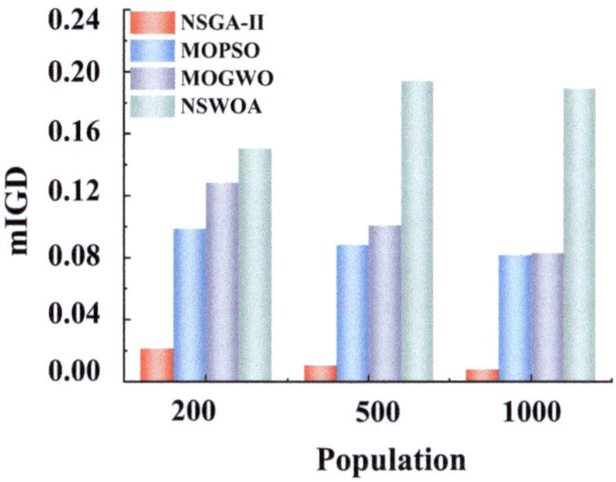

Fig. 11.18 Metric comparison of mIGD with different population sizes

significant increase in computation time. Therefore, in practical applications, it is crucial to strike a balance between algorithm performance and computational cost to achieve efficient solutions to optimization problems.

11.4.6 Design Schemes

The Pareto front obtained using NSGA-II, with a population size of 1000 and 100 generations, is shown in Fig. 11.19. To validate the effectiveness of the optimization cases on the Pareto front, a comparison between the pre-optimization and post-optimization cases is presented in Fig. 11.19a. This comparison highlights the improvements achieved through the multi-objective optimization process, demonstrating the practical utility of the proposed optimization framework.

Sample 5 from Sect. 9.3.1 was selected as the pre-optimization case (i.e., the initial sample), with its geometric parameters and layup scheme listed in Table 9.1. The objectives for this sample were $\overline{u}$ = 2.02 mJ, r_1 = 36.83 mm, f = 18.14 Hz, and m = 3.85E-8 t/mm. Among all the points on the Pareto front, a total of 112 design points were identified that outperformed the initial sample in all four objectives. These design points are primarily located in the upper left region of the Pareto front surface. The detailed optimization schemes for these 112 design points are provided in Appendix F. Figure 11.19a includes a localized zoom-in view of the region containing these 112 design points, where four representative feasible solutions (denoted as Points A, B, C, and D) are highlighted. As shown in Fig. 11.19b, all four objective function values for these representative design points are superior to those of the initial sample.

(a)

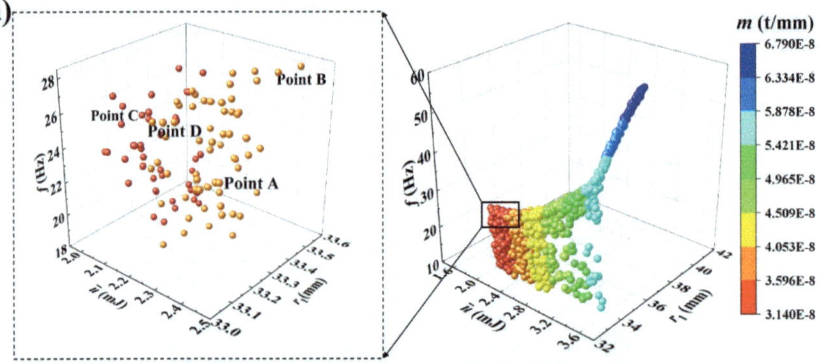

(b)

Designs	Patterns	$\bar{u}$ (mJ)	r_1 (mm)	f (Hz)	m (t/mm)	Values of design variables
Point A	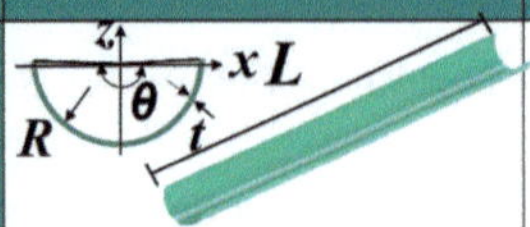	2.45	33.11	24.45	3.84E-8	R=25mm θ=183.4° L=553mm
Point B	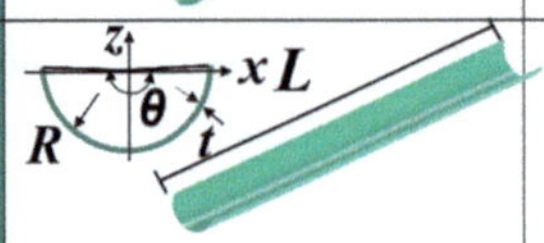	2.43	33.53	28.31	3.85E-8	R=25.1 θ=182° L=504mm
Point C	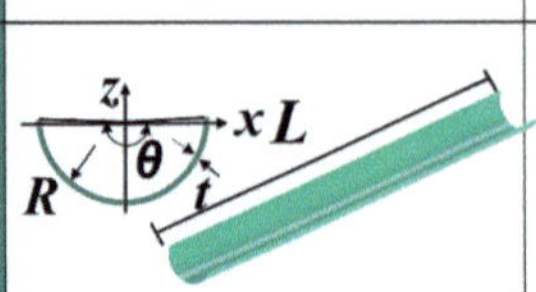	2.03	33.27	24.83	3.19E-8	R=25mm θ=152° L=534mm
Point D	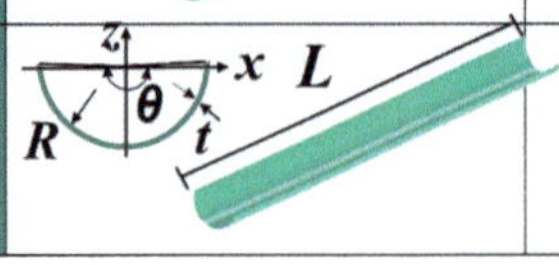	2.35	33.04	27.49	3.69E-8	R=25mm θ=176.4° L=507mm

Fig. 11.19 Optimization results for the bistable deployable composite boom: (**a**) Pareto front. (**b**) Design schemes and geometric configurations

11.5 Results and Discussion

11.5.1 Optimization Performance

Multiple design points were identified on the Pareto frontiers of both cases, showcasing significant optimization effects by surpassing the initial designs in various key performance metrics. However, detailed analysis of these design points reveals that while some points achieved substantial improvement in certain objectives, others showed only minor enhancements or remained unchanged. This underscores the complexity of addressing high-dimensional multi-objective optimization problems. It is challenging to achieve simultaneous optimality across all objectives due to potential trade-offs or conflicts among them. Despite these challenges, optimization algorithms successfully identified design solutions with balanced improvements across multiple objectives. While these solutions may not be optimal for individual objectives, their comprehensive performance makes them valuable for the optimization design of deployable composite structures. These findings highlight the importance of balancing objectives during the design process. In practical applications, it is essential to prioritize objectives and assign appropriate weights based on specific requirements and constraints to achieve the most suitable design solutions.

11.5.2 Significance of Benchmarking

In the optimization field, the "no free lunch" principle is widely accepted, emphasizing that no single algorithm can outperform others across all optimization problems. The optimization of deployable composite structures is a complex task involving multi-objective, multivariable, highly nonlinear, and constrained conditions. Such problems demand not only efficient algorithmic strategies but also a deep understanding of the problem's characteristics and requirements. As a result, benchmarking advanced evolutionary algorithms is a crucial step in research. Through benchmarking, key problem features can be understood, enabling the customization of algorithmic strategies to achieve better optimization results. Although general solvers like NSGA-II often deliver reasonable outcomes, they may not perform optimally for specific problems, as demonstrated by Wang et al. [11, 17]. Therefore, tailoring and optimizing algorithmic strategies to the characteristics of deployable composite structure optimization are essential.

Hyperparameter settings play a pivotal role in the optimization process, directly impacting the search efficiency and quality of results. Practical applications require careful tuning of hyperparameters based on problem-specific characteristics and requirements to ensure high-quality and efficient optimization outcomes. As technology advances and engineering demands grow, more efficient and precise algorithms are expected to emerge, offering solutions for complex problems such as deployable composite structure optimization. Continuous exploration of innovative optimization strategies and techniques will remain a critical direction for future research.

References

1. Forrester AIJ, Keane AJ (2009) Recent advances in surrogate-based optimization. Prog Aerosp Sci 45(1–3):50–79
2. Choi HS, Shin MS (1999) Hydrodynamics study of two different inverse fluidized reactors for the application of wastewater treatment. Korean J Chem Eng 16(5):670–676
3. Malik MH, Arif A (2013) ANN prediction model for composite plates against low velocity impact loads using finite element analysis. Compos Struct 101:290–300

4. Hussain S, Gul S, Khan H et al (2021) Modeling of photolytic degradation of sulfamethoxazole using boosted regression tree (BRT), artificial neural network (ANN) and response surface methodology (RSM); energy consumption and intermediates study. Chemosphere 276:130151
5. Foroutan R, Mohammadi R, Ramavandi B (2021) Waste glass catalyst for biodiesel production from waste chicken fat: optimization by RSM and ANNs and toxicity assessment. Fuel 291:120151
6. Prabhu MV, Karthikeyan R (2018) Comparative studies on modelling and optimization of hydrodynamic parameters on inverse fluidized bed reactor using ANN-GA and RSM. Alex Eng J 57(4):3019–3032
7. Degtyarev VV (2020) Neural networks for predicting shear strength of CFS channels with slotted webs. J Constr Steel Res 177:106443
8. Yang H, Deng ZQ, Liu RQ et al (2014) Optimizing the quasi-static folding and deploying of thin-walled tube flexure hinges with double slots. Chin J Mech Eng 27(2):279–286
9. Willmott CJ, Ackleson SG, Davis RE et al (1986) Statistics for the evaluation and comparison of models. J Geophys Res 33(3):250
10. Onu CE, Nwabanne JT, Ohale PE et al (2021) Comparative analysis of RSM, ANN and ANFIS and the mechanistic modeling in eriochrome black-T dye adsorption using modified clay. S Afr J Chem Eng 36:24–42
11. Wang ZZ, Sobey A (2020) A comparative review between genetic algorithm use in composite optimisation and the state-of-the-art in evolutionary computation. Compos Struct 233:111739
12. Fernandes P, Pinto R, Correia N (2021) Design and optimization of self-deployable damage tolerant composite structures: a review. Compos Part B 221:109029
13. While L, Bradstreet L, Barone L (2012) A fast way of calculating exact hypervolumes. IEEE Trans Evol Comput 16(1):86–95
14. Zitzler E, Deb K, Thiele L (2000) Comparison of multiobjective evolutionary algorithms: empirical results. Evol Comput 8(2):173–195
15. Elarbi M, Bechikh S, Said LB (2021) On the importance of isolated infeasible solutions in the many-objective constrained NSGA-III. Knowl-Based Syst 227:104335
16. Cheng S, Zhan H, Shu Z et al (2019) Effective optimization on bump inlet using meta-model multi-objective particle swarm assisted by expected hyper-volume improvement. Aerosp Sci Technol 87:431–447
17. Wang ZZ, Bai JB, Sobey A et al (2018) Optimal design of triaxial weave fabric composites under tension. Compos Struct 201:616–624
18. Klimm WJ, Ozdemir VB, Kwok K (2020) Multiobjective optimization of deployable visco-elastic tape springs under uncertainty. AIAA Aviation 2020 Forum 2020:3166
19. Su L, Zhang Y, Sun B (2020) Multi-objective optimization of deployable composite cylindrical thin-walled hinges with progressive damage. Struct Multidiscip Optim 61:803–817
20. Jin H, Jia Q, An N et al (2022) Surrogate modeling accelerated shape optimization of deployable composite tape-spring hinges. AIAA J 60(10):5942–5953
21. Li B, Ye H, Zhang Y (2019) Failure analysis of composite tube hinge and optimization design. IOP Conf Seri: Mater Sci Eng 531(1):012062
22. Mallikarachchi H, Pellegrino S (2014) Design of ultrathin composite self-deployable booms. J Spacecr Rocket 51(6):1811–1821
23. https://artes.esa.int/funding/antenna-deployment-arm-integrated-elastic-hinges-artes-ref-5b163-expro
24. Fernandes P, Pinto R, Ferrer A et al (2022) Performance analysis of a damage tolerant composite self-deployable elastic-hinge. Compos Struct 288:115407
25. Yang H, Lu F, Guo H et al (2019) Design of a new N-shape composite ultra-thin deployable boom in the post-buckling range using response surface method and optimization. IEEE Access 7:129659–129665
26. Yang H, Guo HW, Liu RQ et al (2020) Coiling and deploying dynamic optimization of a C-cross section thin-walled composite deployable boom. Struct Multidiscip Optim 61:1731–1738

27. Yang H, Liu L, Guo HW et al (2019) Wrapping dynamic analysis and optimization of deployable composite triangular rollable and collapsible booms. Struct Multidiscip Optim 59:1371–1383
28. Yang H, Fan S, Wang Y et al (2022) Novel four-cell lenticular honeycomb deployable boom with enhanced stiffness. Materials 15(1):306
29. Shi J, Qi J, Wang Y et al (2022) Analysis and optimization of M-shaped boom based on response surface method. Polym Compos 43(7):4327–4338
30. Bessa MA, Pellegrino S (2018) Design of ultra-thin shell structures in the stochastic post-buckling range using Bayesian machine learning and optimization. Int J Solids Struct 139:174–188
31. Zhang Z, Zhou H, Ma J et al (2022) Space deployable bistable composite structures with C-cross section based on machine learning and multi-objective optimization. Compos Struct 297:115983
32. Zhang Z, Liao C, Chai H et al (2021) Multi-objective optimization of controllable configurations for bistable laminates using NSGA-II. Compos Struct 266:113764
33. Wu C, Viquerat A (2017) Natural frequency optimization of braided bistable carbon/epoxy tubes: analysis of braid angles and stacking sequences. Compos Struct 159:528–537
34. Zhang LF, Long ZL, Cai JD et al (2015) Multi-objective optimization design of a connection frame in macro–micro motion platform. Appl Soft Comput 32:369–382
35. Bai JB, Liu TW, Wang ZZ et al (2021) Determining the best practice-optimal designs of composite helical structures using genetic algorithms. Compos Struct 268:113982
36. Liu TW, Bai JB, Fantuzzi N (2023) Folding behaviour of the thin-walled lenticular deployable composite boom: analytical analysis and many-objective optimization. Mech Adv Mater Struct 30(11):2221–2239
37. Bai JB, You FY, Wang ZZ et al (2024) An efficient multi-objective optimization framework for thin-walled tubular deployable composite boom. Compos Struct 327:117713
38. Liu TW, Bai JB, Fantuzzi N et al (2022) Multi-objective optimisation designs for thin-walled deployable composite hinges using surrogate models and genetic algorithms. Compos Struct 280:114757
39. Deb K, Pratap A, Agarwal S et al (2002) A fast and elitist multi-objective genetic algorithm: NSGA-II. IEEE Trans Evol Comput 6(2):182–197
40. Deb K, Jain H (2014) An evolutionary many-objective optimization algorithm using reference-point-based nondominated sorting approach, part I: solving problems with box constraints. IEEE Trans Evol Comput 18(4):577–601
41. Jain H, Deb K (2014) An evolutionary many-objective optimization algorithm using reference-point-based nondominated sorting approach, part II: handling constraints and extending to an adaptive approach. IEEE Trans Evol Comput 18(4):602–622
42. Zitzler E, Laumanns M, Thiele L (2001) SPEA2: improving the strength pareto evolutionary algorithm. 2001:103
43. Cheng R, Jin Y, Olhofer M et al (2016) A reference vector guided evolutionary algorithm for many-objective optimization. IEEE Trans Evol Comput 20(5):773–791
44. Zhang Q, Li H (2007) MOEA/D: a multi-objective evolutionary algorithm based on decomposition. IEEE Trans Evol Comput 11(6):712–731
45. Liu BY, Zhou BW, Yang DS et al (2022) Optimal planning of hybrid renewable energy system considering virtual energy storage of desalination plant based on mixed-integer NSGA-III. Desalination 521:115382
46. Zhang JR, Wang SL, Tang QH et al (2019) An improved NSGA-III integrating adaptive elimination strategy to solution of many-objective optimal power flow problems. Energy 172:945–957
47. Wu K, Zhang DX, Chen ZH et al (2019) Multi-type multi-objective imaging scheduling method based on improved NSGA-III for satellite formation system. Adv Space Res 63(8):2551–2565

48. Wang WW, Dai SJ, Zhao WZ et al (2021) Design optimization of a novel negative Poisson's ratio non-module battery pack system considering crashworthiness and heat dissipation. Compos Struct 275:114458
49. Razmi A, Rahbar M, Bemanian M (2022) PCA-ANN integrated NSGA-III framework for dormitory building design optimization: energy efficiency, daylight, and thermal comfort. Appl Energy 305:117828
50. Cao X, Wen ZG, Xu JJ et al (2020) Many-objective optimization of technology implementation in the industrial symbiosis system based on a modified NSGA-III. J Clean Prod 245:118810
51. Chahardoli S, Hadian H, Vahedi R (2018) Optimization of hole height and wall thickness in perforated capped-end conical absorbers under axial quasi-static loading (using NSGA-III and MOEA/D algorithms). Thin-Walled Struct 127:540–555
52. Wang HY, Xie H, Cheng W et al (2018) Multi-objective optimisation on crashworthiness of front longitudinal beam (FLB) coupled with sheet metal stamping process. Thin-Walled Struct 132:36–47
53. https://github.com/geatpy-dev/geatpy
54. http://geatpy.com/index.php/geatpy%E6%95%99%E7%A8%8B/
55. Karafotias G, Hoogendoorn M, Eiben AE (2015) Parameter control in evolutionary algorithms: trends and challenges. IEEE Trans Evol Comput 19(2):167–187
56. Barbosa HJC, Lemonge ACC, Bernardino HS (2015) A critical review of adaptive penalty techniques in evolutionary computation. In: Evolutionary constrained optimization, pp 1–27
57. Aleti A, Moser I (2016) A systematic literature review of adaptive parameter control methods for evolutionary algorithms. ACM Comput Surv 49(3):1–35
58. Zhang Q, Li H (2008) MOEA/D: a multiobjective evolutionary algorithm based on decomposition. IEEE Trans Evol Comput 11(6):712–731
59. Geatpy2. The Genetic and evolutionary algorithm toolbox for python with high performance. http://geatpy.com/index.php/home/. Accessed 18 May 2021
60. Coello CAC, Pulido GT, Lechuga MS (2004) Handling multiple objectives with particle swarm optimization. IEEE Trans Evol Comput 8(3):256–279
61. Mirjalili S, Mirjalili SM, Lewis A (2014) Grey wolf optimizer. Adv Eng Softw 69:46–61
62. Mirjalili S, Saremi S, Mirjalili SM et al (2016) Multi-objective grey wolf optimizer: a novel algorithm for multi-criterion optimization. Expert Syst Appl 47:106–119
63. Mirjalili S, Lewis A (2016) The whale optimization algorithm. Adv Eng Softw 95:51–67
64. Jangir P, Jangir N (2017) Non-dominated sorting whale optimization algorithm (NSWOA): a multi-objective optimization algorithm for solving engineering design problems. Glob J Res Eng 17:15–42
65. Srinivas N, Deb K (1994) Muiltiobjective optimization using nondominated sorting in genetic algorithms. Evol Comput 2(3):221–248
66. Deb K, Pratap A, Agarwal S et al (2002) A fast and elitist multiobjective genetic algorithm: NSGA-II. IEEE Trans Evol Comput 6(2):182–197

Chapter 12
Application and Outlook

Abstract This chapter reviews the aerospace applications and challenges of deployable composite structures. Since the 1970s, composite materials have replaced metals in deployable structures, such as glass fiber composite masts on Voyager. Deployable composite hinges are used in space telescopes, while booms (lenticular, bipod) feature in solar sails and deorbit sails (e.g., China's Taurus deorbit sail, Pujiang-2 film antenna). Reflectors like taco-shaped models have operated on Mobile Sat-1 and TDRSS. Future challenges include improving folding behavior modeling, enhancing extreme environment adaptability, and advancing multi-objective optimization, requiring breakthroughs in theoretical models, material properties, and auxiliary device design to boost reliability and functionality in space applications.

12.1 Application

In the early days, most of the deployable structures for space applications were made of metal materials such as titanium alloy, high-performance steel, and aluminum alloy. In the 1960s and 1970s, deployable booms were engineered and utilized in flight missions such as Technology Research for Autonomous Aircraft Concepts (TRAAC), Alouette I, and Applications Technology Satellite-6 (AST-6). With technological advancements, the use of space deployable structures has shifted from initially relying on metal materials to lighter-weight composite materials. The technology of continuous longeron mast (CLM), which employs glass fiber composite materials, has been successfully applied in various spacecraft, including Voyager (1977), Galileo (1989), Mars Observer (1992), and the GOES constellation (since 1994 to now). Towards the late 1970s, Lockheed Corporation replaced the aluminum ribs with carbon fiber composite ribs in the design of wrap-ribs for antennas [1]. This section provides a review of the application of deployable composite hinges, booms, and reflectors in the field of space exploration.

J.-B. Bai et al., *Folding Mechanics of Deployable Composite Structures*, Solid Mechanics and Its Applications 283,
https://doi.org/10.1007/978-3-031-99254-4_12

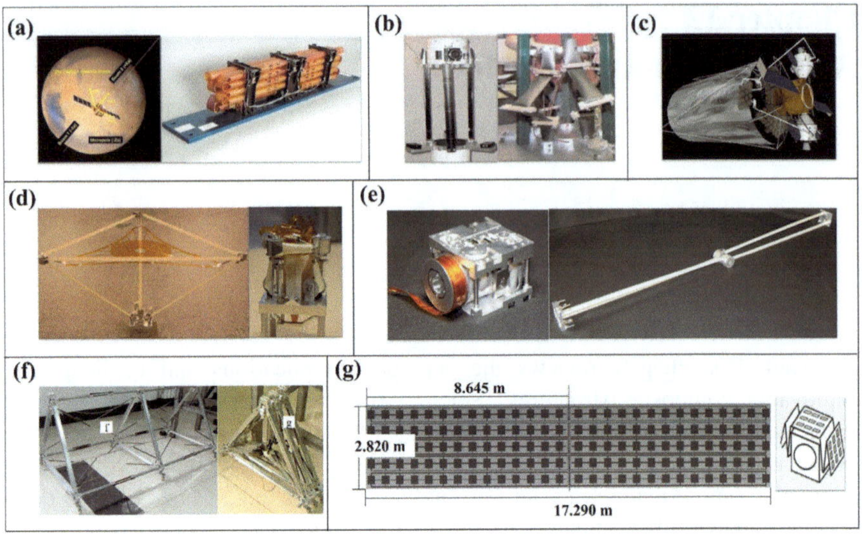

Fig. 12.1 Space applications of deployable composite hinges: (**a**) MARSIS. (**b**) Deployable telescope. (**c**) Notional 20-m telescope. (**d**) Deployable composite telescope support. (**e**) SIMPLE boom. (**f**) Triangular prism mast. (**g**) Large deployable P-band antenna based on tape-spring [3, 6–10]

12.1.1 Deployable Composite Hinges

The concept of using deployable tape springs as deployable structures can be traced back to 1968 [2], serving as an effective alternative to traditional hinge mechanisms. Since 1998, a series of deployable hinges have been developed, with their repeatability initially demonstrated in the integrated hinged optical support structure fabricated at the U.S. Air Force Research Laboratory. Integral deployable composite hinges have found wide-ranging applications in space missions. Examples include the antenna on the Mars Express spacecraft, launched by the ESA in June 2003 [3] (shown in Fig. 12.1a), and the successfully tested deployable optical telescope [4]. Furthermore, deployable composite hinges have also led to various other design concepts and space applications, including astronomical telescopes depicted in Fig. 12.1b–d [5–7], deployable booms depicted in Fig. 12.1e [8], prism masts shown in Fig. 12.1f [9], and antennas illustrated in Fig. 12.1g [10, 11]. Table 12.1 summarizes the various space applications of deployable composite hinges.

12.1.2 Deployable Composite Booms

In early space missions, most deployable booms were made of metal materials [1, 12]: for example, the metal tubular boom used on the Hubble Space Telescope [12] as well as the metal TRAC booms used on NanoSail D2 [13], LightSail 1 [14],

Table 12.1 A summary of space applications of deployable composite hinges

Type	Application	Operator	Material	Status
Integral [3]	Mars advanced radar	ESA	CFRP	In-orbit
Integral [4]	Precision optical telescope mirror	Foster Miller, Inc.	CFRP	In-orbit
Combined [5]	Deployable astronomical telescope	Air force research laboratory	CFRP	Porotype
Combined [6]	20 meter aperture UV-visible-IR telescope	Warren et al.	CFRP	Porotype
Combined [7]	Deployable composite telescope	Footdale and Murphey	CFRP	Concept design
Combined [8]	Self-contained linear meter-class boom	Jeon and Murphey	CFRP	Concept design
Integral [9]	Roll-out solar array	Inc. (DSS)	CFRP	Porotype
Integral [10]	P-band dual-polarization antenna	Soykasap et al.	CFRP	Concept design

LightSail 2 [15], CubeSail [16], and NEA Scout [17], as shown in Fig. 12.2a–f. However, the use of metal materials posed certain challenges due to their heavy weight and high thermal expansion coefficient.

In response to these challenges, many research institutions have begun to focus on developing deployable composite booms with superior performance. Since 1990, the German Aerospace Center (DLR) has been conducting research and development on the lenticular deployable composite boom with funding from the ESA. One successful application of the lenticular deployable composite boom in a flight mission is the jointly developed thin-film solar sail demonstrator by DLR and ESA, which underwent ground deployment tests [18], as shown in Fig. 12.2g. Additionally, the lenticular deployable composite boom was used for the support structure of a solar sail in the GOSSAMER-1 mission [19], as shown in Fig. 12.2h. Furthermore, various deployable structures have been developed based on the lenticular deployable composite boom, including radial ribs, filament-wound ribs, and Astro-tube masts [20, 21], as shown in Fig. 12.2i. These structures have been widely applied in practical space missions, such as the Space Flight Unit (SFU) developed by the Japan Aerospace Exploration Agency (JAXA), Muses-V, Variable Geometry Truss (VGT), solar sails, and pact-truss deployable antennas [22–24]. Some applications are still in the conceptual design [25] and testing stages [26].

Based on the successful development of TRAC booms, the NASA further developed TRAC booms made of composite materials. This concept was proposed by Murphey and Banik and further developed and applied by AFRL. Building upon AFRL's solar sail research, NASA developed the flexible unfurlable lightweight reflector, which utilizes four TRAC booms for the folding and deployment of the solar sail [27, 28], as shown in Fig. 12.2j. In addition, the SIASAIL-I solar sail, launched with the Xiaoxiang-1 07 satellite, utilizes tubular deployable composite booms for the sail deployment [29], as shown in Fig. 12.2k. Other forms of deployable composite booms have also been applied, such as foldable masts [30] and trusses [31], as shown in Fig. 12.2l.

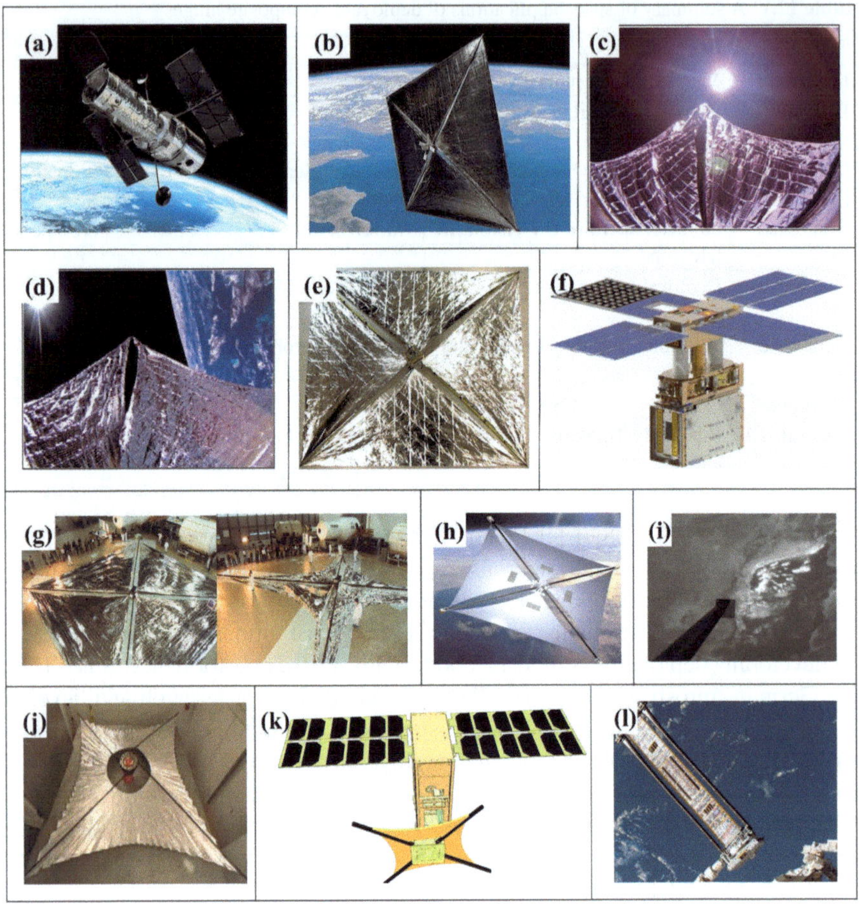

Fig. 12.2 Space applications of deployable booms: (a) Hubble space telescope. (b) NanoSail D2. (c) Lightsail 1. (d) Lightsail 2. (e) CubeSail. (f) NEA Scout spacecraft. (g) ESA-DLR solar sail. (h) Gossamer-1 demonstrator. (i) Astro tube. (j) FURL. (k) SIASAIL-I. (l) ROSA [12–20, 27, 29, 30]

In China, over the past decade, a research group led by the authors in this book has been devoted to the fundamental research and engineering applications of deployable composite structures. They have successfully addressed key technical issues in materials, mechanics, fabrication processes, and experimental validation. The team has developed a series of deployable composite structures with excellent properties, including one of the earliest domestically developed lenticular deployable composite booms. The technological achievements made by the authors' team have been gradually applied in engineering models by institutions such as China Academy of Space Technology (CAST) and Shanghai Academy of Spaceflight

Technology. For example, in 2019, Shanghai Institute of Aerospace Systems Engineering (SIASE) conducted the first technology demonstration of deployable deorbit sail on the Chinese micro-nano satellite "Taurus," achieving significant sail deployment results [32]. Subsequently, deployable deorbit sails were also equipped on commercial satellites such as "Qilu-4," "Foshan-1," "Tianshu-1," and "Jinzijiang-2" [33]. In 2022, a 25-square-meter deorbit sail was installed in the payload section of the Long March 2D Y64 launch vehicle [34]. Additionally, the AIS/VDE thin-film antenna was used on the "Pujiang-2" satellite in 2022 [35], along with other upcoming applications in orbit, such as sun shields and thin-film antennas. Table 12.2 summarizes various space applications of deployable booms.

Table 12.2 A summary of space applications of deployable booms

Type	Mission	Application	Operator	Material	Status
Tubular [12]	N/A	Hubble space telescope	NASA and ESA	Metal	In-orbit
TRAC [13]	Low earth orbit	NanoSail-D2	NASA	Metal	In-orbit
TRAC [14]	Nano-solar sail mission	CubeSail	Surrey Space Centre	Metal	Prototype
TRAC [15]	Low earth orbit	LightSail-1	Planetary Society	Metal	In-orbit
TRAC [16]	Low earth orbit	LightSail-2	Planetary Society	Metal	In-orbit
TRAC [17]	Near-earth asteroid scout	NEA solar sail	NASA	Metal	Prototype
Lenticular [18]	N.A.	ESA-DLR solar sail	DLR and ESA	CFRP	Prototype
Lenticular [19]	Low earth orbit	GOSSAMER-1	DLR and ESA	CFRP	Prototype
Tubular [20]	CubeSat mission	ALSat-1N	Space agencies of the UK and Algeria	CFRP	In-orbit
Lenticular [25]	Risk reduction	ACS3	LaRC	CFRP	Concept design
Lenticular [26]	Near-earth asteroid scout	CS3	LaRC	CFRP	Testing
TRAC [27, 28]	N.A.	FURL solar sail	NASA	CFRP	Prototype
Tubular [29]	Xiaoxiang-1 07 satellite	SIASAIL-I solar sail	Spacety Co., Ltd.	CFRP	In-orbit
Tubular [30]	SpaceX commercial resupply mission	Roll-out solar array	NASA	CFRP	In-orbit
TRAC [34]	Space debris removal	Deorbit sail	SIASE	CFRP	In-orbit
Lenticular [35]	Communication applications	Membrane array antenna	SIASE	CFRP	In-orbit

12.1.3 Deployable Composite Reflectors

Hughes Corporation (now Boeing) utilized carbon fiber composite materials in the design of the taco-shaped deployable reflector antenna and successfully achieved a flight mission on Mobile Sat-1 in 1996 [36]. Subsequently, this concept was expanded to the Tracking and Data Relay Satellite System (TDRSS) in the United States. The second generation of TDRS (TDRS I and TDRS J) was successfully used for missions in the S, Ku, and Ka bands between 2000 and 2002. The same concept was also applied in the third generation of relay satellites, which were launched successfully in 2013 and 2014 [1]. Additionally, CAST conducted ground tests of taco-shaped deployable reflectors [37].

Currently, the application of petal-shaped deployable reflector antennas is limited [38]. Umbrella-shaped deployable reflector antennas are more commonly used, such as the 3-meter-diameter umbrella-shaped deployable antenna carried by technology experiment carrier SAR developed in Israel [39], the 0.5-meter-diameter Ka-band umbrella-shaped deployable antenna developed by the Jet Propulsion Laboratory (JPL) in the United States [40, 41], and the 4.2-meter-diameter umbrella-shaped deployable antenna carried by Chinese relay satellites [42]. However, most reflector antennas currently use metallic materials. With the development of carbon fiber composite material technology, researchers have gradually begun to pay attention to deployable composite reflector antennas. However, there are still few cases of deployable composite reflector antenna applications, and most of them are still in the testing phase [43–45]. Table 12.3 summarizes various space applications of deployable reflectors.

12.2 Outlook

The current research progress and successful engineering application cases have greatly enhanced the understanding of deployable composite structure technology for space applications. However, there are still many urgent challenges that need to be addressed. Through in-depth investigation and analysis of a large number of references, some issues and prospects were discussed. We hope that these prospects will stimulate more research interest to address the current challenges of deployable composite structure technology. Currently, deployable composite structures face the following six challenges:

Table 12.3 A summary of space applications of deployable reflectors

Type	Mission	Application	Operator	Material	Status
Taco-shaped [1, 37]	MSAT	Mobile Sat-1	Stephen A. Robinson	CFRP	In-orbit
	TDRS mission	TDRS-H TDRS-I TDRS-J TDRS-K	NASA	CFRP	In-orbit
	N/A	4.2m flexible spring-back reflector antennae	CAST Xi'an	Polyimide	Testing
Petal-shaped [38]	X-band	Deployable radio frequency reflector antenna	Footdale et al.	CFRP	Concept design
Umbrella-shaped [39–44]	LEO	TECSAR	Israel	CFRP	Prototype
	RainCube mission	Ka-band parabolic deployable antenna	JPL	Metal	Prototype
	Chang'e-4 (CE-4)	Queqiao	China	N/A	In-orbit
	X-band	An unfurlable skin antenna reflector	Astrium GmbH	CFRP	Testing
	N/A	Dual polarization of deployable reflector antennas for SAR	Yoon et al.	CFRP	Testing
	Ku-band	SMART deployable telescope	Murphey et al.	CFRS and CFRP	Testing

12.2.1 Modeling of Folding Behavior

Analytical modeling and finite element modeling are the two most commonly used methods for analyzing deployable composite structures. There are currently multiple analytical models available for predicting the folding behavior of deployable composite structures (including tape springs, hinges, and various types of booms) [46–55], but there are still some issues. For example, the ability to describe geometric configurations in highly nonlinear or large-deformation situations is limited [50], and the influence of boundary conditions is ignored [46, 47], which significantly reduces the prediction accuracy of analytical models. Liu and Bai et al. demonstrated through comparison with various classical analytical models that accurately predicting the folding behavior of tape springs [56], deployable composite booms [57], and bistable deployable composite booms [58, 59] is crucial in determining their geometric configurations during the folding process. Further improving the model, enhancing its accuracy and applicability, will be an important direction for the future. In addition, the failure analysis of deployable composite structures during the folding process is often overlooked, and failure analysis involves the strength properties of materials under different environments and usage conditions, which is a key indicator in the design process of deployable composite structures [60, 61].

Compared with analytical modeling, finite element modeling is a more practical method that discretizes the structure into a finite number of elements and uses numerical calculation methods to simulate the folding process of the structure. Finite element modeling relies on input parameters such as material properties, geometric dimensions, and boundary conditions and predicts structural deformation, stress, and energy changes by solving the equations of the discretized model. This modeling method can provide more detailed and accurate information, which helps evaluate the performance and stability of the structure under different folding states. Due to the nonlinear and deformation characteristics of deployable composite structures, traditional finite element modeling methods may not be able to effectively handle their complex behavior. Therefore, to improve computational efficiency, it is necessary to explore modeling algorithms with stronger applicability for this special structure. These algorithms can include simplification methods based on physical models, model dimensionality reduction techniques, and efficient numerical solution methods. By selecting and optimizing these algorithms reasonably, computation time and resource consumption can be significantly reduced while maintaining the accuracy of simulation results.

12.2.2 Stowage and Deployment Behavior

Deployable composite structures typically go through three different stages: (1) Folding stage: before entering space, deployable composite structures are folded into a compact form. This process is a relatively slow and quasi-static process [55–59]. (2) Storage stage: During launch and transportation to orbit, deployable composite structures are firmly held in a folded state through a locking mechanism. The storage phase can last for a long time, ranging from several months to several years depending on the task requirements. (3) Deployment stage: Once in orbit, deployable composite structures can be deployed by releasing the previous locking mechanism. During the deployment process, the stored strain energy will be rapidly released. The deployment process is a dynamic process that typically completes quickly within a few seconds. The current research mainly focuses on the behavior of deployable composite structures during the folding stage, and the storage and deployment stages are equally crucial for the successful in-orbit operation of the structure. Therefore, in future research, it is necessary to understand and address the challenges faced by deployable composite structures during long-term storage and deployment processes. During the storage stage, prolonged storage may cause changes in the aging properties of materials, such as the evolution of viscoelastic and plastic properties [62–67]. This may lead to changes in material properties, may reduce the stored strain energy, and may affect the deployment ability and shape recovery accuracy of the structure. Therefore, in the design phase, it is necessary to consider and address the aging performance of materials and take corresponding measures to avoid irreversible failure. During the deployment stage, plastic deformation and stress accumulation may lead to a decrease in structural performance, especially at the boundaries or connection points of deployable composite structures. Therefore, in the design phase, it is necessary to fully consider the viscoelasticity of the material and reduce the accumulation of plastic strain through appropriate structural design and connection methods, in order to ensure the reliability and accuracy of the structure during the deployment process [68–71].

Moreover, the utilization of advanced materials and manufacturing processes can effectively enhance the storage and deployment behavior. Firstly, advanced materials exhibit superior physical and chemical properties, such as higher strength, enhanced corrosion resistance, and reduced thermal expansion coefficients. These improvements in performance contribute to reinforcing the stability of deployable composite structures during the storage phase, mitigating material performance degradation triggered by environmental variations. Secondly, cutting-edge manufacturing techniques can optimize the microstructure of composites, minimizing internal defects and stress concentrations. This optimization enhances the overall performance of the materials, imparting superior durability and reliability throughout the storage and deployment cycles.

12.2.3 Impact of Extreme Space Environments

In extreme space environments, deployable composite structures face various challenges, such as special conditions such as high/low temperatures, radiation, atom oxygen, high vacuum, and microgravity. These environmental factors may have a significant impact on the physical and chemical properties of materials, thereby affecting the mechanical properties, stability, and reliability of structures [72, 73]. In response to the impact of extreme space environments, future research directions can focus on the following aspects: (1) Material properties in high/low-temperature environments: In high/low-temperature environments, the mechanical properties of materials such as strength, toughness, and stiffness may change. Therefore, it is necessary to study the stress-strain relationship, fracture behavior, and fatigue life of materials under high/low-temperature conditions to ensure that deployable composite structures can maintain their design target property [74]. (2) The impact of radiation environment on materials: In space environment, radiation can cause damage to materials and may lead to material degradation and property degradation. Therefore, it is necessary to study the radiation failure mechanism, radiation tolerance, and mechanical property changes caused by radiation of materials in radiation environments to ensure the reliability and durability of deployable composite structures [75]. (3) Material properties and stability in vacuum environment: Under vacuum conditions, materials may be affected by factors such as gas escape, volatilization, and adsorption. Therefore, studying the gas escape characteristics, mass loss, and deformation behavior of materials in vacuum environments can help ensure the property stability of deployable composite structures after long-term vacuum condition. (4) Material properties and deployment control in microgravity environments: Under microgravity conditions, the deformation, deployment behavior, and control methods of materials may differ from those on Earth. Therefore, studying the mechanical response, deformation control, and structural dynamics behavior of materials in microgravity environments can help design and optimize the deployment process and behavior of deployable composite structures in space [76–79]. In summary, considering the impact of extreme environments on deployable composite structures is one of the important directions for future research. By thoroughly studying these influencing factors and optimizing the composition, structural design, and preparation process of materials, the performance, stability, and reliability of deployable composite structures in extreme space environments can be improved, thereby achieving more sustainable and reliable space applications.

12.2.4 Structural Design and Space Application

One of the future works is to propose diverse types of deployable composite material structures to meet different space applications. Firstly, deployable composite structures with different shapes and sizes can be designed and developed for different task requirements and space environmental conditions. Through flexible and customizable design, efficient deployment and operation can be achieved in different task scenarios. Secondly, it is necessary to research and develop multifunctional deployable composite structures. For example, by combining technologies such as sensors and actuators, the structure can perceive the external environment and internal state and make corresponding responses or adjustments. This structure can complete specific tasks after deployment, such as shape adjustment, deformation control, and adaptive adjustment [80–83]. New deployment mechanisms can also be explored, such as using shape memory composites to achieve simpler deployment and shape control [84–90]. In addition, new material combinations and processing methods can be explored to improve the behavior and reliability of deployable composite structures: for example, introducing nanomaterials, fiber-reinforced materials, and new resin systems to improve the strength, stiffness, and durability of structures. In summary, future work should focus on proposing new deployable composite structures to meet the needs of different space applications and continuously innovate and optimize in materials, structural design, and manufacturing methods. This will provide more possibilities and opportunities for future space exploration and applications.

12.2.5 Optimization Design

There are three urgent challenges to be addressed in the multi-objective optimization design of deployable composite structures. (1) Choosing appropriate surrogate models and evolutionary algorithms: Different surrogate models (such as RSM,

Kriging model, EBF, RBF, ANN, SVR, and RF) and evolutionary algorithms (such as GA, SA, PSO, and ACO) have been widely used, but there is still a lack of analytical basis for selecting appropriate surrogate models and evolutionary algorithms for specific optimization problems [91]. In most cases, surrogate models and evolutionary algorithms are usually selected based on experience. Therefore, establishing a guide with analytical basis for selecting appropriate surrogate models and evolutionary algorithms can greatly improve the efficiency of deployable composite structure optimization. It is necessary to develop new surrogate models and evolutionary algorithms suitable for specific deployable composite structure optimization problems or further improve existing surrogate models and evolutionary algorithms to adapt to the current challenges. (2) Optimization of hyperparameters for surrogate models and evolutionary algorithms: The prediction accuracy and robustness of surrogate models are closely related to their hyperparameters. Evolutionary algorithms implement random global search in a defined optimization space, so the definition of hyperparameters in evolutionary algorithms directly affects the efficiency and convergence of optimal search in optimization problems. However, the selection of hyperparameters for surrogate models and evolutionary algorithms is often based on previous research or experience. This may result in the inability to identify the optimal hyperparameter settings for surrogate models and evolutionary algorithms [92–94]. Therefore, there is a great demand in the future to develop hyperparameter optimization methods suitable for deployable composite structures. (3) The number of design variables, objectives, and constraints: Deployable composite structure optimization is a complex engineering problem. Therefore, the number of design variables and objectives also plays a crucial role in determining the final optimization scheme. In most published literature, geometric parameters and layering schemes are the most commonly considered design variables [95, 96]. Folding and deployment behavior (such as the required load for folding, stiffness, and frequency of deployment state) are the most commonly considered objective functions. Non-failure, geometric dimensions, and weight are the most commonly considered constraints. Generally speaking, the more design variables, objectives, and constraints are considered, the closer the optimization solution is to the practical application scenarios of deployable composite structures. Therefore, considering more design variables, objectives, and constraints in future optimization models, and designing suitable surrogate models and evolutionary algorithms, can significantly improve the quality of optimization solutions.

Furthermore, the emergence of advanced design methodologies, such as generative artificial intelligence, will provide robust support for the optimization of deployable composite structures. By integrating artificial intelligence algorithms with numerical simulation techniques, the design space can be efficiently explored, revealing novel structural forms and connection methods. This integration is expected to enhance the performance of structures while simultaneously reducing design costs.

12.2.6 Auxiliary Devices for Folding and Deployment

For practical engineering applications, auxiliary devices used in deployable composite structures can achieve specific task objectives. A large amount of engineering practice has shown that the smooth deployment based on deployable composite structures is the key to the success of the entire system. In other words, faults are often prone to occur during the deployment process, leading to the inability to achieve complete tasks or even complete failures. Auxiliary devices aim to support, drive, and control the movement process of deployable composite structures, enabling them to smoothly achieve folding and deployment functions [97–99]. These auxiliary devices include mechanical structures, mechanisms, drivers, controllers, etc. The most representative ones include folding mechanisms, release devices, etc. When designing auxiliary devices, engineers not only focus on the behavior of the deployable composite structure, but also pay special attention to the reliability and deployability of the auxiliary device from the perspective of the success of the entire task. The ideal solution is to maximize the efficiency of deployable composite structures while minimizing the impact on the entire system's tasks and ensuring the achievement of critical capabilities, even in extreme cases where partial deployment or failure to deploy is possible. In addition, specific requirements and working environment conditions for deployable composite structures need to be considered, such as overload and vibration during launch, extreme high/low temperatures during orbit, vacuum, weight loss, and radiation. These devices must have sufficient rigidity and stability to support and control the movement process of deployable composite structures [100]. Meanwhile, precise position control capability is also essential to ensure that the structure can be accurately deployed to the predetermined position and retracted back into a compact form when needed [101, 102]. In addition, auxiliary devices also need to meet the requirements of lightweight and compactness to adapt to the mass and volume limitations of deployable composite structures. With the continuous development of technology, future research can focus on the following aspects to further improve auxiliary devices. Firstly, according to the needs of different application scenarios, more advanced materials and manufacturing technologies can be explored to improve the property and reliability of auxiliary devices. Secondly, intelligent materials and intelligent structures can be used to achieve adaptive deformation and functional adjustment, enabling auxiliary devices to autonomously adjust according to environmental changes and operational requirements. In addition, for long-term space missions, the durability and maintainability of auxiliary devices are also fields that need attention. In summary, auxiliary devices play an important role in the research of deployable composite structures for space applications. Future research should focus on technological improvement, material innovation, new concept design, optimization, and other aspects to further enhance the functionality and reliability of auxiliary devices [103].

References

1. Murphey TW, Francis W, Davis B et al (2015) High strain composites. In: 2nd AIAA spacecraft structures conference, Kissimmee, Florida
2. Vyvyan WW. Self-actuating, self-locking hinge. U.S. Patent 3,386,128.1968-6-4
3. Marks GW, Reilly MT, Huff RL (2002) The lightweight deployable antenna for the MARSIS experiment on the mars express spacecraft. In: 36th aerospace mechanisms symposium, Cleveland, Ohio
4. Domber JL, Hinkle JD, Peterson LD et al (2002) Dimensional repeatability of an elastically folded composite hinge for deployed spacecraft optics. J Spacecr Rocket 39(5):646–652
5. Warren PA, Silver MJ, Dobson BJ Jr et al (2013) Experimental characterization of deployable outer barrel assemblies for large space telescopes. In: UV/optical/IR space telescopes and instruments: innovative technologies and concepts VI, San Diego, California
6. Maji AK, Harris M, Garcia D et al (2011) Feasibility assessment of deployable composite telescope. J Aerosp Eng 24(1):12–19
7. Footdale J, Murphey T (2011) Structural design of a CubeSat-based diffractive optic telescope. In: 52nd AIAA/ASME/ASCE/AHS/ASC structures, structural dynamics and materials conference, Denver, Colorado
8. Jeon S, Murphey T (2011) Design and analysis of a meter-class CubeSat boom with a motorless deployment by bi-stable tape springs. In: 2nd AIAA/ASME/ASCE/AHS/ASC structures, Denver, Colorado
9. Yang H, Guo HW, Wang Y et al (2018) Design and experiment of triangular prism mast with tape-spring hyperelastic hinges. Chin J Mech Eng 31(1):1–10
10. Soykasap Ö, Pellegrino S, Howard P et al (2008) Folding large antenna tape spring. J Spacecr Rocket 45(3):560–567
11. Footdale J, Banik J, Murphey T (2010) Design developments of a non-planar deployable structure. In: 51st AIAA/ASME/ASCE/AHS/ASC structures, structural dynamics, and materials conference, Orlando, Florida
12. Foster CL, Tinker ML, Nurre GS et al (1995) Solar-array-induced disturbance of the Hubble space telescope pointing system. J Spacecr Rocket 32(4):634–644
13. Vulpetti G, Johnson L, Matloff GL et al (2015) The nanoSAIL-D2 NASA mission. In: Solar sails: a novel approach to interplanetary travel, pp 173–178
14. Ridenoure RW, Munakata R, Wong SD et al (2016) Testing the lightsail program: advancing solar sailing technology using a cubesat platform. J Small Satell 5(2):531–550
15. Spencer DA, Betts B, Bellardo JM et al (2021) The LightSail 2 solar sailing technology demonstration. Adv Space Res 67(9):2878–2889
16. Lappas V, Adeli N, Visagie L et al (2011) CubeSail: a low cost CubeSat based solar sail demonstration mission. Adv Space Res 48(11):1890–1901
17. Lockett TR, Johnson C, Stewart ER (2019) Lessons learned from the flight unit testing of the near earth asteroid scout flight system. In: International symposium on solar sailing, Aachen, Germany
18. Sickinger C, Herbeck L, Breitbach E (2006) Structural engineering on deployable CFRP booms for a solar propelled sailcraft. Acta Astronaut 58(4):185–196
19. Seefeldt P, Spietz P, Sproewitz T et al (2017) Gossamer-1: mission concept and technology for a controlled deployment of gossamer spacecraft. Adv Space Res 59(1):434–456
20. Reveles JR, Lawton M, Fraux V et al (2016) In-orbit performance of AstroTube: AlSat Nano's low mass deployable composite boom payload. In: 31st annual AIAA/USU conference on small satellites, Logan, Utah
21. Chen W, Fang G, Hu Y (2017) An experimental and numerical study of flattening and wrapping process of deployable composite thin-walled lenticular tubes. Thin-Walled Struct 111:38–47
22. Lemak M, Banerjee A (1994) Comparison of simulation with test of deployment of a wraprib antenna. In: Guidance, navigation, and control conference, Scottsdale, Arizon

23. Greschik G, Park K, Natori M (1994) Unfurling of a rolled-up strip into a helically curved tube. In: 35th Structures, structural dynamics, and materials conference, Hilton Head, South Carolina
24. Tanaka H (2006) Design optimization studies for large-scale contoured beam deployable satellite antennas. Acta Astronaut 58(9):443–451
25. Fernandez JM, Rose G, Stohlman OR et al (2018) An advanced composites-based solar sail system for interplanetary small satellite missions. In: 2018 AIAA spacecraft structures conference, Kissimmee, Florida
26. Fernandez JM, Rose GK, Younger CJ et al (2017) NASA'S advanced solar sail propulsion system for low-cost deep space exploration and science missions that use high performance rollable composite booms. In: International symposium on solar sailing, Kyoto, Japan
27. Murphey TW, Turse D, Adams L (2017) TRAC boom structural mechanics. In: 4th AIAA spacecraft structures conference, Grapevine, Texas
28. Banik J, Ardelean E (2010) Verification of a retractable solar sail in a thermal-vacuum environment. In: 51st AIAA/ASME/ASCE/AHS/ASC structures, structural dynamics, and materials conference, Orlando, Florida
29. Liu J, Zhao P, Wu C et al (2022) SIASAIL-I solar sail: from system design to on-orbit demonstration mission. Acta Astronaut 192:133–142
30. Chamberlain MK, Kiefer SH, LaPointe M et al (2021) On-orbit flight testing of the roll-out solar Array. Acta Astronaut 179:407–414
31. Takano T, Natori M, Ohnishi A et al (2000) Large deployable antenna with cable net composition for satellite use. Electron Commun Jpn (Part I: Commun) 83(8):105–115
32. Fu Y, Peng F, Zhang C et al (2021) Multi-faceted simulation of atomic oxygen erosion of deorbit sail for cleaning space debris. Acta Astronaut 187:61–69
33. https://m.thepaper.cn/baijiahao_18982027
34. Yun WD, Fang GQ, Fu YL et al (2021) The progress and applications of deployable membrane sail for space debris removal. Space Debris Res 21(3):20–28
35. Xie C, Zhang E, Yan B et al (2023) Configuration design and demonstration of a spaceborne deployable membrane array antenna. J Mech Eng 60(3):11–27
36. Baier H, Datashvili L, Nathrath N et al (2004) Technical assessment of high accuracy large space borne reflector antenna. In: Final Report of ESA Contract
37. Ma XF (2002) Design and modeling of flexible spring-back antenna. China Academy of Space Technology, Xi'an
38. Footdale JN, Banik J (2016) Design and deployment testing of the multi-arm radial composite (MARCO) reflector antenna. In: 3rd AIAA spacecraft structures conference, San Diego, California
39. Sharay Y, Naftaly U (2006) TECSAR: design considerations and programme status. IEE Proc Radar Sonar Navigat 153(2):117–121
40. Sauder JF, Chahat N, Hirsch B et al (2017) From prototype to flight: qualifying a Ka-band parabolic deployable antenna (KaPDA) for CubeSats. In: 4th AIAA spacecraft structures conference, Grapevine, Texas
41. Chahat N, Sauder J, Thomson M et al (2017) Ka-band deployable mesh reflector antenna compatible with the deep space network. In: 11th European conference on antennas and propagation, Paris, France
42. Tang Y, Wu W, Qiao D et al (2017) Effect of orbital shadow at an earth-moon Lagrange point on relay communication mission. Sci China Inf Sci 60:1–10
43. Yoon SS, Lee JW, Lee TK et al (2018) Insensitivity characteristics in the dual polarization of deployable CFRP reflector antennas for SAR. IEEE Trans Antennas Propag 66(1):88–95
44. Datashvili L, Baier H, Wehrle E et al (2010) Large shell-membrane space reflectors. In: 51st AIAA/ASME/ASCE/AHS/ASC structures, structural dynamics, and materials conference, Orlando, Florida

45. Datashvili L, Baier H, Schimitschek J et al (2007) High precision large deployable space reflector based on pillow-effect-free technology. In: 48th AIAA/ASME/ASCE/AHS/ASC structures, structural dynamics, and materials conference, Honolulu, Hawaii
46. Wuest W (1954) Einige anwendungen der theorie der zylinderschale. ZAMM-J Appl Math Mech 34(12):444–454
47. Panovko YG, Gubanova II (1965) Stabilities and oscillations of elastic systems: paradoxes, fallacies and new concepts. Consultants Bureau, New York
48. Seffen KA, Pellegrino S (1999) Deployment dynamics of tape-springs. Proc R Soc Lond Seri A: Math Phys Eng Sci 455(1983):1003–1048
49. Mansfield EH (1973) Large-deflexion torsion and flexure of initially curved strips. Proc R Soc Lond A Math Phys Eng Sci 334(1598):279–298
50. Yao XF, Ma YJ, Yin YJ et al (2011) Design theory and dynamic mechanical characterization of the deployable composite tube hinge. Sci China Phys Mech Astron 54:633–639
51. Iqbal K, Pellegrino S, Daton-Lovett A (1998) Bistable composite slit tubes. In: IUTAM-IASS symposium on deployable structures, Cambridge, UK
52. Galletly DA, Guest SD (2004) Bistable composite slit tubes. I. A beam model. Int J Solids Struct 41(16–17):4517–4533
53. Galletly DA, Guest SD (2004) Bistable composite slit tubes. II. A shell model. Int J Solids Struct 41(16–17):4503–4516
54. Guest SD, Pellegrino S (2006) Analytical models for bistable cylindrical shells. Proc R Soc A: Math Phys Eng Sci 462:839–854
55. Wang S, Schenk M, Jiang S et al (2020) Blossoming analysis of composite deployable booms. Thin-Walled Struct 157:107098
56. Liu TW, Bai JB, Fantuzzi N (2024) New insights into folding tape-spring. Compos Struct 341:118225
57. Liu TW, Bai JB, Fantuzzi N (2023) Analytical models for predicting folding behaviour of thin-walled tubular deployable composite boom for space applications. Acta Astronaut 208:167–178
58. Liu TW, Bai JB, Wang ZZ et al (2023) Determining folded stable state of bistable deployable composite boom: an improved two-parameter analytical model. Acta Astronaut 213:507–515
59. Liu TW, Bai JB, Fantuzzi N (2023) Analytical model for predicting folding stable state of bistable deployable composite boom. Chin J Aeronaut 37(8):460–469
60. Liu TW, Bai JB, Lin QH et al (2021) An analytical model for predicting compressive behaviour of composite helical structures: considering geometric nonlinearity effect. Compos Struct 255:112908
61. Liu TW, Bai JB (2022) Folding behaviour of a deployable composite cabin for space habitats-part 2: analytical investigation. Compos Struct 297:115929
62. Chen CP, Lakes RS (1993) Analysis of high-loss viscoelastic composites. J Mater Sci 28:4299–4304
63. Kang JH, Hinkley JA, Gordon KL et al (2021) Viscoelastic characterization of polymers for deployable composite booms. Adv Space Res 67(9):2727–2735
64. Roh JH, Kim HI, Lee SY (2015) Viscoelastic effect on unfolding behaviours of shape memory composite booms. Compos Struct 133:235–245
65. Kwok K, Pellegrino S (2011) Viscoelastic effects in tape-springs. In: AIAA/ASME/ASCE/AHS/ASC structures, structural dynamics and materials, p 2022
66. Hamillage MY, Leung C, Kwok K (2022) Viscoelastic modeling and characterization of thin-ply composite laminates. Compos Struct 280:114901
67. Khan AI, Borowski EC, Soliman EM et al (2017) Examining energy dissipation of deployable aerospace composites using matrix viscoelasticity. J Aerosp Eng 30(5):04017040
68. Borowski EC, Soliman EM, Khan AI et al (2018) Stowage and deployment of a viscoelastic orthotropic carbon-fiber composite tape spring. J Spacecr Rocket 55(4):829–840
69. Adamcik B, Firth J, Pankow M et al (2020) Impact of storage time and operational temperature on deployable composite booms. In: AIAA Scitech 2020 Forum, Orlando, Florida

70. Guo R, Jin X, Jia Q et al (2023) Folding, stowage, and deployment of composite thin-walled lenticular tubes. Acta Astronaut 213:567–577
71. Deng J, An N, Jia Q et al (2024) Deployment analysis of composite thin-walled lenticular tubes with effect of storage time and temperature. Chin J Aeronaut 37(1):162–172
72. He Y, Suliga A, Brinkmeyer A et al (2018) Durability of composite materials in deployable structures for space applications. In: ECCM 2018-18th European conference on composite materials, Athens, Greece
73. Grossman E, Gouzman I (2003) Space environment effects on polymers in low earth orbit. Nucl Instrum Methods Phys Res, Sect B 208:48–57
74. De Zanet G, Viquerat A (2020) Thermal response of CFRP deployable tubes in the space environment. In: AIAA Scitech 2020 Forum, Orlando, Florida
75. Rojdev K, O'Rourke MJE, Hill C et al (2014) Radiation effects on composites for long-duration lunar habitats. J Compos Mater 48(7):861–878
76. Amamoto T, Sakamoto H, Furuya H et al (2018) Modeling of composite booms' deployment dynamics under microgravity based on ground tests. In: 2018 AIAA spacecraft structures conference, Kissimmee, Florida
77. Furuya H, Yashima K, Yokomatsu T et al (2017) Micro-gravity testing of boom-membrane integrated deployable structures for micro-satellites. In: 4th AIAA spacecraft structures conference, Grapevine, Texas
78. Sarantinos N, Loginos P, Charlaftis P et al (2019) Behaviour of photopolymer fiber structures in microgravity. SN Appl Sci 1:1–10
79. Chae S, Oh YE, Lee SY et al (2020) Deployment behaviours of CFRP reflector under zero-gravity environment. Int J Aerosp Syst Eng 7(1):1–6
80. Wang W, Rodrigue H, Ahn SH (2016) Deployable soft composite structures. Sci Rep 6(1):20869
81. Daton-Lovett AJ, Compton-Bishop QM, Curry RG (2000) Deployable structures using bistable reeled composites. In: Smart structures and materials 2000: active materials: behaviour and mechanics, Newport Beach, California
82. Tarazaga P, Inman D, Wilkie W (2006) Control of a space rigidizable-inflatable boom using embedded piezoelectric composite actuators. In: 47th AIAA/ASME/ASCE/AHS/ASC structures, structural dynamics, Newport, Rhode Island
83. Laurenzi S, Rufo D, Sabatini M et al (2019) Characterization of deployable ultrathin composite boom for microsatellites excited by attitude maneuvers. Compos Struct 220:502–509
84. Liu Y, Du H, Liu L et al (2014) Shape memory polymers and their composites in aerospace applications: a review. Smart Mater Struct 23(2):023001
85. Santo L, Quadrini F, Accettura A et al (2014) Shape memory composites for self-deployable structures in aerospace applications. Procedia Eng 88:42–47
86. Margoy D, Gouzman I, Grossman E et al (2021) Epoxy-based shape memory composite for space applications. Acta Astronaut 178:908–919
87. Li F, Liu Y, Leng J (2019) Progress of shape memory polymers and their composites in aerospace applications. Smart Mater Struct 28(10):103003
88. Mu T, Liu L, Lan X et al (2018) Shape memory polymers for composites. Compos Sci Technol 160:169–198
89. Xia Y, He Y, Zhang F et al (2021) A review of shape memory polymers and composites: mechanisms, materials, and applications. Adv Mater 33(6):2000713
90. Xin X, Liu L, Liu Y et al (2019) Mechanical models, structures, and applications of shape-memory polymers and their composites. Acta Mech Solida Sin 32:535–565
91. Liu TW, Bai JB, Fantuzzi N (2022) Folding behaviour of the thin-walled lenticular deployable composite boom: analytical analysis and many-objective optimization. Mech Adv Mater Struct:1–19
92. Wang ZZ, Sobey A (2020) A comparative review between genetic algorithm use in composite optimisation and the state-of-the-art in evolutionary computation. Compos Struct 233:111739

93. Bai JB, You FY, Wang ZZ et al (2023) An efficient multi-objective optimization framework for thin-walled tubular deployable composite boom. Compos Struct 2023:117713
94. Bai JB, Liu TW, Wang ZZ et al (2021) Determining the best practice-optimal designs of composite helical structures using genetic algorithms. Compos Struct 268:113982
95. Ferraro S, Pellegrino S (2021) Topology and shape optimization of ultrathin composite self-deployable shell structures with cutouts. AIAA J 59(9):3696–3709
96. Wu J, Li J, Yan S (2014) Design of deployable bistable structures for morphing skin and its structural optimization. Eng Optim 46(6):745–762
97. Zhang D, Liu L, Xu P et al (2023) Ancient papyrus scroll-inspired self-deployable mechanism based on shape memory polymer composites for Mars explorations. Compos Struct 304:116391
98. Bai JB, Bu GY (2022) Progress in 4D printing technology. J Adv Manufact Sci Technol 2(1):2022001–2022001
99. Hu J, Zhu Y, Huang H et al (2012) Recent advances in shape-memory polymers: structure, mechanism, functionality, modeling and applications. Prog Polym Sci 37(12):1720–1763
100. Chu ZY, Lei YA (2014) Design theory and dynamic analysis of a deployable boom. Mech Mach Theory 71:126–141
101. Chu ZY, Hu J, Yan SB et al (2015) Experiment on the retraction/deployment of an active-passive composited driving deployable boom for space probes. Mech Mach Theory 92:436–446
102. Santiago-Prowald J, Baier H (2013) Advances in deployable structures and surfaces for large apertures in space. CEAS Space J 5:89–115
103. Liu TW, Bai JB, Fantuzzi N (2024) Thin-walled deployable composite structures: a review. Prog Aerosp Sci 146:100985

Appendixes

1.1 Appendix A. 22 Optimal Design Schemes

No.	$r_{1,0}$ (mm)	$r_{2,0}$ (mm)	a (mm)	R_{ultimate} (mm)	T (N m)	S (g/cm)	I_y (cm^4)	I_z (cm^4)
1	20.02	22.55	5.01	60.835	1.153	1.231	1.024	2.878
2	24.73	28.29	5.01	47.164	1.164	1.496	1.933	5.502
3	23.93	39.64	5.33	48.729	1.259	1.597	1.781	7.145
4	33.35	20.63	5.59	58.183	1.118	1.772	4.679	8.513
5	35.77	20.51	5.20	58.687	1.120	1.859	5.774	10.053
6	38.35	20.88	5.26	57.253	1.134	1.970	7.109	12.033
7	35.42	35.03	5.20	36.840	1.248	2.028	5.654	14.261
8	38.10	29.75	5.32	40.630	1.213	2.076	7.003	14.799
9	40.94	25.33	5.08	46.123	1.184	2.130	8.661	15.754
10	41.88	25.57	5.76	45.741	1.201	2.189	9.271	16.731
11	44.53	21.63	5.01	54.650	1.171	2.225	11.124	17.762
12	37.61	47.58	5.19	35.558	1.383	2.266	6.823	21.215
13	43.24	30.37	5.33	40.075	1.250	2.295	10.219	20.107
14	45.37	25.22	5.23	46.309	1.213	2.312	11.776	20.157
15	46.91	22.98	5.63	50.857	1.207	2.355	13.003	20.852
16	47.45	25.04	5.01	46.621	1.223	2.389	13.466	22.417
17	44.08	37.30	5.04	35.695	1.321	2.410	10.861	24.307
18	41.97	46.90	5.07	33.892	1.404	2.439	9.440	26.332
19	49.07	25.23	5.23	46.303	1.238	2.463	14.888	24.454
20	44.43	41.56	5.13	34.029	1.368	2.479	11.143	26.895
21	48.76	28.43	5.05	41.966	1.266	2.488	14.618	25.691
22	48.08	30.94	5.32	39.589	1.289	2.500	14.033	26.136

J.-B. Bai et al., *Folding Mechanics of Deployable Composite Structures*, Solid Mechanics and Its Applications 283,
https://doi.org/10.1007/978-3-031-99254-4

1.2 Appendix B. 48 Optimal Design Schemes

No.	$r_{1,0}$ (mm)	$r_{2,0}$ (mm)	a (mm)	$R_{ultimate}$ (mm)	T (N m)	S (g/cm)	I_y (cm^4)	I_z (cm^4)
1	34.58	34.79	5.53	37.093	1.246	1.998	5.263	13.458
2	34.58	34.79	5.53	37.093	1.246	1.998	5.263	13.458
3	34.81	34.90	5.38	36.960	1.246	2.005	5.370	13.695
4	34.81	34.90	5.38	36.960	1.246	2.005	5.370	13.695
5	34.82	34.84	5.49	36.959	1.247	2.008	5.371	13.677
6	34.82	34.79	5.53	36.972	1.247	2.008	5.373	13.665
7	34.82	34.79	5.53	36.972	1.247	2.008	5.373	13.665
8	34.82	34.79	5.53	36.972	1.247	2.008	5.373	13.665
9	34.82	34.91	5.53	36.956	1.248	2.010	5.374	13.703
10	34.82	34.91	5.53	36.956	1.248	2.010	5.374	13.703
11	35.22	35.22	5.16	36.736	1.248	2.020	5.559	14.147
12	35.22	35.22	5.16	36.736	1.248	2.020	5.559	14.147
13	35.22	35.22	5.16	36.736	1.248	2.020	5.559	14.147
14	35.22	35.22	5.16	36.736	1.248	2.020	5.559	14.147
15	35.22	35.22	5.16	36.736	1.248	2.020	5.559	14.147
16	35.22	35.22	5.16	36.736	1.248	2.020	5.559	14.147
17	35.22	35.22	5.16	36.736	1.248	2.020	5.559	14.148
18	35.28	35.46	5.00	36.705	1.249	2.022	5.588	14.279
19	35.31	35.03	5.22	36.838	1.248	2.023	5.599	14.160
20	35.31	35.03	5.22	36.838	1.248	2.023	5.599	14.160
21	35.31	35.03	5.22	36.838	1.248	2.023	5.599	14.160
22	35.31	35.03	5.22	36.838	1.248	2.023	5.599	14.160
23	35.39	35.46	5.00	36.641	1.250	2.027	5.643	14.382
24	35.39	35.46	5.00	36.641	1.250	2.027	5.643	14.382
25	35.42	35.03	5.20	36.840	1.248	2.028	5.654	14.261
26	35.42	35.03	5.20	36.840	1.248	2.028	5.654	14.261
27	35.42	35.03	5.20	36.840	1.248	2.028	5.654	14.261
28	35.42	35.03	5.20	36.840	1.248	2.028	5.654	14.261
29	35.42	35.03	5.20	36.840	1.248	2.028	5.655	14.261
30	35.42	35.03	5.20	36.839	1.248	2.028	5.655	14.262
31	35.39	35.22	5.16	36.734	1.249	2.028	5.642	14.301
32	34.37	39.12	5.09	37.214	1.280	2.030	5.189	14.703
33	34.37	39.12	5.09	37.214	1.280	2.030	5.189	14.703
34	35.39	35.46	5.16	36.641	1.252	2.031	5.643	14.382
35	35.39	35.46	5.17	36.641	1.252	2.031	5.643	14.382
36	35.69	34.99	5.00	36.864	1.247	2.033	5.781	14.479
37	35.69	34.99	5.00	36.864	1.247	2.033	5.781	14.479
38	34.38	40.09	5.01	37.211	1.288	2.039	5.195	15.032
39	35.92	34.97	5.04	36.871	1.248	2.044	5.893	14.680
40	35.89	35.03	5.20	36.840	1.251	2.047	5.880	14.675
41	35.59	36.68	5.33	36.535	1.266	2.058	5.744	14.977

1.2 Appendix B. (continued)

No.	$r_{1,0}$ (mm)	$r_{2,0}$ (mm)	a (mm)	R_{ultimate} (mm)	T (N m)	S (g/cm)	I_y (cm^4)	I_z (cm^4)
42	35.59	36.73	5.33	36.535	1.266	2.059	5.744	14.995
43	35.59	36.80	5.33	36.535	1.267	2.059	5.744	15.018
44	36.53	34.20	5.10	37.319	1.244	2.061	6.191	14.954
45	36.53	34.22	5.10	37.301	1.245	2.061	6.191	14.964
46	36.53	34.23	5.10	37.300	1.245	2.061	6.191	14.965
47	36.55	34.22	5.10	37.301	1.245	2.062	6.206	14.990
48	36.55	34.22	5.10	37.301	1.245	2.062	6.206	14.990

1.3 Appendix C. Optimal Design Points for Ply Scheme 1

No.	R (mm)	θ (°)	$\bar{T}$ (N mm)	$E_x I$ (N mm^2)
1	40.00	270.00	88.70	1.62E+09
20	40.00	269.29	88.57	1.61E+09
40	40.00	268.72	88.38	1.59E+09
60	40.00	268.14	88.19	1.58E+09
80	40.00	267.57	87.99	1.57E+09
100	40.00	267.00	87.78	1.56E+09
120	40.00	266.43	87.57	1.55E+09
140	40.00	265.85	87.34	1.53E+09
160	40.00	265.28	87.10	1.52E+09
180	40.00	264.13	86.86	1.50E+09
200	40.00	263.56	86.60	1.49E+09
220	40.00	262.99	86.34	1.47E+09
240	40.00	261.84	86.06	1.46E+09
260	40.00	261.27	85.76	1.44E+09
280	40.00	260.12	85.46	1.42E+09
300	40.00	258.98	85.13	1.40E+09
320	40.00	257.83	84.79	1.38E+09
340	40.00	257.26	84.44	1.36E+09
360	40.00	255.54	84.06	1.34E+09
380	40.00	254.39	83.67	1.32E+09
400	40.00	253.25	83.24	1.30E+09
420	40.00	252.10	82.80	1.27E+09
440	40.00	250.38	82.33	1.25E+09
460	40.00	249.24	81.83	1.22E+09
480	40.00	247.52	81.30	1.19E+09
500	40.00	245.80	80.74	1.16E+09
520	39.99	244.08	80.14	1.13E+09
540	40.00	241.79	79.48	1.10E+09
560	40.00	240.07	78.79	1.06E+09
580	40.00	237.78	78.06	1.03E+09

(continued)

1.3 Appendix C. (continued)

No.	R (mm)	θ (°)	$\overline{T}$ (N mm)	E_xI (N mm^2)
600	40.00	234.91	77.24	9.87E+08
620	39.99	232.62	76.38	9.46E+08
640	39.99	229.76	75.44	9.02E+08
660	39.99	226.32	74.42	8.57E+08
680	40.00	222.88	73.31	8.08E+08
700	40.00	219.44	72.09	7.57E+08
720	40.00	215.43	70.76	7.04E+08
740	40.00	210.85	69.30	6.48E+08
760	40.00	206.26	67.70	5.89E+08
780	40.00	200.54	65.93	5.29E+08
800	40.00	194.81	63.98	4.68E+08
820	40.00	188.50	61.84	4.06+E8

1.4 Appendix D. Optimal Design Points for Ply Scheme 2

No.	R (mm)	θ (°)	$\overline{T}$ (N mm)	E_xI (N mm^2)
840	40.00	270.00	52.90	3.73E+08
860	40.00	269.86	52.87	3.73E+08
880	40.00	267.57	52.35	3.61E+08
900	40.00	254.39	49.83	3.05E+08
920	40.00	240.64	47.07	2.48E+08
940	40.00	226.32	44.3	1.97E+08
960	40.00	210.85	41.3	1.5E+08
980	40.00	195.38	38.3	1.1E+08
1000	40.00	180.00	35.2	7.6E+07

1.5 Appendix E. Finite Element Analysis Dataset

No.	R (mm)	θ (°)	L (mm)	f (Hz)
1	25	150	500	27.378
2	25	150	625	20.415
3	25	150	750	16.158
4	25	150	875	13.305
5	25	150	1000	10.266
6	25	180	500	28.125
7	25	180	625	20.229
8	25	180	750	15.611
9	25	180	875	12.632

1.5 Appendix E. (continued)

No.	*R* (mm)	θ (°)	*L* (mm)	*f* (Hz)
10	25	180	1000	10.554
11	25	210	500	31.722
12	25	210	625	21.777
13	25	210	750	16.256
14	25	210	875	12.815
15	25	210	1000	10.503
16	25	240	500	37.850
17	25	240	625	25.134
18	25	240	750	18.175
19	25	240	875	13.939
20	25	240	1000	11.155
21	25	270	500	45.836
22	25	270	625	29.927
23	25	270	750	21.226
24	25	270	875	15.959
25	25	270	1000	12.526
26	26.25	150	500	26.985
27	26.25	150	625	20.036
28	26.25	150	750	15.815
29	26.25	150	875	12.997
30	26.25	150	1000	10.783
31	26.25	180	500	28.259
32	26.25	180	625	20.140
33	26.25	180	750	15.474
34	26.25	180	875	12.476
35	26.25	180	1000	10.398
36	26.25	210	500	32.497
37	26.25	210	625	22.129
38	26.25	210	750	16.396
39	26.25	210	875	12.856
40	26.25	210	1000	10.490
41	26.25	240	500	39.251
42	26.25	240	625	25.937
43	26.25	240	750	18.656
44	26.25	240	875	14.234
45	26.25	240	1000	11.337
46	26.25	270	500	47.779
47	26.25	270	625	31.143
48	26.25	270	750	22.026
49	26.25	270	875	16.505
50	26.25	270	1000	12.912
51	27.5	150	500	26.678
52	27.5	150	625	19.716
53	27.5	150	750	15.517

(continued)

1.5 Appendix E.(continued)

No.	R (mm)	θ (°)	L (mm)	f (Hz)
54	27.5	150	875	12.726
55	27.5	150	1000	10.746
56	27.5	180	500	28.488
57	27.5	180	625	20.137
58	27.5	180	750	15.382
59	27.5	180	875	12.353
60	27.5	180	1000	10.268
61	27.5	210	500	33.357
62	27.5	210	625	22.549
63	27.5	210	750	16.597
64	27.5	210	875	12.940
65	27.5	210	1000	10.510
66	27.5	240	500	40.694
67	27.5	240	625	26.787
68	27.5	240	750	19.180
69	27.5	240	875	14.566
70	27.5	240	1000	11.550
71	27.5	270	500	49.721
72	27.5	270	625	32.390
73	27.5	270	750	22.855
74	27.5	270	875	17.085
75	27.5	270	1000	13.324
76	28.75	150	500	26.442
77	28.75	150	625	19.444
78	28.75	150	750	15.256
79	28.75	150	875	12.485
80	28.75	150	1000	10.525
81	28.75	180	500	28.798
82	28.75	180	625	20.193
83	28.75	180	750	15.334
84	28.75	180	875	12.262
85	28.75	180	1000	10.161
86	28.75	210	500	34.281
87	28.75	210	625	23.027
88	28.75	210	750	16.845
89	28.75	210	875	13.062
90	28.75	210	1000	10.560
91	28.75	240	500	42.165
92	28.75	240	625	27.165
93	28.75	240	750	19.738
94	28.75	240	875	14.93
95	28.75	240	1000	11.797
96	28.75	270	500	51.632
97	28.75	270	625	33.636

1.5 Appendix E. (continued)

No.	R (mm)	θ (°)	L (mm)	f (Hz)
98	28.75	270	750	23.702
99	28.75	270	875	17.677
100	28.75	270	1000	13.754
101	30	150	500	26.279
102	30	150	625	19.219
103	30	150	750	15.029
104	30	150	875	12.272
105	30	150	1000	10.328
106	30	180	500	29.190
107	30	180	625	20.039
108	30	180	750	15.329
109	30	180	875	13.300
110	30	180	1000	10.079
111	30	210	500	35.258
112	30	210	625	23.553
113	30	210	750	17.134
114	30	210	875	13.219
115	30	210	1000	10.638
116	30	240	500	43.650
117	30	240	625	28.587
118	30	240	750	20.327
119	30	240	875	15.321
120	30	240	1000	12.057
121	30	270	500	53.501
122	30	270	625	34.895
123	30	270	750	24.566
124	30	270	875	18.291
125	30	270	1000	14.203

1.6 Appendix F. 112 Optimal Design Points Superior to Initial Sample

No.	R (mm)	θ (°)	L (mm)	$\overline{u}$ (mJ)	r_1 (mm)	f (Hz)	m (t/mm)
1	25.01	164.16	511.71	2.19	33.34	26.50	3.44E-08
2	25.00	159.76	582.06	2.13	33.44	21.96	3.35E-08
3	25.00	173.80	615.50	2.32	33.49	20.33	3.64E-08
4	25.00	166.27	646.55	2.22	33.14	18.92	3.48E-08
5	25.00	176.26	580.21	2.35	33.24	22.31	3.69E-08
6	25.01	165.75	531.31	2.21	33.17	25.13	3.47E-08
7	25.01	153.07	504.96	2.04	33.14	26.77	3.21E-08
8	25.00	179.75	606.62	2.40	33.28	20.93	3.77E-08

(continued)

1.6 Appendix F. (continued)

No.	R (mm)	θ (°)	L (mm)	$\overline{u}$ (mJ)	r_1 (mm)	f (Hz)	m (t/mm)
9	25.02	155.86	541.34	2.08	33.38	24.32	3.27E-08
10	25.00	151.69	554.01	2.02	33.11	23.60	3.18E-08
11	25.01	160.38	628.95	2.14	33.41	19.73	3.36E-08
12	25.00	159.76	583.06	2.13	33.25	21.91	3.35E-08
13	25.00	174.57	651.11	2.33	33.35	18.72	3.66E-08
14	25.00	152.14	533.62	2.03	33.27	24.83	3.19E-08
15	25.00	160.36	630.90	2.14	33.41	19.65	3.36E-08
16	25.00	154.97	523.62	2.07	33.46	25.48	3.25E-08
17	25.02	182.94	523.75	2.44	33.16	26.62	3.83E-08
18	25.00	156.31	549.97	2.08	33.30	23.79	3.27E-08
19	25.01	165.87	605.04	2.21	33.38	20.77	3.47E-08
20	25.01	165.02	509.82	2.20	33.24	26.68	3.46E-08
21	25.00	171.06	523.59	2.28	33.36	25.89	3.58E-08
22	25.00	174.89	523.51	2.33	33.36	26.10	3.66E-08
23	25.00	176.36	507.27	2.35	33.04	27.49	3.69E-08
24	25.01	153.07	504.96	2.04	33.14	26.77	3.21E-08
25	25.00	183.55	552.83	2.45	33.11	24.45	3.84E-08
26	25.00	153.33	630.89	2.04	33.32	19.84	3.21E-08
27	25.01	165.41	605.10	2.20	33.38	20.77	3.47E-08
28	25.00	178.56	554.84	2.38	33.40	24.03	3.74E-08
29	25.02	162.98	531.98	2.17	33.08	25.01	3.42E-08
30	25.01	178.14	552.42	2.37	33.11	24.17	3.73E-08
31	25.00	181.14	535.24	2.41	33.28	25.59	3.79E-08
32	25.00	154.44	502.12	2.06	33.43	27.00	3.23E-08
33	25.00	158.49	630.90	2.11	33.41	19.69	3.32E-08
34	25.01	170.64	550.61	2.27	33.30	23.97	3.57E-08
35	25.02	178.23	530.26	2.37	33.17	25.78	3.74E-08
36	25.01	165.58	552.26	2.21	33.11	23.73	3.47E-08
37	25.00	155.61	598.00	2.07	33.27	21.21	3.26E-08
38	25.02	168.16	516.11	2.24	33.35	26.33	3.52E-08
39	25.02	169.78	500.24	2.26	33.42	27.66	3.56E-08
40	25.01	157.35	595.00	2.10	33.46	21.33	3.3E-08
41	25.00	173.25	507.89	2.31	33.04	27.23	3.63E-08
42	25.00	175.37	533.15	2.34	33.27	25.40	3.67E-08
43	25.02	154.18	535.17	2.05	33.28	24.71	3.23E-08
44	25.00	177.91	562.83	2.37	33.12	23.46	3.73E-08
45	25.00	156.05	554.87	2.08	33.21	23.50	3.27E-08
46	25.00	177.88	606.62	2.37	33.28	20.87	3.73E-08
47	25.00	158.89	618.97	2.12	33.40	20.19	3.33E-08
48	25.00	182.61	552.83	2.43	33.11	24.39	3.82E-08
49	25.01	164.03	599.81	2.19	33.17	21.03	3.44E-08
50	25.02	166.29	516.11	2.22	33.35	26.25	3.48E-08
51	25.02	174.22	501.94	2.32	33.33	27.79	3.65E-08

1.6 Appendix F. (continued)

No.	R (mm)	θ (°)	L (mm)	$\overline{u}$ (mJ)	r_1 (mm)	f (Hz)	m (t/mm)
52	25.00	173.24	507.89	2.31	33.04	27.23	3.63E-08
53	25.01	158.27	664.52	2.11	33.46	18.42	3.32E-08
54	25.08	161.94	502.12	2.15	33.43	27.16	3.4E-08
55	25.02	166.74	646.55	2.22	33.24	18.92	3.5E-08
56	25.00	173.75	512.63	2.32	33.34	26.88	3.64E-08
57	25.01	168.38	543.21	2.24	33.09	24.39	3.53E-08
58	25.00	156.98	649.00	2.09	33.25	19.02	3.29E-08
59	25.02	162.75	531.98	2.17	33.08	25.01	3.41E-08
60	25.00	164.74	606.62	2.20	33.28	20.70	3.45E-08
61	25.01	165.87	542.53	2.21	33.39	24.37	3.47E-08
62	25.00	176.82	508.13	2.36	33.24	27.45	3.7E-08
63	25.00	170.16	552.76	2.27	33.11	23.82	3.56E-08
64	25.00	155.58	566.43	2.07	33.22	22.84	3.26E-08
65	25.00	176.97	562.77	2.36	33.12	23.42	3.71E-08
66	25.00	154.97	590.04	2.07	33.16	21.61	3.25E-08
67	25.00	156.31	550.45	2.08	33.20	23.76	3.27E-08
68	25.00	169.06	512.18	2.25	33.24	26.67	3.54E-08
69	25.00	159.76	583.04	2.13	33.44	21.91	3.35E-08
70	25.00	158.02	551.78	2.11	33.11	23.68	3.31E-08
71	25.00	155.43	523.62	2.07	33.46	25.48	3.26E-08
72	25.01	170.42	508.16	2.27	33.04	27.05	3.57E-08
73	25.00	181.71	586.63	2.42	33.16	22.14	3.81E-08
74	25.02	169.66	613.59	2.26	33.29	20.37	3.56E-08
75	25.00	171.22	564.61	2.28	33.22	23.10	3.59E-08
76	25.00	176.69	554.84	2.35	33.31	23.94	3.70E-08
77	25.00	153.39	629.18	2.04	33.51	19.91	3.21E-08
78	25.02	160.16	575.01	2.13	33.24	22.34	3.36E-08
79	25.02	167.92	516.11	2.24	33.35	26.32	3.52E-08
80	25.00	158.91	618.48	2.12	33.40	20.21	3.33E-08
81	25.00	176.97	564.78	2.36	33.32	23.29	3.71E-08
82	25.00	182.20	576.23	2.43	33.14	22.81	3.82E-08
83	25.01	167.46	559.10	2.23	33.41	23.34	3.51E-08
84	25.00	165.33	512.67	2.20	33.24	26.47	3.46E-08
85	25.00	152.59	552.83	2.03	33.11	23.65	3.2E-08
86	25.00	178.44	554.84	2.38	33.40	24.02	3.74E-08
87	25.00	162.70	511.94	2.17	33.15	26.44	3.41E-08
88	25.00	152.45	532.77	2.03	33.18	24.88	3.19E-08
89	25.00	155.20	511.94	2.07	33.15	26.29	3.25E-08
90	25.00	157.45	653.02	2.10	33.35	18.86	3.3E-08
91	25.00	154.44	502.12	2.06	33.43	27.00	3.23E-08
92	25.01	168.32	542.97	2.24	33.29	24.41	3.53E-08
93	25.02	155.86	545.24	2.08	33.39	24.08	3.27E-08
94	25.02	170.04	500.48	2.27	33.23	27.65	3.56E-08

(continued)

1.6 Appendix F. (continued)

No.	R (mm)	θ (°)	L (mm)	$\overline{u}$ (mJ)	r_1 (mm)	f (Hz)	m (t/mm)
95	25.01	180.64	550.20	2.41	33.30	24.46	3.78E-08
96	25.01	176.94	555.73	2.36	33.31	23.89	3.71E-08
97	25.00	173.01	523.51	2.31	33.36	25.99	3.62E-08
98	25.04	180.32	517.24	2.40	33.25	26.95	3.78E-08
99	25.00	157.08	660.19	2.09	33.46	18.61	3.29E-08
100	25.00	160.64	644.61	2.14	33.24	19.09	3.36E-08
101	25.08	158.21	507.82	2.10	33.24	26.62	3.32E-08
102	25.00	158.49	631.99	2.11	33.32	19.65	3.32E-08
103	25.00	180.95	535.24	2.41	33.28	25.58	3.79E-08
104	25.02	181.57	501.30	2.42	33.42	28.40	3.81E-08
105	25.01	167.70	654.05	2.23	33.35	18.61	3.51E-08
106	25.08	183.02	503.93	2.43	33.53	28.31	3.85E-08
107	25.10	181.57	501.30	2.41	33.33	28.42	3.82E-08
108	25.00	153.33	628.94	2.04	33.31	19.92	3.21E-08
109	25.01	180.64	550.20	2.41	33.30	24.46	3.78E-08
110	25.13	176.38	500.01	2.34	33.42	28.11	3.71E-08
111	25.01	163.19	552.41	2.17	33.11	23.68	3.42E-08
112	25.01	171.48	606.23	2.28	33.18	20.76	3.59E-08

MIX
Papier aus verantwortungsvollen Quellen
Paper from responsible sources
FSC® C105338

If you have any concerns about our products,
you can contact us on
ProductSafety@springernature.com

In case Publisher is established outside the EU,
the EU authorized representative is:
Springer Nature Customer Service Center GmbH
Europaplatz 3, 69115 Heidelberg, Germany

Printed by Libri Plureos GmbH
in Hamburg, Germany